Stalin's Aviation Gulag

Smithsonian History of Aviation Series
Von Hardesty, Series Editor

On December 17, 1903, human flight became a reality when Orville Wright piloted the *Wright Flyer* across a 120-foot course above the sands at Kitty Hawk, North Carolina. That awe-inspiring 12 seconds of manned, powered flight inaugurated a new technology and a new era. The airplane quickly evolved as a means of transportation and a weapon of war. Flying faster, farther, and higher, airplanes soon encircled the globe, dramatically altering human perceptions of time and space. The dream of flight appeared to be without bounds. Having conquered the skies, the heirs to the Wrights eventually orbited the Earth and landed on the Moon.

Aerospace history is punctuated with many triumphs, acts of heroism, and technological achievements. But that same history also showcases technological failures and the devastating impact of aviation technology in modern warfare. As adapted to modern life, the airplane—as with many other important technological breakthroughs—mirrors both the genius, as well as the darker impulses of its creators. For millions, however, commercial aviation provides safe, reliable, and inexpensive travel for business and leisure.

This book series chronicles the development of aerospace technology in all its manifestations and subtlety. International in scope, this scholarly series includes original monographs, biographies, reprints of out-of-print classics, translations, and reference materials. Both civil and military themes are included, along with a systematic study of the cultural impact of the airplane. Together, these diverse titles contribute to our overall understanding of aeronautical technology and its evolution.

Advisory Board

Stalin's Aviation Gulag

A Memoir of Andrei Tupolev and the Purge Era

L. L. KERBER
EDITED BY VON HARDESTY

Smithsonian Institution Press
Washington and London

Edited by Initial Cap Editorial Services
Production Editor: Jenelle Walthour

Library of Congress Cataloging-in-Publication Data
Kerber, L. L. (Leonid L'vovich), 1903–1993.
[Tupolevskaĭa sharaga. English]
    Stalin's aviation gulag : a memoir of Andrei Tupolev and the Purge Era / L. L.
Kerber ; edited by Von Hardesty.
        p.   cm.—(Smithsonian history of aviation series)
    Translation of: Tupolevskaĭa sharaga.
    Includes bibliographical references (p.).
    ISBN 1-56098-640-9 (alk. paper)
    1. English ed. of work previously published under different pseudonyms. 2.
Aeronautics—Soviet Union—History. 3. Tupolev, Andreĭ Nikolaevich, 1888–1972—
Imprisonment. 4. Kerber, L. L. (Leonid L'vovich), 1903–1993—Imprisonment. 5.
Aeronautical engineers—Soviet Union. 6. Concentration camps—Soviet Union—
History. I. Hardesty, Von, 1939– . II. Title. III. Series.
    TL526.R9K47 1996   96-17488
    629.13'00947—dc20

British Library Cataloguing-in-Publication Data is available

Manufactured in the United States of America
03 02 01 00 99 98 97 96   5 4 3 2 1

CONTENTS

FOREWORD

The Soviet Union of Stalin's day is a world still hard for Westerners to imagine. Seldom has any nation endured such a massive, self-inflicted genocide. And seldom, if ever, has a major state had for decades so high a proportion of its population in prison. The society of which *Stalin's Aviation Gulag* shows us such a fascinating corner is one that seems bewildering. History has seen plenty of tyrannies. But why would even a particularly harsh one throw into jail one of its leading aircraft designers, its leading rocket scientist, and tens of thousands of its most skilled engineers and technicians? The vast wave of arrests and executions that rolled across the Soviet Union in the late 1930s had, in the usual sense of the word, no logic.

By the mid-1930s, the Communist Party had complete power over Soviet society, and Josef Stalin had dictatorial control over the Party. His archenemy, Leon Trotsky, was in exile abroad. Surviving monarchists had long since found new lives in Paris or New York. Political opponents on the left—and there once had been many who had hoped the Russian Revolution would turn into something more humane and democratic—had all been shot, imprisoned, or exiled. At a huge cost in human lives, all private farmers had been forced onto collective farms. All the major Western powers had at last recognized the Soviet state.

Within the Party, there had not been the slightest whisper of open dissent for several years. Unthreatened by internal enemies, Soviet power—and Stalin's own power—triumphed. Strangely, it was only at this point that there began the massive wave of arrests and killings known as the Great Purge.

Between the beginning of 1935 and the German invasion of Russia in mid-1941, according to Russian security officials today, more than 19 million Soviets were arrested. Roughly 7 million of these people were executed. The vast majority of the remainder were scattered with long sentences into the far-flung prison camps of the *Gulag* system. And there, eventually, most died: of malnutrition and exposure, in industrial and mining accidents, and sometimes in further waves of executions that swept the camps themselves.

During this period, the secret police seized more than one out of eight Soviet men, women, and children. From the writings of eloquent survivors such as Nadezhda Mandelstam, Eugenia Ginzburg, Victor Serge, and many others, we know the human toll the Great Purge wrought. The constant fear. The nervous, extravagant, frightened praise of Stalin that flowed forth on all occasions. The homes torn apart when a parent was taken away in the middle of the night and never seen again. Because the toll was so high, few extended families did not lose someone to that midnight knock on the door. In many ways, Russia today is still suffering the moral aftermath of all this.

But there was another consequence to the wave of arrests, and that is what *Stalin's Aviation Gulag* gives a window on. With more than 10 percent of its population in custody or just executed, what happens to a country's economy?

The effects of the Great Purge for Russian industry were even greater than the arrest statistics imply, for a peculiar blend of paranoia fueled the arrests. One part was Stalin's own personal madness. Another part was Russia's centuries-old xenophobia. And a third, clearly visible among the secret police, was an ancient, suspicious hostility toward the educated. This meant that the police dragnet disproportionately scooped up scientists, technicians, and engineers. They were blamed for the inevitable bottlenecks and glitches in an economy frantically struggling to industrialize. It was lethal, for example, to be head of any kind of scientific institute that failed to solve immediately some pressing problem. During this period, the secret police arrested *thirteen* successive directors of the Academy of Sciences in Kiev.

The xenophobia meant that the police dragnet also swept up a category of people who heavily overlapped with the scientists and engineers: anyone with a foreign connection. You were at grave risk if you had a family member living abroad, had lived or traveled abroad, spoke a foreign language, or even had a foreign-sounding name. The head of protocol at the Foreign Ministry was arrested for having "connections with foreigners"—even though that was, of course, part of his job.

Throughout the peak purge years of 1936, 1937, and 1938, Black Marias roared through the streets of every Soviet city, and secret police interrogators and executioners worked overtime. Then, with a large part of its scientific and technical elite either dead or behind barbed wire, the Soviet Union began to feel the results. So many railroad personnel had been arrested that some trains failed to run; so many engineers were gone that factories suffered breakdowns. And these disasters in turn provoked new arrests of those supposedly responsible.

By 1939, the Great Purge started to wind down. It ended in peculiarly Stalinist fashion, with the arrest of secret police chief Nikolai Yezhov and some 20,000 of his officers, many of whom ended up in the same camps as people they had been interrogating several years earlier. Arrests continued, but they were now mostly of a different sort of people—inhabitants of eastern Poland and the Baltic states, which Russia had taken over under the Hitler-Stalin pact of 1939.

Despite the ending of the Purge, the Soviets did not release many of the people they held in prison and labor camps. Statistics for *Gulag* population and mortality are more in dispute than those for arrests and executions, and scholars' calculations vary as to just how many live prisoners the system held on the eve of World War II, the period of which L. L. Kerber writes so much here. Almost everyone agrees, though, that the total number of prisoners was well into the millions. Many estimates place the number higher than 5 million, which would be a significant part of the Soviet labor force. In any event, this army of prisoners was carefully integrated into the economy.

The Soviets had begun doing this some years earlier. In 1931–33 the White Sea Canal, connecting the Arctic Ocean and the Baltic Sea, was built almost entirely by prisoners—more than 250,000 of them all told. The canal itself was an expensive failure: it was too small for big ships, it soon started filling with silt, and linking the Baltic and Arctic looked impressive on a map but put no food on anyone's table. Nonetheless, Maksim Gorkiy and other Soviet writers published a special volume praising the way this grand project was built and "enemies of the people" rehabilitated at the very same time.

And so all long-term prisoners were put to work. The needs of the Soviet economy dictated where the secret police placed the major islands of the *Gulag* archipelago. One huge group of labor camps, for example, sprang up in the district around Karaganda, in Kazakhstan. The region was rich in coal; prisoners mined the coal and built the railroads to send it to Russia's industries. Alexander Solzhenitsyn served part of his sentence in this area, and it is the setting for his novel, *One Day in the Life of Ivan Denisovich*. Traveling near Karaganda today, you can still see the aftereffects of the *Gulag* years: where prisoners once dug coal far under the earth, the ground above the mine has gradually sunk, and fields become pools of water when the spring snow melts.

Another large group of labor camps, legendary for their harsh climate and high death rate, sprang up in Kolyma, near where Russia faces Alaska across the Bering Strait. The bulk of Russia's gold lies in Kolyma, and more than 2 million prisoners were sent there to dig it out of the frozen ground. Here, too, a visitor can still see the snowbound ruins of stone punishment cells, rusting barbed wire, rotting wood barracks, and other debris from the area's vast network of more than a hundred camps. Other groups of camps were clustered around mines near Vorkuta in the Arctic, other mines in the Urals, and throughout the vast Siberian *taiga,* or conifer forest. Cutting timber, in fact, was probably the most common prisoner occupation of all.

Besides putting prisoners to work as unskilled laborers, the construction of the White Sea Canal set another precedent. Any engineering project calls for blueprints and plans, and even at that early date many of the USSR's best engineers were in custody. Hence the secret police developed the *sharaga* (often referred to by its diminutive, *sharashka*), or special prison, where engineers and technicians were put to work, usually in collaboration with free employees. One of the earliest *sharagas* on record was the Design Bureau for the White Sea Canal, Solzhenitsyn tells us in *The Gulag Archipelago*. It was in the warren of offices owned by the secret police just behind their headquarters in the famous Lubyanka building near Red Square. The prisoner-designers, however, were not allowed to inspect the terrain their canal had to pass through. Not surprisingly, this caused problems.

Also at the beginning of the 1930s, prisoner-engineers in another *sharaga* designed a new steel mill. There was, Solzhenitsyn says, an entire department of the secret police whose job was to "solve scientific problems using prisoners." He calls the *sharagas* the "paradise islands" of the *Gulag,* because the half-starved prisoners in the ordinary labor camps so envied the *sharagas'* superior food and working conditions. Rumors of Tupolev's paradise island spread through the *Gulag*.

To most readers, the best-known *sharaga* is the one where Solzhenitsyn himself spent some years, which he made the setting of his novel *The First Circle*. Solzhenitsyn's fellow-prisoner, the distinguished literary critic Lev Kopelev, wrote a nonfiction account of the same place, *Ease My Sorrows*. The *sharaga* was in the Moscow suburb of Marfino, housed in what had once been a church called Our Lady of Ease My Sorrows. Kopelev and Solzhenitsyn were both there between the end of World War II and Stalin's death in 1953. The *sharaga* had an acoustics laboratory and the prisoners were mathematicians, engineers, and scientists. They were put to work developing a scrambling device for secret telephone communications, and a "voice-print" system for identifying the voices on tapped telephone lines.

Much of what Kopelev describes about life at Our Lady of Ease My Sorrows echoes L. L. Kerber's account of the *sharaga* in *Stalin's Aviation Gulag*. There was good food, a library of technical books and journals, a workforce of free civilians who came in each day, and incessant demands for new inventions, to be produced immediately. These orders came from secret police commanders, afraid for their skins, who knew nothing of science or technology. But the prisoners of Our Lady of Ease

My Sorrows had one consolation: with a laboratory full of the country's most advanced audio equipment, they were able to quietly build themselves tiny radios to receive the BBC.

Three particular features of Soviet society during the terror of the Great Purge years stand out, and reflections of each of them appear in the life of A. N. Tupolev's *sharaga*. One is the odd way almost everyone arrested was accused of spying for a foreign power. The Russian mania about spies and foreign influence is an old one: the 1914 Baedeker for travelers to Russia, for example, urges visitors not to wrap objects in their suitcases in foreign newspapers, or else they will be confiscated. The Tsars were always suspicious of Russian travelers bringing home subversive ideas from Western Europe, and often with reason. The aristocratic Decembrist rebels of 1825, to take just one example, were mostly officers who had picked up liberal notions when they had been in the West after defeating Napoleon.

Under Stalin, this xenophobia reached new heights. Anyone who was arrested was, by definition, guilty. And most people judged guilty were, by definition, foreign spies. Hence the confessions that almost all prisoners broke down and signed after long, harsh interrogations usually involved espionage. If the prisoner had the slightest connection to another country, or had a foreign journal around the house when arrested, that might determine which country he had been spying for. But almost any country would do. Tupolev was accused of selling airplane blueprints to Germany. The executed writer Isaac Babel was accused of giving military secrets to France. One loyal Old Bolshevik in southern Russia was, to his utter mystification, accused of spying for Argentina.

The practice of calling any political enemy a foreign spy persisted for many years. After Stalin died in 1953, his Politburo colleagues soon arrested and executed the much-feared secret police chief Lavrenti Beria. This was the first step in dismantling the Stalin-era terror system. But when they announced Beria's death, he was accused not of brutality in running the secret police for 14 years, and not of sending millions of innocent people to needless deaths, but of being a British spy!

A second feature of the time, seen in these pages in the poignant deathbed confession of the aircraft designer Sergei Ilyushin, is the role of denunciations. Unlike the marching orders of the Gestapo, who went after Jews, Gypsies, and other clearly recognizable groups, there was little pattern to how Stalin's NKVD chose its victims. Its agents knew that they had to make huge quantities of arrests—often to meet numerical quotas—or else they themselves might be shot for insufficient vigilance. If ethnicity or actual evidence of wrongdoing (which was seldom there) was not the criterion for arresting someone, how, then, could you know you were seizing an enemy? One way was by denunciations.

The NKVD set up special mailboxes, and encouraged good citizens to denounce confidentially "enemies of the people." Millions, possibly tens of millions, of Soviets did so. Often a denunciation was a convenient way of getting rid of a rival at work, or of someone who occupied an apartment you had your eye on, or someone you had a quarrel with. And if you were in a position of authority and never denounced anyone, why, that might be an indication that you were not vigilant enough, and maybe even were deliberately harboring spies. Many people denounced others out of fear for their own lives.

More denunciations came from prisoners. When a person was arrested, he or she was often beaten or deprived of sleep for days at a time, and was always asked: Who were your confederates in espionage? Interrogators seldom stopped until they had beaten at least a few names out of their unlucky victims. Brave and clever prisoners tried to name people who had left the country or were dead. Many Stalin-era secret police files are today open at last, and they are filled with the names of people who denounced a particular prisoner, and the names of those he or she was forced to denounce in turn. Within a wildly irrational system, this methodical recording of denunciations was a weird kind of proof, to the police, that they were being rational. What do you mean we've arrested an innocent person? See? This dangerous spy Ivanov has been denounced by three people! How can you say he's innocent?

Finally, closely connected to the spy and denunciation mania is one more aspect of the Purge years: the assumption that for anything that went wrong, someone was at fault. Again, it is a perverse exaggeration of something long part of Russian life, summed up in the common phrase *"kto vinovat?"* "Who is the guilty one?" The same habit of thinking underlay other historical outbreaks of mass paranoia, whether the Spanish Inquisition, witchcraft scares, or (in a vastly milder form, to be sure) McCarthyism in the United States. If something has gone wrong—a factory breakdown in Russia, an epidemic of disease in medieval Spain, the "loss" of China to the United States—*someone* must be guilty. And, of course, that someone must be punished.

The men in *Stalin's Aviation Gulag* worried constantly about this: if a new aircraft had a accident on its maiden flight, on whom would the axe fall—the test pilot? The designer? One of the engineers? In any bureaucracy, of course, someone's head may roll when a costly mistake is made. But we use the phrase figuratively; in Stalin's Russia the head often got a bullet in the neck. The director of a factory that didn't meet its production quota, the manager of a mine that didn't mine enough ore, an inventor whose invention failed—all could end up dead, or, if lucky, sent for ten years to the *Gulag*. Sometimes the crime was merely producing an embarrassing piece of information. When, for instance, the census showed that the purges and famines were decreasing the Soviet Union's population, Stalin had the members of the census board shot. The new board quickly came up with higher figures.

This was the society, then, in which Tupolev and his fellow engineers worked. The complex teamwork of designing new aircraft had to take place in a country undergoing an endless inquisition. Most amazing of all is that Tupolev and the talented men around him had to do some of the most creative work of their lives while in prison. To find anything fully analogous in history one must go back to the courts of empires in the distant past, where sometimes certain slaves were highly trained craftsmen. *Stalin's Aviation Gulag* offers an unparalleled insider's look at such an empire in our own time.

Adam Hochschild

A NOTE FROM THE AUTHOR

We are not that of which little remains, we already practically do not exist.
Words heard during a conversation between two imprisoned scientists in a camp.

The recollections I have offered about one of the most prominent aviation designers of the twentieth century, A. N. Tupolev, do not fit within a purely biographical framework. Owing to his gifts and breadth of interests, Tupolev not only designed new airplanes, but also assisted N. Ye. Zhukovskiy in studying the problems of aero- and hydrodynamics at the Central Aero-Hydrodynamics Institute, and helped G. K. Ordzhonikidze to reorganize or, more correctly, to develop the country's aviation industry.

I could not ignore these aspects of his activity, just as I could not keep silent about his meticulous care in selecting and assigning his design bureau employees. He brought in specialists from the most varied fields. Along with "pure" airplane builders—specialists on construction mechanics, strength of design, and aviation engines—there were (heretofore somewhat unusual) specialists on instruments, radio communications, electromechanics, and other related disciplines. I was one of those employees.

I should explain how I arrived in the midst of Soviet aviation events, and how eventually I turned out to be a deputy of Tupolev.

I came from a suspect social class for those who led the October Revolution. My father was a naval officer and ended his life as a vice admiral. Soon after the Bolshevik revolution, his wife and children were resettled out of the capital. We made our way to our mother's sister in Luga. In order to survive, I went to work for the Luga Artillery Range as a telephone operator. Soon the artillery unit of the XI Petrograd Division was formed out of the range employees, and was subsequently sent to the front. After the battles against the White Poles, our division was quartered in Petrograd. In the spring of 1921 the Kronstadt sailors rebelled against the communist dictatorship. The 32nd Brigade of our division was sent to storm the fortress. The telephone operators pulled communication lines from Martyshkino Village across the ice to the fortress. After burying the fallen Red Army soldiers and numerous sailors, we returned to the city. I was demobilized and went to live with my brother, V. L. Kerber-Korvin, in Taganrog, where he served at an aviation factory.

In 1922 Taganrog Factory engineers M. M. Shishmarev, N. G. Mikhel'son, and V. L. Kerber-Korvin took first place in a competition for a naval fighter airplane. They were called to Moscow to build this airplane. We settled near Khodynskiy Airfield, where I also began to work installing radio apparatuses on planes and developing radio communications.

In the early 1930s the TsAGI designed the ANT-25 airplane to compete in world competition for long-distance flight. A radio station was ordered for its long-range flight headquarters (party leaders G. K. Ordzhonikidze and K. Ye. Voroshilov, along with Air Force commander Ya. I. Alksnis) at the Morse Factory. I was assigned to test the radio on a flight to Chita on a TB-1 (piloted by Gorodilov). The tests went excellently, and chief designer Tupolev wanted to become familiar with them in detail. That was when I met this outstanding man for the first time.

Soon after that test, I was assigned to install radio direction finders on TB-3 bombers. The tense situation in the Far East did not exclude conflict with Japan, and it was not possible to fly over the sea without radio navigation equipment. Engaged in this work, I understood that aviation navigators were not familiar with radio navigation equipment. They were already widely in use at sea. Relying on the works of Sakelar, a professor at the Naval Academy, I wrote the work *Aerial Means of Radio Navigation and Their Operation* for aviation navigators. After my return to Moscow I was again summoned to Tupolev. He assessed my work highly and ordered that the book be printed. This was the first publication in our country in the field of aviation radio navigation. The entire edition was purchased by the Air Forces to train navigators. Probably these two successfully accomplished assignments inclined him to use me more. The flight headquarters brought me in to instruct ANT-25 crews preparing to cross the North Pole to the United States. At this time I became acquainted with, and later became close friends with their participants, G. F. Baidukov and S. A. Danilin.

Gradually Tupolev involved me more and more in discussing his future flying machines. I felt that I could be myself around him. One thing alone frightened me, my lack of necessary education. As a person alien to the working class, I was not allowed to study at a higher educational institution. I had to engage in self-education, which I did for the rest of my life.

In parallel with the work on the ANT-25, I was assigned to take part in equipping the DB-A airplane (N-209) of chief designer V. F. Bolkhovitinov, on which Sigismund Levanevskiy and N. Kastanayev prepared a commercial crossing over the North Pole to Alaska. One of Bolkhovitinov's assistants, Professor M. M. Shishmarev, my wife's father, recommended that he include me in the crew as a radio operator and radio navigator.

On 21 October 1937, the harmonious work with Tupolev's group all fell apart when A. N. Tupolev was arrested. A year later, I was also sent to prison, and ended up in a work camp near Archangel cutting timber.

When the authorities organized a special NKVD prison around Tupolev, TsKB-29, for the development of new combat aircraft, Tupolev requested me as well as other aviation specialists from the various camps. This decision saved my life. Having arrived in Moscow, I ended up as chief of the equipment construction team.

In 1952 Tupolev named me his assistant for aircraft equipment. I worked in this post until 1972, the year that Andrei Nikolayevich Tupolev died. I was responsible for developing the instrumentation of all of his airplanes from the Tu-2 to the Tu-154.

ACKNOWLEDGMENTS

Stalin's Aviation Gulag: A Memoir of Andrei Tupolev and the Purge Era represents a collective effort of many people in both Russia and the United States. The idea of an English translation of L. L. Kerber's classic account of Andrei Tupolev and the purges arose in 1989, when the Smithsonian History of Aviation book series began as a collaborative effort of the National Air and Space Museum (NASM) and the Smithsonian Institution Press. Funding for a translation became a reality through the publication of *Milestones of Aviation,* edited by John T. Greenwood in association with Trish Graboske, publications chief at NASM. A meeting I had with L. L. Kerber in September 1991 set the stage for the Smithsonian Institution Press English version. Kerber had been interested in finding an English-language publisher for many years and agreed to work with us. Two decades had passed since he wrote the original manuscript, so he was encouraged to revise the text in minor ways, if necessary, in light of new information.

To prepare the new Russian-language manuscript for translation, several people worked with Leonid Kerber, contributing significantly to the final work. Paul Mitchell prepared the initial translation, which became a baseline for all our subsequent translation and editing work. Maksimilian Saukke, a close associate of Kerber, provided technical and historical assistance. He and N. V. Grigorev provided written drafts on Kerber's life. No less important was the assistance of Mikhail L. Kerber, Leonid Kerber's son. Together with other members of the Kerber family these individuals worked to identify historical photographs for *Stalin's Aviation Gulag*. For this translation, we used the transliteration system of the United States Board of Geographic Names (BGN), except for the names that have acquired a familiar transliterated form—for example, "Sikorsky" instead of "Sikorskiy."

Preparing the translation for publication required considerable patience because there were many delays. In Moscow Olga Tyuleneva worked with us to move the project forward. It is difficult to imagine the project reaching fruition without her expert assistance. She provided liaison with the Kerber family and assistance on many requests, major and minor. There were others who assisted as well from the Russian side, in particular Yuri Salnikov and Aleksei Drozhilov. A graduate of the Moscow Aviation Institute, Aleksei Drozhilov provided consultation on a whole range of technical questions. I appreciate the many kindnesses and assistance offered to me by my Russian collaborators.

I am grateful as well to those on the American side who shepherded this book at various stages. Felix Lowe, former director of the Smithsonian Institution Press, was an early enthusiast for the Kerber book. His successor at the Press, Dan Goodwin, and Mark Hirsch, the acquisitions editor for aerospace history at the Press, also promoted the project in numerous ways. Therese Boyd, the editor for the book, deserves special mention for her impressive work in the careful adaptation of the English-language translation for publication.

Finally, there are others who offered advice and support at various stages of the project: Adam Hochschild, Harold Shukman, Mark Kulikowski, Tom Alison, Lisa Jankoski, Stephen Hardesty, Michael Gorn, and Patricia Hardesty.

INTRODUCTION

Stalin's Aviation Gulag is a curious title for a book. The world of aeronautics, especially for Americans, seems quite separate from prisons or the secret police. For those familiar with the Stalinist legacy, in particular the purge era of 1936–39, the intersection of aviation and political repression is well known—in fact, this era became a dramatic example of Stalin's ambivalent posture toward his technological elite. At the very epicenter of this story is Andrei Tupolev (1889–1972), the Soviet Union's premier aircraft designer in the twentieth century.

When Tupolev died on 23 December 1972, the Soviet press gave considerable place and emphasis to his career. No one equaled Tupolev in the sheer magnitude of his accomplishments as an aircraft designer. The Soviet media chronicled the Tupolev legacy in detail, noting his many triumphs in both civil and military aviation. At the time, the Cold War dominated East-West relations, and Tupolev-designed strategic bombers—specifically the Tu-16 Badger, Tu-95 Bear, and Tu-22 Blinder—projected Soviet military power. Tupolev was heralded as a genius, the legendary chief designer, the builder of Soviet aviation. Fittingly, Tupolev had spent his last years working on the futuristic Tu-144, the Soviet Union's supersonic transport built to challenge the

Anglo-French Concorde. The appearance of the Tu-144 confirmed Tupolev's place at the cutting edge of Soviet aviation.[1]

Not revealed in the obituaries and retrospective articles were the shadowy circumstances of Tupolev's arrest in 1937 and his subsequent work in a *sharaga* (special prison workshop) during the years 1939–41. Even at the time of Tupolev's death the very existence of the prison workshops remained a state secret, something outside official history and the narrow confines of public discourse. During Nikita Khrushchev's years in power, roughly the decade following Stalin's death in 1953, an effort was made to expose the worst excesses of Stalin and to condemn the "Cult of Personality." Leonid Brezhnev, coming to power in 1964, preserved de-Stalinization in its broad outlines, but reemphasized certain forms of censorship. The official organs of the Soviet Union could embrace Tupolev as a national hero, but not acknowledge his victimization during the purge era. This same reflex to guard the past would force Alexander Solzhenitsyn into exile later in the decade, once Solzhenitsyn engaged in a systematic effort to expose the dark side of the Stalin years.

Tupolev's story, as events proved, could not be censored. A year before his death, an anonymous memoir began to circulate in the "samizdat," or underground press, in the Soviet Union. Given the title *Tupolevskaya sharaga* (translated literally as *Tupolev's Special Prison Workshop*), the typewritten manuscript with its focus on an aviation design bureau differed markedly from the typical samizdat fare of banned poetry, literature, and philosophical writings. Yet it moved quickly across the underground network. Bold in tone and filled with many anecdotes, *Tupolevskaya sharaga* soon attracted a large readership, becoming in the 1970s a classic in the literature of dissent. For members of the Soviet aviation community, the anonymous memoir represented a candid account of how Tupolev and many other specialists endured political repression under Stalin. For them, the memoir was also timely, coming as it did when the official press simply ignored Tupolev's arrest and imprisonment under Stalin. Seen from this perspective, Tupolev's achievements assumed heroic proportions because he had worked in an atmosphere of arbitrary arrest and the often mindless interference of Lavrenti Beria and the police bureaucracy.

After the passage of a quarter-century this samizdat classic is now available to the English reader, under the title *Stalin's Aviation Gulag: A Memoir of Andrei Tupolev and the Purge Era*. Behind this publication event, however, is a long and complicated story. The first typewritten version of *Tupolevskaya sharaga* circulated among a small circle of aviation professionals, mostly friends and former colleagues of Andrei Tupolev. The manuscript entered the clandestine world of samizdat literature around 1971. Rumors circulated that Tupolev himself had approved the move. *Tupolevskaya sharaga* was copied, avidly read, and passed on to an ever-widening group of readers in the Soviet Union. Ultimately it would find its way to West Germany, to be read by another circle of interested readers. This sequence of events passed swiftly, spontaneously, and outside the control of the author, still unidentified.

For Soviet dissidents, *Tupolevskaya sharaga* exposed the excesses of the Stalin years. Tupolev's plight in the purge era had been revealed for the first time in dramatic detail. Historical memory had triumphed over the concerted efforts of the state to control images of the past. No less important was the fact that *Tupolevskaya sharaga* gave historical weight to Alexander Solzhenitsyn's novel, *The First Circle,* which had employed imaginative literature to capture the peculiar world of the prison workshops run by the NKVD, the Soviet secret police.[2]

When *Tupolevskaya sharaga* reached West Germany in 1971, the book acquired an eager readership among emigré Russians and observers of the Soviet Union. That same year a West German publishing house, Possev-Verlag, decided to publish the Russian-language memoir unabridged. They adopted the name "A. Sharagin" as the author. The transparent adaptation of the word *sharaga* to fashion the surname "Sharagin" only added to the atmosphere of mystery surrounding the book. At the eleventh hour, even as the Russian-language version went to press in West Germany, news arrived that the author was the late G. A. Ozerov, an aeronautical engineer and one-time associate of Tupolev.[3] The West German publishers stopped the presses to superimpose the following breathless note at the bottom of the title page: "Even as the book went to press we discovered the name of the author— Georgiy A. Ozerov."[4] The sudden revelation that Ozerov was the author prompted skepticism in the Soviet Union, where some asked, could this be a ruse to deflect the attention of the police authorities? Their doubts stemmed from the fact that Ozerov was dead. Moreover, there were no descendants of Ozerov to suffer the consequences of his alleged authorship of the controversial memoir. Whatever the reasons, the police did not actively investigate the matter or seek out the actual author.

By 1973 public interest in *Tupolevskaya sharaga* reached another milestone. That year a French translation appeared under the title, *En Prison avec Tupolev*.[5] The French edition also identified Ozerov as the author of the book, while preserving "A. Charaguine" as the pseudonym for the author. The French translation gave "Ozerov's" memoir wide exposure outside the narrow confines of the Russian emigré community and Western specialists who monitored Soviet affairs. Coincidental to the German and French publications, the BBC gave *Tupolevskaya sharaga* exposure on several shortwave programs. The BBC broadcast excerpts from the book as early as 1971, showcasing the controversial samizdat publication as a fresh and insightful memoir on the Stalin purges. The BBC took special interest in *Tupolevskaya sharaga* because of its coverage of Sergei Korolyov (1907–66), former head of the Soviet space program. Korolyov had spent time with Tupolev in his special prison workshop. He had been rescued from the dreaded Kolyma gold mines in Siberia just in time, it appeared, for few escaped the Kolyma camps alive.[6]

Throughout this period the real author of *Tupolevskaya sharaga*, Leonid L'vovich Kerber, had managed to escape notice. His anonymity, in fact, lasted until the late 1980s. Only then, during the Gorbachev years, did his identity became well known. The Gorbachev government, with its unprecedented openness and desire to liquidate the worst features of Soviet police repression, had established a secure environment in which Kerber ran no risk of retribution for his authorship of this celebrated book in the samizdat.

Kerber published an official biography of his former boss in 1973, *Tu—Chelovek i samolet* (*Tupolev—The Man and His Aircraft*). While never lacking detail or nuance, this official biography makes only one fleeting reference to Tupolev working in a "special design bureau."[7] The reader is left with little knowledge of the design bureau's purpose, location, and administration. Kerber's biography was widely distributed in the Soviet Union at the very moment his samizdat version of Tupolev's life appeared in West Germany and France. The existence of the official biography, with its studied restraint and polite bows to state censorship, did provide a convenient cover for Leonid Kerber. Few people, it appeared, suspected him of being the author of *Tupolevskaya sharaga.*

Throughout his long career Kerber enjoyed a reputation as a highly talented engineer. He had won several medals for his service to the state, including Hero of Socialist Labor and the Lenin Prize. His close ties with Andrei Tupolev had allowed him to advance to positions of responsibility in the post–World War II context when Tupolev once again stood at the pinnacle of Soviet aviation.

Kerber's retirement years coincided with the relaxed censorship of the Gorbachev era. Glasnost'—Gorbachev's deliberate campaign to deal openly and honestly with the history of the Soviet Union—allowed Kerber many opportunities to write about Tupolev and Soviet aviation. Before his death in October 1993 Kerber made full use of this new ethos to produce a series of articles on Tupolev, the purge era, and Soviet aviation.[8] Always a zealous defender of Tupolev, Kerber did much to chronicle the many hidden dimensions of Soviet aviation development. Together with *Tupolevskaya sharaga,* these writings constitute an important legacy of historical materials on aviation in the Soviet Union.

Among Kerber's writings one will find frequent reference to A. S. Yakovlev, Tupolev's great rival. Yakovlev belonged to a younger generation of aircraft designers who vied with Tupolev for Stalin's favor in the late 1930s, a group that also included Sergei Ilyushin, A. I. Mikoyan, and Vladimir Myasishchev, to name a few. Yakovlev criticized the inordinate emphasis on large aircraft, in particular the production of bombers at the expense of fighters and ground-attack aircraft. After Tupolev's arrest in 1937 and the calamitous involvement of Soviet aviation in the Spanish Civil War, Yakovlev's star was on the rise. Stalin's reorganization of the Air Force in 1940 mirrored Yakovlev's priorities. During World War II Yakovlev served as Deputy Commissar for Aircraft Production, even as Tupolev spent time in a *sharaga.*[9]

In the Stalinist world of arbitrary shifts of policy, Yakovlev eventually stumbled. Toward the end of World War II, as Stalin began to remove many of his wartime chieftains, Yakovlev made an ill-advised speech questioning the future of jet engines. This misstep came just as Tupolev, through reverse engineering, ably copied the American B-29 Superfortress, to give the Soviet Union its first real strategic bomber, the Tu-4. In the post-1945 world Tupolev's star rose again: he assumed his customary position as the Soviet Union's premier designer. However, Kerber—at Tupolev's side during the war—never forgot the humiliation of the prison workshop. He resented Yakovlev's numerous books and writings, which highlighted his own achievements at the expense of Tupolev. Kerber used his retirement years to engage in a polemic against Tupolev's rival.

I met Kerber in Moscow in September 1991. He took great delight in talking about aviation. His eyes conveyed vitality and wisdom. Surrounded by family and friends that evening, he told many anecdotes about his life, moving with verbal images across the vast landscape of the Soviet era. Soviet history was nearly co-extensive with his own life; he was born in the reign of Nicholas II and he lived to see the fall of communism. His recollections of the Stalin years were detailed and told with great élan, but never with any outward bitterness or rancor. Based on this brief encounter, I sensed that Kerber, if patriotic, was essentially apolitical in his approach to life. At heart, he was an aeronautical engineer. He appeared to contend with politics as the Russian peasant had traditionally viewed the weather: if the sun shines, you rejoice, if ill winds come your way, you wait patiently for a better day. Kerber possessed more than an ample measure of resiliency in the face of life's vagaries, the ability to work hard, and to see the humorous and the absurd in life, even at moments when others capitulated to despair.

That same evening in Moscow I asked him to explain the circumstances surrounding his decision to write *Tupolevskaya sharaga*. He said that he began writing his reminiscences as early as 1951, two years before Stalin's death. He explained that his simple goal was to record for his grandchildren some of the extraordinary events he had observed. At that time, he confessed, there was no more grandiose motive at play. He was not a professional writer, but informal, handwritten notes on his life were a way to preserve the history of those years for his family.

Once he completed a handwritten draft, the manuscript remained a well-guarded secret of the Kerber family for several years. Eventually Kerber decided to show his manuscript to a close friend, who strongly urged him to prepare a typewritten copy so it could be shared with others. Kerber agreed. This fateful step, taken around 1971, moved the memoir outside the narrow and secure family circle to the clandestine world of samizdat. This step was risky because Kerber's recollections contained a powerful indictment of the Soviet regime. His motives were no doubt complex. He had a compelling desire to recount history honestly, as he remembered it. Kerber was not content merely to tell stories or be remembered as a contributor to oral tradition.

What is known about Kerber's own life?[10] Kerber was born on 16 June 1903 in St. Petersburg. His father served as a naval officer in the Imperial Russian Navy. At the time of Kerber's birth the Russian autocracy was still intact, but the Russo-Japanese War and the Revolution of 1905 quickly followed, setting the stage for a more tumultuous political crisis. In 1917 Kerber decided to follow the family tradition and enter the naval cadet corps. The revolutionary events that year abruptly ended Kerber's career in the navy. Being from the upper classes, he found many doors closed to him as the Bolsheviks took political power in November 1917.

As a young man, Kerber had not been a radical or revolutionary activist, although in the 1917 revolution he cast his lot with the Bolsheviks. From 1918 to 1921 he served as a common soldier in the ranks of the Red Army. He participated in one of the most controversial episodes associated with the Bolshevik regime, the storming of the Kronstadt Naval Base by the Red Army. Kerber also served with the Red forces in the campaigns against the White armies of Kolchak and Denikin. Following the bloody Civil War Kerber moved to the south in 1921, to work as a fitter in an aviation plant in Taganrog. Here he became enamored with aviation. Despite his upper-class background Kerber managed in the following years to pursue an engineering career in the Red Army, specializing in radio communications, electronics, and navigation. He published an army manual on navigation in 1934, which clearly established his reputation as a specialist in his field.

The lives of Kerber and Tupolev intersected in the 1930s. When the Tupolev design bureau developed the long-range ANT-25 for the record-breaking transpolar flights of 1937, Kerber served as an army communications specialist assigned to the project. That same year Tupolev was arrested. Kerber suffered the same fate in 1938, but, unlike Tupolev, who was held in a Moscow prison, the NKVD shipped Kerber off to a distant spot in the Gulag archipelago, a labor camp in the forests of the North near Archangel. They would be reunited in 1939, when Tupolev requested that Kerber be summoned from the labor camp to join him at the newly established special prison workshop on Radio Street in Moscow, adjacent to TsAGI (the Central State Aero-Hydrodynamics Institute). This prison experience forged a lifelong friendship between Tupolev and Kerber.

Stalin's Aviation Gulag, at its core, is a biography of Andrei Tupolev. It differs from Kerber's official biography of Tupolev in tone and content. To pass the censors, Kerber wrote his account of Tupolev's life in a highly stylized fashion, omitting much, and never conveying any of his deeply held convictions. For his unofficial account, however, Kerber allowed himself more freedom to express his own understanding of the past. Kerber had enjoyed unique access to Tupolev, allowing for in-depth interviews. The opening chapter of *Stalin's Aviation Gulag*, for example, reflects one of Kerber's informal conversations with Tupolev, one in which the famed designer spoke candidly about his formative years. Kerber's research was enriched as well by other colleagues' reminiscences and his own intimate knowledge of Tupolev's career.

Kerber begins his story with Andrei Tupolev's fateful encounter with aviation in 1909, the year he entered the Moscow Higher Technical School. Highly talented and ambitious, Tupolev became a student of the famed aerodynamist N. Ye. Zhukovskiy, who at the time was a prominent faculty member at the school. Zhukovskiy—often called the "Father of Soviet Aviation"—made substantial contributions to aeronautical theory in the decade following the Wright Brothers' invention of the airplane in 1903. Later Lenin mobilized Zhukovskiy to found TsAGI, which became the major aeronautical research institute in the Soviet Union. As a protégé of Zhukovskiy, the young Tupolev found himself at the very center of Russia's embryonic aviation establishment. Through Kerber's retelling of Tupolev's early years we catch fleeting glimpses of Russia's lively aeronautical community in the last days of the tsarist regime.[11]

For Russia's small aeronautical community, the year 1917 brought great disruption and internal conflict. Pilots, engineers, aircraft designers, aviation workers—all had to decide whether or not to embrace the Bolshevik cause. Some aeronautical figures such as Zhukovskiy and the young Tupolev saw in Lenin and his fledgling regime modernity and technological progress. Other aviation figures refused to cooperate with the Bolsheviks, the most notable being Igor Sikorsky, who emigrated to the United States in 1919 to pursue a career in aircraft and helicopter design equal in magnitude to his contemporary, Tupolev. Both began their careers in the twilight of tsarism, going on to attach their names to some of the most important aircraft designs of the twentieth century. Sikorsky died just a few weeks before Tupolev in 1972. It would not be until 1989 that the Soviet Union would fully recognize Sikorsky.[12]

Reading *Stalin's Aviation Gulag,* one acquires a deeper appreciation for the peculiarities of Soviet aviation. The October Revolution of 1917, as Soviet propaganda organs routinely portrayed it, was an epic moment in history, a watershed that divided the primitive past from the future. Leading the Soviet Union into the future, the Communist Party embraced the goal of rapid technological change, often from a highly visionary posture and frequently backed up by ruthless force. All workers, including highly trained specialists and engineers, were subordinated to this process. Under Stalin, beginning in 1927, the Soviet Union inaugurated a series of Five-Year Plans which committed society to the herculean task of building a modern industrial state. The script called for vast projects of electrification, the mechanization of agriculture, and industrial expansion—all to be launched in one generation. As the world looked on in awe, Stalin presided over many dramatic events such as the building of the great Dnieper hydroelectric plant, the new steel city of Magnitogorsk, the Moscow subway, and the White Sea Canal. The airplane—one of the most powerful symbols of modernity—would play a dramatic role in this new ethos.[13]

Stalin's need for engineers was manifest. The technological intelligentsia would, by necessity, be at the center of any industrial transformation. Progress was inconceivable without them. This reality was most apparent in the sphere of aviation, where trained aircraft designers, pilots, engineers, and highly skilled workers were in short supply. Men such as Tupolev possessed rare skills essential to the aviation sector. While the prerevolutionary legacy in aviation had been strong and enduring, the Soviet Union in the interwar years enjoyed uneven progress in aeronautics. Aero engines, in particular, were a chronic weakness, forcing Moscow to spend hard currency to purchase Western-designed powerplants. Still, the Soviet Union was not without some talented designers and possessed the institutional base to expand the aviation industry in the 1930s. Accordingly, Stalin showered many favors on the aviation establishment and allocated for it an enormous share of the Soviet Union's national resources.

The 1930s proved to be an unusually fertile period for aviation development worldwide. This context for rapid technological change made the Soviet Union's quest to become a major air power more difficult. Change came rapidly, in ways that literally transformed the appearance and design of the modern airplane. At the dawn of the decade the airplane still looked much as it did at the end of World War I: most aircraft were still biplanes, fashioned of wood and fabric, and equipped with open cockpits, externally braced wings, and fixed landing gear. Such aircraft typically were slow, lacked range, and possessed only primitive instrumentation. A few short years passed and the airplane projected a much different look: now most aircraft were monoplanes, characterized by all metal construction, enclosed cockpits, streamlined cowlings and fairings, and retractable landing gear. As World War II approached, these sleek new aircraft established a new standard of performance. Airplanes could fly higher, faster, and farther. Air records were quickly set and then broken as nations vied with one another for dominance in this rapidly evolving arena of technology. Valery Chkalov, the Soviet Union's most celebrated flier, flew a Tupolev-designed ANT-25 aircraft over the North Pole to Vancouver, Washington, in 1937, one of many aerial spectaculars of "Stalin's Falcons."[14] Dependable passenger aircraft such as the DC-3 allowed airlines to achieve their long-cherished goal of transporting large numbers of travelers safely across vast stretches of the globe. All these changes implied that the next war would be decided by air power—no doubt with the bomber at the cutting edge.

Tupolev and the growing Soviet aeronautical community faced enormous challenges in their attempt to achieve parity with the West. The regime's mania for large, highly visible projects influenced Soviet aviation as it did other sectors of industry. Kerber makes reference to the Tupolev-designed *Maksim Gorkiy,* an eight-engined behemoth that made a dramatic flyby over Red Square. This red-winged aircraft became the ultimate expression of "gigantomania" with its unprecedented power and range. The *Maksim Gorkiy* flew across the vast Soviet land mass in the mid-1930s to showcase Soviet technical achievement. On board was a printing press, photo lab, and cinema. The aircraft operated with a crew of 20 and, depending on the interior configuration, could transport 43 to 76 passengers. Such aerial spectaculars gave the regime recognition abroad and legitimacy at home. Ultimately, disaster befell the *Maksim Gorkiy,* on 18 May 1935 when the famed aircraft crashed after a midair collision. But there were always high risks in the international competition for air records.

Flying "higher, faster, and farther" also suggested that the airplane possessed considerable potential as a weapon in any future war. Stalin expressed a willingness in the 1930s to test his military designs in several minor wars and skirmishes. In 1937–38, Moscow deployed aircraft to Spain to defend the Republic in the Spanish Civil War. Soviet I-15 and I-16 fighters with SB-2 bombers greatly assisted the Republic in the air war against the Nationalists under Francisco Franco. Initially, Republican air units, flying these late-model aircraft, established air superiority. This dominance abruptly ended, however, when Germany deployed the Luftwaffe's latest aircraft, in particular the Messerschmitt Bf-109E, in support of the Nationalists.

Soviet aircraft in Spain demonstrated an inability to match the technical and tactical expertise of the enemy. The humiliation of the Spanish war forced Moscow to abandon military intervention. By 1940, it was apparent that Soviet aviation had slipped behind that of Nazi Germany, its potential rival in the next war. During the 1930s the Soviet Union had placed emphasis on aerial stunts, record-breaking flights, and giant flying machines. With the approach of war these activities, pursued largely for propaganda purposes, would prove foolhardy.

Stalin moved to reorganize the Soviet aeronautical establishment in 1940, a campaign that dovetailed with the purges. Already Tupolev was at work in the *sharaga* on Radio Street in Moscow. The favoritism that had been afforded aviation in general and Tupolev in particular ended abruptly. Pilots, Air Force leaders, designers, aircraft plant managers, the entire edifice of the aeronautical establishment, fell victim to the purges. Air commanders Ya. I. Alksnis and Ya. V. Smushkevich were arrested and executed, along with a large percentage of the air staff and officer corps. Stalin elevated a new generation—one less experienced and often less talented—to replace them. As World War II approached, the Soviet Union found its aviation establishment profoundly weakened.[15]

Tupolev knew firsthand the arbitrary aspects of Soviet rule, the political whims and shifting party priorities that could undermine a project or end a career. But as a young man Tupolev had learned the skills of survival; he would not be consumed in the purge process. The Communist Party sought to exercise control over all aspects of Soviet life. Aviation had enjoyed an exalted status because of its obvious technological and propaganda value. Accordingly, the enthusiasm of Soviet authorities for aviation often meant meddling by party functionaries with little, if any, technical knowledge. Kerber describes Tupolev's forceful personality at work in these dangerous times. He managed to walk a tightrope between his independent frame of mind and the demands of his police overseers.

How the purge mechanism victimized Tupolev is told in detail in *Stalin's Aviation Gulag*. Kerber recounts how the NKVD, the Soviet secret police, arrested Andrei Tupolev on 21 October 1937. Three agents from the NKVD, following the typical script, came to Tupolev's apartment at 11 o'clock in the evening. They ordered a stunned Tupolev into his study while they thoroughly searched the apartment. Finally, the NKVD led Tupolev away, later to charge him with selling blueprints to Germany to build the Messerschmitt Bf-110 aircraft.

Tupolev's arrest traumatized the aeronautical establishment. No one was safe. The bizarre accusation that Tupolev had been in league with

Nazi Germany only added to the sense of vulnerability among the technical intelligentsia. As events unfolded, Tupolev was spared execution. He was never tortured, although he was compelled to stand for hours during interrogation.

Kerber soon fell into the purge dragnet as well. He described his arrest and imprisonment during my interview with him in September 1991. The first sign of trouble, he told me, came on the morning of 21 October 1938. On that fateful day, precisely a year after Tupolev's arrest, Kerber's boss asked him to surrender his pass. He dutifully turned it over and left the design bureau. On his way home Kerber stopped to visit his pregnant wife, who had entered a hospital to give birth to their second child. He told her of the ominous request to surrender his pass. Returning home, Kerber anticipated the inevitable. At 11 o'clock there was a knock at the door. A police unit consisting of two officers and three soldiers abruptly entered his apartment and searched each room. Prior to the search they asked a man from the neighborhood to accompany them as a witness. This polite bow to legality became an empty gesture; the same man later gave "evidence" to the police that Kerber had engaged in counterrevolutionary activity. The thorough search of the Kerber apartment, however, revealed little: a toy gun and photo development equipment. Although Kerber was an amateur photographer, the NKVD accused him of using his developer for espionage work. Kerber was taken to Lubyanka, joining countless others as victims of Stalin's repression.

Beginning in 1939, Kerber joined Tupolev and 150 other specialists in the *sharaga* located in the KOSOS building on Radio Street. Located on the edge of the TsAGI complex and near the Yauza River, the *sharaga* gave no outward signs of being a prison. The KOSOS structure (KOSOS was the Russian acronym for the Design Department for Aircraft Construction) appeared to be a normal, modern building with many windows and rounded lines. To conceal its true character, the police had barred the windows from the inside. All the prisoners lived on the top three floors.[16]

The prison workers lived according to a strict routine, arising early and working a ten-hour day, with dinner served at 1 P.M. Most prisoners discovered the food to be excellent, especially for a prison—two-course dinners with dessert served by waitresses with aprons. A small shop provided cigarettes, razor blades, and other amenities. The police constructed a caged enclosure on the roof in which the prisoners could exercise.

Kerber's insider's account of the *sharaga* suggests a prison reality reminiscent of the French Bastille. The *sharaga,* as with the Bastille, offered many luxuries unheard of in the camps. But all prisoners lived and worked without any clear idea of when, if ever, they would be released. This uncertainty gave the *sharaga* an oppressive atmosphere. There was no rule of law defining the length of imprisonment or guaranteeing rudimentary human rights. Drunkenness or unauthorized contact with the outside brought immediate banishment to the camps. Kerber recounts the chilling reminders of the purge mechanism that controlled their lives: once the *sharaga* library received a collection of books with the nameplate "From the Library of Nikolai Bukharin." Bukharin was well known to the prisoners as Lenin's one-time collaborator who had been executed after a Moscow show trial.

Kerber resented the grotesque and brutal side of communist rule, although he had endorsed the larger ideological goals of the Revolution and remained throughout his entire life a loyal citizen of the Soviet Union. Still he possessed a profound appreciation for traditional Russian culture—often condemned in official Soviet ideology. He found the destruction of churches and historic monuments, in particular, to be painful, a reflection of the low level of cultural attainment evident in the party bureaucracy. The NKVD, the secret police, held sway over his life and countless others. Kerber lived restively under this regime of subordination. In retrospect, it is difficult to understand the pervasive influence of men such as Beria over Soviet cultural life. Nikolai Yezhov, Beria's immediate successor, was eulogized in *Pravda* in 1937 as a heroic figure who had destroyed "poisonous snakes" and "smoked vipers from their lairs and dens." This "devoted friend of the mighty Stalin" would be the last victim of his own tenure as NKVD chief, to be replaced by Beria. His campaign against "Trotskyist bands," "Bukharinites," and "spies," had swept up both Tupolev and Kerber. Such men as Yezhov and Beria were to be feared, but never respected.[17]

Work in the *sharaga* focused on three major designs: Project 100, a high-altitude fighter; Project 102, a long distance bomber; and Project 103, a dive bomber. Kerber records that many important specialists were mobilized for this *sharaga,* a group that included talented aircraft designers such as V. M. Petlyakov and Vladimir Myasishchev. Even Sergei Korolyov, the future Soviet space leader, joined the *sharaga* after being imprisoned in Kolyma, one of Siberia's most harsh prison complexes. The specialists at KOSOS helped to develop two major Soviet aircraft of World War II, the Pe-2 and the Tu-2. How Tupolev and his specialists were able to circumvent the interference of Soviet authorities, especially Beria, is revealing, demonstrating the commitment of the prisoners to quality work. Kerber identifies Tupolev as a primal force, a leader who stressed duty, and a strong-willed figure who sustained the morale of the *sharaga* specialists.

The legacy of the *sharaga* would be the Pe-2 and the Tu-2 aircraft. Both would see considerable action in World War II, suggesting that the *sharaga* made a substantial contribution to the war effort. Kerber gives little coverage to these aircraft outside of the design work associated with their production. Tupolev's design, the Tu-2 attack bomber, entered operational service in 1943, earning the Stalin Prize that same year. The Tu-2, a twin-engined bomber with its crew of three (sometimes four), participated in many of the air offensives against the Germans in 1944–45. The Tu-2 cruised at 340 MPH, being powered by two 1850-hp engines. Well armed and fast, the Tu-2 fit well into the tactical combat role the Soviet Air Force preferred. By Soviet standards the Tu-2 reached the front only in small numbers: a mere 5,256 were manufactured, compared to over 36,000 of the Il-2 Shturmovik ground-attack aircraft.[18]

Stalin freed Tupolev in 1941 and then pressed his design bureau into a frenetic schedule of wartime projects. Perhaps Tupolev's most extraordinary challenge came in 1945, when he was ordered to design and build the Tu-4, a replica of the U.S. Air Force's B-29 Superfortress— the most advanced long-range bomber in the world. To copy the B-29, the Soviets used three interned B-29s that had made emergency landings in the Soviet Far East after bombing missions over Japan.[19] *Stalin's*

Aviation Gulag provides an insider's perspective on this unique chapter in the history of Soviet aviation.

The Boeing B-29 offered many challenges to the Tupolev team. The American designers equipped the high-altitude, cigar-shaped bomber with a pressurized cabin, four powerful R-3350 engines (2,220 hp), the latest radar and navigation equipment, remote-controlled armament, and sophisticated electrical systems. Tupolev and his engineers faced enormous problems copying such a complex aircraft, a task complicated by the fact that the Soviets had to adapt American measurements to the metric system. Kerber gives considerable space to this unusual story of "technology transfer."[20]

The Tu-4 project signaled the Soviet commitment to building a strategic bomber. In the context of the postwar years, with the advent of nuclear weapons and the Cold War, the Soviet Union required a genuine long-range bomber. The air triumph over the Luftwaffe during World War II had ironically left the Soviet Air Force ill-equipped for the Cold War. Simply stated, the Soviets lacked a strategic bomber. Soviet wartime priorities dictated the mass production of fighters and ground-attack aircraft, allowing for a meager force of less than 80 four-engined Pe-8 bombers. Overtures to the United States to ship B-17 bombers as part of Lend-Lease had been politely refused. The engineering required to complete this high-priority project was immense. Kerber possessed a unique vantage point from which to comment on Tupolev's extraordinary feat of manufacturing the Soviet copy of the B-29.

Stalin's Aviation Gulag also provides an overview of Tupolev's career in the postwar years. The Tu-4 program set the stage for a sequence of Tupolev-designed strategic bombers. These designs dovetailed into the jet age, as the Soviet Union endeavored to maintain parity with the United States. Before Tupolev's death in 1972 his design bureau produced the large civilian passenger aircraft that allowed Aeroflot to become the largest airline in the world. As Kerber points out, the postwar years would be an extraordinarily creative period for Tupolev. The memory of the *sharaga* waned as Tupolev reemerged as the Soviet Union's preeminent aircraft designer.

For L. L. Kerber, the telling of the Tupolev story was an important personal mission. His candid account of his own imprisonment as part of the Tupolev *sharaga* conveys little bitterness. Kerber viewed his life with detachment. An underlying optimism and patriotism animated his outlook. Yet he lamented the absurdities that punctuated Soviet rule, the lack of sophistication among communist functionaries, and the loss of historic monuments to socialist wrecking balls.

L. L. Kerber died on 8 October 1993, aged 90 years. *Stalin's Aviation Gulag* will endure as his personal legacy. His family and friends remember him as a decent man, a survivor of a turbulent period of history, and a wonderful storyteller. By reading his book we capture a sense of the man.

Von Hardesty
Smithsonian Institution
Washington, D.C.

FORMATION OF THE MAN | 1

On a dreary February day in 1972, as a light snow was falling and fog concealed the treetops of the old forest in the village of Nikolina Gora outside Moscow, seated on the divan looking chilly and morose, I sat with elderly Andrei Nikolayevich Tupolev as he reminisced: "Our roots are in Siberia. Father recounted that my great-grandfather (for whom I am named, by the way) was elected ataman[1] of some portion of the Siberian Cossack forces quartered in Krasnoyarsk. In the main, the commanders serving in these forces were poorly educated officers who aspired to romance and adventures, and in their own way continued the work of the Russian pioneer explorers in Siberia. These were the people who elected our grandfather their Ataman. He must have in some way distinguished himself and thereby merited their trust. It may have been his desire for knowledge. He not only taught his six children to read and write, but sent them to continue their education in Tomsk. Quite some distance from Krasnoyarsk! But in those years Tomsk was the administrative and cultural capital of Siberia."

At this moment, Lyalya, one of Tupolev's daughters, came in. For some time the old man had been ill, and she had assumed the role of in-home doctor.

"Is that you again, Lyalya, with your medicines? It is unbearable." He began to get angry. "How much of this filth can I swallow, and why is it always so bitter?"

Once the medicine was administered, he returned to his family's story. "After they finished their education, Great-grandfather's children scattered throughout Russia: one was a doctor, one built railroads, one taught. My grandfather, Ivan Andreyevich, settled in Tomsk and taught at the *gymnasium* [secondary school] there. He dedicated his energies to giving his ten children a complete education. As a consequence, the Tupolev family became members of the Russian educated class.

"Now about my father, Nikolai Ivanovich Tupolev [1842–1911], who was graduated from the *gymnasium* at Tobolsk in 1860. For a short interlude he taught at the Berezovskiy Teachers School. He then entered Moscow University. A radical, he participated in the student agitation against the university authorities, an activity that denied him his diploma. In 1870 Father took the position of notary at the Tver Region Court in the city of Korchev.

"Mama, Anna Vasilyevna [née Lisitsina, 1850–1928], was the daughter of a court investigator, and was born in Tiflis. There were many children in the Lisitsin family and, like my father's family, they valued education. All the children pursued formal education. Anna Vasilyevna completed the Tver women's *gymnasium*. She had a facility for languages, and spoke German and French freely." Tupolev unexpectedly burst out laughing. "When I went to France for the first time, I was embarrassed for so long, afraid to speak the language, but then I went ahead and spoke, and . . . it worked! How can I not remember her with a good word?

"My parents settled on the land. The political conditions and the material side of life played a part in their decision, but there was also the desire to live with the people, through their own labor—the movement that gripped the Russian educated class following the freeing of the peasants from serfdom in 1861. With their scant savings and my mother's dowry, they bought a small parcel of land in Pustomazovo, located in Korchevskiy District, Tver Province, on the Luninka River, and moved into the wooden, single-story 'manor' house, as the local peasants called it.

"On 10 November 1888, in that house I entered the world. I was their sixth and next-to-last child.

"Sergei, the oldest, later fought in the Balkans to free the Slavs from the Turkish yoke. The other five—Tatyana, Maria, Vera, Nikolai and Natalya—thanks to our parents' influence went to work among the people and taught in Tver, Torzhok, and Chernigov. I was the only one to choose a technical field.

"Our family did not live richly; to the contrary we lived rather poorly. This is understandable. Our small plot of land had four dessiatinas [approximately 11 acres] of arable land, and one more that was sparsely tree-covered. We kept a horse, two cows, ten sheep, and two pigs. Well, of course, we also had our chickens and ducks. The 'estate' did not bring in any income. We ran a natural economy—the large family ate the entire crop.

"As a youth I looked for mushrooms and berries, and went to the Luninka to fish with a line, and sometimes also a drag-net. In the summer I played knucklebones or nail toss with the boys, and played with toys when it rained or in the winter. The toys were not store-bought, but homemade, fashioned out of chips of wood, twigs and pieces of wire. I was fascinated with them because they could be sawed, shaved, punctured, and bent, things I loved to do.

"For their education in schools and teachers' seminaries, the four oldest children moved to Tver, where they lived with Mother's brother, Uncle Vasiliy. When it was my turn to go off to study I took the examination at the Tver *gymnasium* and failed. That summer I had to study; in the fall I passed the examination and went to the *gymnasium*. My sister Natalya still has the application in which our mother wrote: 'N. I. Tupolev, having seven children, is in his declining years and in ill health, lives in extreme poverty, and through his labor is not able to maintain his family, and even less so to pay for their higher education.' Payment for school was sent in every half-year in advance. Those who did not deposit their payments at the designated time were considered to have left the institution.

"After she arrived in Tver to get me into the *gymnasium* and Natalya into preparatory school, Mother was asked to sign an obligation, which in part stated, 'See that your son does not handle firearms.' [Truly the leadership's zeal knows no bounds: Andrei had just turned seven!]

"By determination of the pedagogical council, I was enrolled in the first class at Tver Province *gymnasium*. When Mamma brought this joyous news home, my father and mother decided that they could not burden Uncle Vasiliy with two more 'lodgers,' and that they should look for an apartment or little house in Tver where Mother with all six children would live during the school year. Father had to remain in Pustomazovo because it was where he worked. Together with an acquaintance who also had a large family, he purchased a tiny, single-story wooden house in quiet Solodovaya Street. It stands today, a little ramshackle house, with its small windows and low ceilings.

"We occupied one-half of that house. Of the five rooms, two overlooked the street, two the alley, and the last the courtyard terrace. The kitchen was in the basement. In back of our landlord Vrasskiy's houses was a huge orchard with fruit trees and raspberry bushes, which extended down to the Lazur Stream. Vrasskiy was uncommonly stingy, did not allow anyone to pick apples, and dragged persons caught redhanded by the ear—including me.

"In Tver our way of life changed little from what it had been at home. In the morning we went after water to the well, there was no electricity, we cut firewood, and the toilet was made of wood. The household duties were distributed among the six children, and we fulfilled our duties faithfully, trying in every way possible to free Mother from all of the heavy work.

"The new expenses put a strain on our parents. In October my mother had to submit a petition to the *gymnasium:* 'I most humbly request deferment of the payment for the study of my son, first-year student Andrei Tupolev, until the first days of December of this year.' The school director answered the petition: 'Refused.' It was necessary to go into debt.

"We had few acquaintances. Uncle Vasiliy, a military doctor at the local garrison, was the only person to visit regularly. On Sundays he brought us delicacies, until my mother told him, 'Enough of this, you are ruining the children's teeth. If you must bring something, then bring something more substantial.' From then on Uncle Vasiliy sent us each two loaves of the soldiers' still warm, uncommonly tasty rye bread in the mornings. He came very early and started the samovar. After the samovar had boiled, the tea had been made, and the milk and butter brought from the cellar, around 7 o'clock we were awakened to a delightful breakfast. How tasty those breakfasts were!

"For the holidays and in the summer we traveled on carts to Pustomazovo to visit Father. We could have taken a steamboat on the Volga, but that was more costly. At the start of the school year we moved back to Tver, traveling on several carts, burdened down with provisions from the country. In the city products were more expensive, and for this reason we tried to be self-sufficient. There was a cellar at the house, and we filled it up with goods from the country.

"Finally Father reached his goal and was able to obtain an external university degree. Work in the court, however, was closed to him, and for the rest of his life he worked as a village notary. Between his work as a notary and that as a farmer, Father was always on trips or out in the field, and had no time to spend with us. I was brought up by my mother, my older sisters, and a nanny. We lived easily, simply, and well with our parents. We had a firmly established relationship of mutual love and friendship. Father and Mother tried in every way possible to inculcate in us love for our people."

Andrei's Early Education

Andrei's grades that first year of the *gymnasium* were uneven. He loved physics, mathematics, geography; he was indifferent toward other subjects, and could not stand singing, dancing and religious studies. "They got all over me for penmanship. My problem was that I had terrible handwriting, which I simply could not correct. No doubt, this is why for my whole life I could not stand to write letters, a fact that distressed both my mother and my wife.

"The *gymnasium* was located in a grand, three-story building on Millionnaya Street. The governor-general lived across the street in the Putevoi Palace, and next door was the municipal garden, where the young people walked in the evenings. At the *gymnasium*, I recall the luxurious staircase and wide corridors where we rushed about during breaks, and the very high ceilings in the classrooms, in which there was always fresh air and it was easy to breathe.[2]

"In winter a skating rink was cleared on the Lazur River, at which the brass band of the Tver cavalry school played. We had only to hear the first bars of 'Danube Waves,' or 'In the Hills of Manchuria,' and we strapped our skates onto our felt boots and ran headlong onto the ice. But we were allowed to skate only after we had studied our lessons. The music was a great incentive. I retained my fondness for skating in my college years, and, imagine, I met my future wife at the skating rink."

Andrei was self-willed. He greatly valued his independence and personal freedom. Frequently and unexpectedly he would retreat into his shell, and, absorbed in his own thoughts, would become unsociable. He loved to play practical tricks, but at the same time found it difficult to make friends with his contemporaries, and opened up completely to almost no one. He read a great deal without any particular selectivity.

His circle of fellow-students, especially in grades 6 and 7, often discussed which fields of study to enter. As in his own family, the majority were inclined toward the humanities. The recent populist movement prompted many to enter medicine, teaching, veterinarian service, and the law, all practical careers where they might help the people. It is difficult not to go along with one's friends, but to take one's own path. This is even more true when the others have loftier ideals, which a young man, frankly speaking, has no way of countering. Nonetheless, Andrei's attraction to technology outweighed everything.

"My love for physics," Tupolev recalled, "remained with me my whole life. Nikolai Fedorovich Platonov, my mentor, who taught with unusual clarity and beauty, inculcated this love in me. He did not limit himself to the *gymnasium* course, but organized an astronomy circle, took us on excursions to the railroad, textile factories, many of which were in the city, to Volga steamboats, and organized intricate and captivating experiments in mechanics, optics and electricity. I remember how, with our own hands from on-hand materials, we mastered lift cranes, a model steam engine, dynamo, and even tried to make a powered carriage. We all loved him greatly.

"Platonov was one of the teachers who, little by little, also opened up students' eyes to the negative sides of life. Once he organized for us an excursion on the Volga to Rybinsk. Here I was confronted with reality for the first time, and my eyes were literally opened. Reality differed rather greatly from what we observed in rural life, which at the time was still patriarchal. Everything that our parents had told us at home about the lack of rights for the working people I saw with my own eyes. I saw the artels of barge haulers, raftsmen and loaders, sweat pouring down their faces, loading, turning over logs, and towing. They were extremely poor people, dressed in rags, but were physically very strong. I remember how we leaned over the railing and watched them when they sprawled out to eat. Sitting on logs, quietly, not hurrying, they ate bread with Caspian roaches and watermelons. And a little higher, above us on the first-class deck, there flowed an entirely different life, a life without care. Not thinking about money, the merchant class passed the time gaily in the buffet, enjoying caviar, grouse and champagne.

"Although I was still young, what I saw forced me to think. Platonov never forced any conclusions upon us, but the questions begged for answers. Why is everything so unfair? Gradually the negative aspects of life found their way into my young consciousness, the difference between wealth and poverty, between gluttony and hunger. And I formed my attitude toward wealth for once and for all. Platonov helped us by this trip, so that when, much later, after the revolution, none of us questioned which side to join."

An Introduction to Aviation

While Tupolev was formulating his beliefs, as well as developing the foundation for his future calling, steps were being taken to make human flight a reality. In 1885, Russian naval officer A. F. Mozhayskiy tested a heavier-than-air flying apparatus. It is true that it did not fly but, characteristically, Mozhayskiy anticipated in his machine all of what would become the main parts of an aircraft.

In 1895, when Andrei was 7 years old, German engineer Otto Lilienthal began flying in a glider that he had built himself. N. Ye. Zhukovskiy observed Lilienthal's flights at the Wasserkuppe in Germany. He shared with the inventor his thoughts about ways to conquer the air. Touched by the scientist's attention, the inventor agreed to sell him one of his gliders. It had little in common with modern gliders. It had two wings, similar to birds' wings, made of bamboo and wooden rods covered with material. The glider pilot would shove his arms into the straps, run down the hill with this apparatus on his back, and, once he left the ground, fly up to 100 meters. The Lilienthal glider is now preserved in the N. Ye. Zhukovskiy Scientific Memorial Museum in Moscow.

In 1903, when Andrei was 15, the Wright brothers, Orville and Wilbur, built an airplane with a gas engine. At Kitty Hawk, they powered their flying machine into the air, flying 35 meters.

The prophetic words of N. Ye. Zhukovskiy came true: "Man will fly, relying not on the strength of his muscles, but on the strength of his intellect." Reports about these exciting events reached even far-off Tver and made for much discussion among the *gymnasium* students. Although the young men could not resist the excitement of these developments, general opinion held that the sciences should be considered the most worthy pursuit. Against the advice of his father, his family, and his school friends, Tupolev decided to dedicate himself to the natural sciences, and to enter the Imperial Moscow Technical Institute.

His mother, Anna Vasilyevna, went with him to Moscow because it was the boy's first time in a large city, and she wanted to see how he would live and eat. They stayed with acquaintances while in the city. In the fall of 1908 Andrei passed the competitive examinations for the mechanical factory of the Imperial Moscow Higher Technical School, this "alma mater" of Russian technical thought, which gave Russia, and then the Soviet Union, so many glorious names.

Since he came from the provinces and had no family in the city, he was given a place in a dormitory, at the corner of Brigadirskiy Lane and Koroviy Brod Street, next to the school. There was a cookhouse on the first floor where he could eat. Seemingly everything was well. But Andrei did not like the fact that there were five other students in the room. The still-provincial lone wolf avoided strangers. Anna Vasilyevna came by to see how her Andryusha was living. They sat down before parting, she cried, made the sign of the cross over him, and went back to the family.

And so Tupolev was enrolled in the best technical educational institution in Russia, but he was still not sure about what direction he should take. Tupolev's inclination toward aviation took shape somewhat later, and not from long reflections by the young man, but from two external promptings. But let Tupolev himself tell about this: "N. Ye. Zhukovskiy, the renowned Russian aerodynamist, first became aware of me in a rather curious way. I was studying in the first class. I did not have any particularly deep interest in flying, although its novelty attracted me. A flying exhibition was organized once at the university. One day I went there. I saw some glider pilot pulling a cable. I began to help, and found myself next to Delone, who was a student of N. Ye. Zhukovskiy, and later became the famous mathematician. He immediately introduced me to Zhukovskiy. That is how, by grabbing the cable, I became involved in this business."

Tupolev began to attend Zhukovskiy's lectures. He fell under the spell of the professor's talent, erudition, and personal charm. In order to understand why so many young men fell under that same spell, it is only necessary to listen to one or two of his lectures. No one else could so beautifully and resourcefully set forth the physical essence of mechanical processes. It seemed that abstract, entirely theoretical phenomena took on visible, concrete images. Tupolev recalled, "Nikolai Zhukovskiy was the first to provide a scientific explanation for the lifting force of a wing, and he deduced the famous theorem that defined the amount of this force. By combining the mathematical development of precise theory with experimental observations, he demonstrated the fruitfulness of a new methodology of scientific inquiry. He was not only a great scientist, but an engineer 'of the first rank.' Zhukovskiy understood that scientists in the twentieth century would be required to combine theory with practice."

Zhukovskiy inspired Tupolev to pursue aeronautics. Observing Tupolev's first, but not timid, steps toward what he very quickly understood was his calling, Zhukovskiy soon became convinced of his new student's exceptional capabilities. While still a second-year student, Tupolev brilliantly completed the professor's assignment and independently built a wind tunnel for the school. Gradually this student, not at all accidentally, became Zhukovskiy's closest student and assistant. A scientist as great as Zhukovskiy would not succumb simply to Tupolev's charm; the master understood that out of all his apprentices this one was worthy to become his successor.

An aeronautics exhibition was held in Moscow from 18 to 25 April 1910. In issue 9 of the *Biblioteka Vozdukhoplavaniya* [Aeronautics Library] magazine it was noted: "If the visitors to the exhibition do not see the model airplane, 'Antoinette,' built by the student Tupolev, with all of its tiny details, they will miss a lot." The student was efficient and stubborn, and, if he took up something, did it honestly.

Tupolev's school years coincided with the revolutionary enthusiasm following the revolution of 1905. There were student uprisings in 1910–1911, an agitation that attracted students from Moscow State University and the Imperial Moscow Technical School. Tupolev decided to use his apartment address "for dealings among the city coalition committees of the higher educational institutions in Petersburg and Moscow, for the purpose of uniting these institutions in holding strikes." On 8 March 1911, the Moscow city governor, Major General Ardianov, signed decree no. 536 on arresting a student, and on 14 March Tupolev was sent to the Arbat police station.

Regarding Tupolev's absence at lectures after this incident, one can find a curious document in the archives of the Zhukovskiy Museum: "Case No. 8 of the Office of the Imperial Moscow Technical School for 1911. Dismissal of students for various reasons." Thirty-eight people were dismissed. Of these, thirty-four submitted requests for dismissal themselves, three died, and one was expelled for one year—student A. Tupolev. In vain Zhukovskiy, citing the student's uncommon abilities, attempted to defend him. However, political loyalty took precedence over any reasonable arguments in defense of Tupolev's character and he remained in custody.

On 19 April 1911 Tupolev's father, Nikolai Ivanovich, died. All of the relatives came for the funeral. Tupolev was released on 21 April to take part in the funeral. His father was buried in the village churchyard, some kind words were said, and all went their own way. Sergei, Tupolev's oldest brother, had been studying for the priesthood. But under the influence of the revolutionary ferment and the "new" ideas of the time, he lost his religious faith and left the clergy. At that time war was going on in the Balkans to free the Slavs from the Turkish yoke. Sergei set off for the war as a volunteer and was killed in Bulgaria.

The Tupolev home soon came to life again. Tupolev returned to Pustomazovo and, for almost a year and a half, he took care of the farm alone. He plowed, gathered the harvest, did carpentry and metal work, worked at the forge, repaired the house, built stoves, recovered the roof, built a well, drained the cellar. In winter, when it grew dark early, he spent much time in self-education. He diligently studied textbooks and read. When the books he needed were not at the farm he walked to Korchevo or Kimry, each a distance of some 20 kilometers, and bought what he needed. It seems that in these years his moral and aesthetic views took shape, greatly influenced by the nineteenth-century radical writers Belinskiy, Chernyshevskiy, Dobrolyubov, and Pisarev.[3]

When Tupolev was permitted to return to Moscow in 1912, he resumed his formal studies. He not only listened to lectures, but willingly took up all manner of practical work. While World War I was in its second year, he designed a wind tunnel for a Petrograd laboratory. In that same year, despite the fact that he had not yet defended his diploma work, he was invited to be chief of the hydroplane department at the Duks Factory. In 1916 N. Ye. Zhukovskiy lured him to work in the first aviation design and testing bureau in Russia. Their first published work, under the editorship of Zhukovskiy, Lukyanov, and Tupolev, united a number of articles under a common theme, "Aerodynamic Calculation of the Airplane." At the end of the collection it states that sections I ("Determination of Propeller Setting Angle"), VI ("Constructing Power Balance Depending on Flight Regime"), VII ("Determination of Flight Altitude"), and VIII ("Constructing a Curve of Greatest Vertical Velocity for a Given Power") were penned by A. N. Tupolev.

Around the same time, and also at Zhukovskiy's initiative, a commission was created to develop something similar to modern strength standards for airplanes. Tupolev joined this commission, along with

Professor S. P. Timoshenko. Thus, Zhukovskiy gradually inculcated in young Tupolev an interest in science and theoretical studies.

Of course, the student understood quite well how lucky he was. Daily communication with the professor, his mentor, and work under his guidance in the laboratories enriched the young engineering student. Tupolev remembered well many of Zhukovskiy's statements his whole life. In 1919, Zhukovskiy, at a session of the commission for the promotion of aviation of the Air Forces Directorate, said, "For military needs the existence of heavy aviation seems necessary, since the tasks of bombing cannot be accomplished successfully by light aviation. In the field of peaceful applications, heavy aviation will play a very important role, all the more so in that in many cases it will be preferable to light aviation." The young Tupolev shared these two fundamental postulates and retained them throughout his long life.

During World War I, military pilots trained at the Imperial Moscow Technical School. Their course on flight dynamics included Zhukovskiy and Tupolev as lecturers. Zhukovskiy was advanced in age. As he took tests from the young officers, he usually sat in a divan with his head drooped wearily. After one of the examinations he met Tupolev, and, slightly confused, said, "You know, this is surprising, but I assure you that today the very same pair of shoes gave me the theory of bombing three different times!"

"Nadashkevich!" Tupolev burst out laughing. He knew it had to be A. V. Nadashkevich, who in the near future would become Tupolev's closest assistant on aircraft armaments. Wishing to help his friends, he had taken the bombing examination for them. "Do not be angry at them. They all want to get to the front as soon as possible. They are all drawn by the romance of aerial clashes!"

When the revolution occurred in October 1917, Zhukovskiy was among those older scientists who immediately pledged loyalty to the new Soviet government, and gave to it all of their talents and all of their knowledge. In 1947, at a session dedicated to the 100th anniversary of Zhukovskiy's date of birth, Tupolev noted, "Zhukovskiy believed in the new forces of our country and wanted to go with these new forces. This is to his great credit. He was a great patriot and loved the homeland deeply. He grieved about her misfortunes and tribulations and always wanted to be useful to her. When Soviet power came and he had the opportunity to work together with the collective that formed around him, to work on a broader scale and be useful to our country, not one man, no one out of our collective, vacillated for a single minute."

And this was so. The collective included A. A. Arkhangel'skiy, V. P. Vetchinkin, G. M. Musinyants, G. Sabinin, B. S. Stechkin, Tupolev, Ushakov, B. Yuryev along with the professor. Tupolev understood for the first time that he belonged to aviation completely, with every fiber in his body, and that he could not live without it.

His first, perhaps naive, step was construction of his own, independently calculated, heavier-than-air flying apparatus—a biplane glider made of wood and cloth. In it Tupolev made his first flight across the Yauza. "Our glider was controlled by movements of the pilot, who hung between the two wings. And he gathered speed through the physical force of another man. Yuryev 'harnessed' himself into a strap and ran. I felt the ground departing from beneath my feet, and . . . flew. Someone was able to take a photograph at this moment. I flew to the other bank of the Yauza and fell to the ground. . . . This flight confirmed our calculations, only to a certain extent it is true, for the glider crumpled up beyond repair upon landing. By the way, the pilot, as you see, survived."

On 11 June 1918, Tupolev defended his diploma thesis. By this time he was already a full-fledged engineer-inventor. His work at the Duks Factory, and at the design bureau of the Air Forces (Chief) Directorate, when he had already been given true responsibility, had some influence on him. He became capable of taking responsibility for his own decisions as an engineer. Zhukovskiy praised Tupolev's diploma work, "Hydroplane Design Calculation." Speaking at the Second All-Russian Aviation Congress, he recalled, "One of these diploma projects—a hydroplane—presented by our engineer Tupolev, is a prominent study on how a hydroplane takes off from the water and how it lands on the water, and thanks to the studies of the young scientist this matter was fully clarified. Had these studies been published, they would have brought fame to Russian aviation science." In the conclusion to this project Zhukovskiy wrote, "The hydroplane design presented by student Tupolev is an excellent indication of the maturity of his engineering thought."

Beginnings of the TsAGI

When the new school year began, the Moscow Higher Technical School Academic Council decided to leave Tupolev at the school to prepare to become a professor in Zhukovskiy's department. Tupolev was flattered, but he also thirsted for practical engineering activity. But where could he begin? After four years of an exhausting war and the associated ruin, the aviation industry was dying. Factories were halting production one after another, in part because of the disruption in the supply of metal, fuel, and electric power. Many workers had abandoned the factories to participate in the Civil War. Industry leaders were emigrating after rejecting the ideas of the October Revolution. Igor Sikorsky, the creator of outstanding four-engine Ilya Muromets airplanes, and engineer V. Lebedev left the country. Factory owners Meller, Shchetinin, Anatra, and a number of others also left.

World War I had shown aviation to be a formidable and promising weapon. Like any great power, the young Soviet state could not get by without a strong air force. For the systematic restoration and development of Soviet aviation, however, a solid scientific center was needed. United by their interest in creating such a center, world-renowned scientist Zhukovskiy and young engineers A. N. Tupolev and I. A. Rubinskiy spent their evenings, under the dim light of a kerosene lamp in the scientist's unheated apartment, formulating their views. Zhukovskiy's other students, members of the aeronautics circle, also contributed. They proposed the creation of a single center in the fields of aerodynamics and hydrodynamics for all of the country's departments. In this center would be conducted scientific studies, design, construction and testing of airplanes, speedboats, motors, and windmills—in short, everything capable of flying or moving at high speeds.

On 23 November 1918, Tupolev sent the plan for the future Central Aero-Hydrodynamics Institute (TsAGI) to the Scientific and Technical

Department of the Supreme Council of the National Economy (SCNE), which was the highest body created to guide the developmental prospects of Soviet science and industry. It was a rather thick plan. Everything was included: the structure of the institute, number of laboratories and activity of each of them, and a listing of the initial construction projects. In short, everything was done with a broad sweep and perspective.

By then Zhukovskiy was in his eighth decade. He had aches and pains, and could not walk through the institutions to delve into the progress of a project. The majority of the responsibilities lay on Tupolev's shoulders. Officials in the RSFSR [Russian Federation of the Soviet Union] administrative agencies in no way strove for quick solutions to problems. The young and energetic Tupolev had no patience for these bureaucratic delays. Relating the results to Zhukovskiy, he would recall his clashes with "officials, scribblers and iron-headed bureaucrats." The elderly professor was troubled. "Alas, please, somehow be a bit milder and more delicate, heaven forbid that you offend someone, and they bury our whole idea."

One of the people Tupolev encountered was the chief of the Scientific and Technical Department of the Supreme Council of the National Economy, Nikolai Petrovich Gorbunov. Gorbunov had previously worked as secretary of the Council of People's Commissars, and was a close associate of V. I. Lenin. He immediately understood the value of the report. At the first opportunity, he told Lenin of the scientists' proposal. Lenin assessed the benefits of uniting the disparate scientific and engineering forces into a single organism, and requested that the decision be drawn up as soon as possible.

On 1 December 1918 the Scientific and Technical Department (SCNE) decreed that the new scientific center was to be developed using the laboratories of the Moscow Higher Technical School and the personnel of the design bureau of the Air Forces (Chief) Directorate.

During his lifetime Tupolev frequently recalled Gorbunov: "He was an amazing man. Inclined toward science, he possessed the gift of uncommon foresight. In the period of economic ruin following the many years of war, when even the trains ran sporadically, he well understood not the romance, but the real future of aviation. When Gorbunov said that the decision to create the TsAGI had been approved, Zhukov-skiy and I set off from Myasnitskaya, where the Supreme Council of the National Economy was located in the building of a former religious consistory, to his home on Mylnikov Lane. I no longer remember exactly where, but we found the miracle of a surviving café. It had nothing to offer except for sour clotted milk. We raised glasses of the clotted milk in a toast."

There are few left who know what took place on 12 January 1917, during the solemn commencement on the occasion of the 162nd anniversary of Moscow University. Rector M. K. Lyubavskiy, as usual, named the assembled new honorary members of Moscow University, who had earned this title in the past year. Among others was pronounced the name of "Father of Russian Aviation," honored university professor N. Ye. Zhukovskiy. Lyubavskiy had not managed to finish the phrase when the entire auditorium began to applaud Zhukovskiy, who had appeared in the hall.[4] From that time on, and quite justifiably for our country's science, N. Ye. Zhukovskiy became the "Father of Russian Aviation."

On this overcast December day, as huge snowflakes settled slowly to the ground, it was very difficult to comprehend the greatness of the moment. In a country where there was nothing to eat, nothing with which to heat the houses, the creation of a truly world-class scientific aviation center began on an empty spot of ground. And this was done at a time when the White armies of Kolchak, Dutov, Denikin, Skoro-padskiy, Yudenich, Mannerheim, Miller, Chaikovskiy and Unger completely encircled Soviet Russia. The noose around the industrial heart of the European part of the country, with its center in Moscow, was drawing ever tighter.[5]

TsAGI Becomes a Reality

On 3 December, those who held the creation of the institute closest to their hearts met in Zhukovskiy's apartment on Mylnikov Lane. Engineers N. V. Krasovskiy and I. A. Rubinskiy, long acquainted with the professor, and Andrei Tupolev, the closest of his students, came.

"The small iron stove was started in the low rooms of the apartment. We stoked it with damp wood, our eyes stung, but it was possible to work tolerably well. Zhukovskiy was ill, and he lay on the divan, covered up in a rug. He apologized, but said that he could not get up. In order not to burden the professor we decided to limit ourselves to one question, which could not be put off: How would the TsAGI be managed? We agreed that it should be done collegially. We had no doubt that the chairman of the collegium would be Zhukovskiy. After thinking a little, he said softly, 'I agree, but you must choose a young energetic comrade to assist me.'"

They parted after dark, wet snow was falling, the streets were not lighted, and they splashed through the puddles. "Yes, it was cold in the professor's apartment, our boots were wet through, we did not exactly know what we would eat tomorrow, but just the same the TsAGI was born!" Tupolev smiled when he recalled that night.

In the morning, the collegium chairman's comrade was chosen at a general meeting of the scientific associates. It was generally assumed that it would be Tupolev. He was closest to Zhukovskiy, and had put much effort and work into preparing the proposal to create the institute. But his strong will and authoritative nature frightened people. Of course, forcefulness, firmness of purpose, intellectual curiosity, and complete lack of intimidation toward the highest authorities were extremely important qualities for the collegium chairman's comrade. Nevertheless, his demanding and uncompromising nature, and his not-always-restrained irritability, and, at times (as it seemed to them), even ruthlessness, made them wary. It was not without reason that behind his back they called him "the Fist" or "the Petty Proprietor."

When the votes were counted, the majority supported V. A. Arkhangel'skiy, a mathematician and pianist who was a good administrator, but who did not possess the necessary authoritative manner. Soon, however, they saw that he was not suitable, and by general decision of the collective Tupolev replaced his predecessor.

The school was located in two buildings: a free-standing house, previously a dormitory for student pilots attending classes at Moscow Higher Technical School, and a neighboring building, the former Rayek Inn, which was well suited for shops. The Technical School courses had already been moved to Khodynskiy Airfield and the owner of the house had gone into hiding somewhere, so both the house and the inn were vacant. To get the unauthorized establishment registered, it was necessary to present a plan to the Moscow Soviet. When the contours of the two small buildings were drawn on tracing paper, Tupolev asked, "You know, why don't we draw the entire block, and not limit ourselves to these two mangy little houses? It is clear what we will begin to build." Today the entire block is full of the large and beautiful buildings of the institute.

In the beginning, several departments were organized within the TsAGI: general theoretical, aviation, motor, wind engines, and materials study. No complications arose with selecting the scientific staff. First, many of the employees of the design and testing bureau of Moscow Higher Technical School shifted to the institute, as did almost all the members of the aeronautics circle, and a number of professors and docents from the school. Many skilled workers were without work in those years, making it possible to select outstanding mechanics and carpenters.

Since the TsAGI had not yet begun to receive assignments, the staff worked to obtain materials and supplies. Mandates in hand, they combed through Moscow establishments and warehouses for drawing equipment, instruments, machine tools, and tools. Even today TsAGI veterans do not mind recounting such legends as how Tupolev, the Arkhangel'skiy brothers, and Stechkin "abducted" an uncommon machine tool from the Kursk Railroad. Another successful acquisition was found to be recorded the archives of the collegium: "On behalf of the service, we express gratitude to Vorogushin, Musinyants and Ushakov for work they accomplished with exceptional energy." Having spied unclaimed property at the Moscow Customs Office, they took much that was valuable and greatly needed by the TsAGI.

After a month the courtyard and inn were heaped with an assortment of property. The empty rooms of the house and halls of the inn filled with drawing tables, measuring apparatus, carpenters' benches, and even machine tools. Occasionally something truly valuable was recovered.

The staff also began to work on the buildings themselves. With their own hands they pulled electrical wire, installed glass in broken windows, changed pipes that had burst from the cold, patched the roof, and set up iron stoves for heating.

The cells of the future departments began to take shape. The general theoretical department developed more quickly than the others. The young scientists completed studies begun earlier at the Moscow Higher Technical School, and then ran new wind-tunnel tests and conducted strength calculations of a cantilever wing without braces or tie rods.

In those days colleagues at the TsAGI had to address the most unexpected matters. Here is an extract from some of the protocols:

21 January 1919. Order No. 1. In view of the disarray of city transportation, the household committee is asked to allocate one room for the overnight stay of employees detained at work.

2 December 1919. Para. 5—On Developing the Theory of Wave Resistance of Ships. It is resolved to ask Professor Zhukovskiy to take up this theory.

Para. 6—On Replacement of the Position of Groundskeeper in the TsAGI buildings. It is resolved, in view of the refusal of the former groundskeeper, to name a new one.

24 February 1920. Para. 5—On Reading Scientific Reports. It is resolved to ask Professor Zhukovskiy to read the report on wind engines; Engineer Yuryev that on heavy aviation; and Engineer Vetchinkin that on airplane testing and flight dynamics.

Para. 6—On Granting a Large Room to the Theoretical Department. It is resolved to grant them the former office, and they are to move all property themselves.

Para. 7—On the Horse Called "Grom." It is resolved, in view of the total lack of fodder, that the household committee is to transfer it to the Baumanovskiy District Council, and to grease the horse buggy and put it under the shed.

Everyone in the school collective understood that creating some new, better airplane lay in the future. However, they also knew that the homeland needed not only theoretical research, but also practical help.

In those days, everything began with a commission. The Defense Council assigned the TsAGI the task of replacing the machine-gun carts that became stuck in deep snow in the winter with propeller sleds, powered by a motor with propeller, and capable of sliding across ice-covered snow on skis. This commission was called KOMPAS (Commission for the Construction of Propeller Sleighs), and was headed by Moscow Higher Technical School Professor N. N. Brilling, with Tupolev as his deputy.

Domestic industry was not producing suitable engines for such a sled. B. S. Stechkin was designated to search the depots for imported engines. He found foreign-made rotary aviation engines, for which the commission designed propellers using NYeZh-type blades. N. Ye. Zhukovskiy, with his student V. T. Vetchinkin developed the theory for the family of blades.

Employees of the commission, together with those of the Moscow Higher Technical School and the TsAGI, designed and built the sleds themselves. It was difficult to understand which of these people worked out of necessity, and which out of pure enthusiasm. No rigid dividing line had yet arisen between the technical school and the new institute. Even Tupolev and Arkhangel'skiy, prompted by Stechkin, did not define which group they belonged to and, therefore, worked in both places.

The intense work, evenings spent thinking out his future airplanes, sometimes staying up half the night, the poor food, and the damp, unheated room took their toll on Tupolev. He began coughing and became feverish with tuberculosis, but stubbornly refused to go to a hospital. Nevertheless, his friends insisted. Like all the hospitals in those years, the city hospital was very inadequate. The patients were underfed, medicines were not delivered. To save firewood, the wards were barely heated and not ventilated at all. Treatment consisted of the doctor tapping on Tupolev and explaining to his students in Latin the course of the illness. Tupolev understood that he would not get well there, and fled.

When Zhukovskiy learned that Tupolev had left the hospital, he was alarmed and told the other professors that the talented, young engineer must be saved. They placed Tupolev in a good pulmonary sanitarium, Vysokiye Gory [High Mountains], located on a bank of the Yauza near

Sadovoy Ring. Experienced doctors, qualified treatment, intensified feeding, and an attentive attitude helped the young patient cope with the illness. Nevertheless, Tupolev was in treatment for about a year.

In spring 1920, the chief physician at the sanitarium, M. Kondorskiy, who was sent on a trip to the Ukraine by the People's Commissar of Health, invited Tupolev to go with him, to fatten up and gain strength. But then threatening events developed. The army of General Wrangel had shifted to its final offensive. The initial successes of the White Guards forced Tupolev and Kondorskiy to hasten back to Moscow.[6]

"I do not like to remember this trip," said Tupolev, "overcrowded heated cars, lice, typhus. I was just lucky that I did not catch anything."

Nonetheless, he returned from the south refreshed, having gained weight and strength.

The Advent of Duralumin

Tupolev had no doubts that it was time to build a real airplane, but he did not know what materials or design to use. Not having any experience in building airplanes, but possessing a curious and critical mind, Tupolev thoroughly familiarized himself with the construction of numerous imported Farmans, Voisins, and Nieuports, which had been delivered to us in the war. A critical understanding of other designers' decisions is useful, but Tupolev went further and engaged in theory and conclusions. The designs he studied had much in common and were dictated by the available materials—wood with fabric covering.

The lack of an orderly calculation method also had an impact. Before 1924 the general belief was that "an aircraft is not a machine. One cannot calculate it like an engineering structure, and it will never be possible to determine precisely its air load." Therefore, aircraft builders would increase the strength of the wing and fuselage framework by any means possible. They inserted tie rods made of steel wire everywhere between their spars and struts, where they were necessary and where they were not. The innovation helped, but it complicated calculations because it engendered statically indeterminate conditions, and it became impossible to determine how certain construction changes would alter flying performance.

At Khodynskiy Airport, Musinyants, one of the members of the aeronautics circle, watched as two mechanics pulled a Morane Parasol fighter across the field. When both landing wheels plunged into a small ditch simultaneously, one of the tie rods connecting the wing to a strut on the fuselage broke.

"I watched the mechanic take a piece of copper piping, lead the ends of the broken tie rod into it, straighten them, and bandage it with tie wire," Musinyants later told his colleagues. "I asked, 'And will it hold?' They replied, 'Of course, see how many there are here!'"

Tupolev smiled sadly. The mechanic was right. The large number of tie rods clearly illustrated the airplane designer's inability to calculate the wing precisely.

Will these conditions change? Tupolev asked himself. The future is pounding on the door. Tomorrow large load-carrying aircraft will be needed. And will we just guess, and increase the number of wooden struts, tie rods and braces, doubting as before whether they will hold? No, I am convinced that it is necessary to shift to metal. Only by using it will we be able to move the process of creating an airplane from the sphere of intuition into the sphere of the exact sciences, that is, normal engineering activity. One does not have to go far for examples. Girders for bridges, including cantilevered ones, are calculated precisely, as is an aircraft wing. Just take the draw span of the new Okhtenskiy Bridge in Petrograd. Next to the abutment to which it is fastened, the span girder is about twice as thick as it is at the end. It is quite obvious that if we increase the aircraft wing section at the root in this way we will be able to apply a much stricter method of calculation. But of what should such a wing be made? Wood is not suitable, and steel is heavy. A strong, light metal is needed. Such a metal exists; it is simply necessary to set up its production in our country. Will it be difficult? Undoubtedly. But if it will be the future of aircraft construction for many years, there is reason to overcome the difficulties.

What was this strong, light metal Tupolev was considering, where was it created, and how could it be used? Before World War I, the German designer Count Zeppelin developed a project for a huge, rigid dirigible. Its framework was made out of light, skeletal trusses, within which were fitted soft, hydrogen-filled silk balloons. It was not possible to make the dirigible out of wood; it was too heavy and lacked strength. In terms of weight, aluminum was best, but it was too soft. A search began for an alloy that would combine the lightness of aluminum with the strength of steel. The Germans conducted this research in the city of Duren, and the new metal was called "duralumin" or "dural."

During the war the Germans used the dirigible to bomb Paris, London, and Petrograd. One of the Zeppelins was shot down by Russian artillery and fell near Riga. Its remnants were collected and brought to Petrograd (St. Petersburg). Here they discovered that "duralumin" was an alloy of aluminum with copper and manganese. Later, pieces of the duralumin frame of a shot-down Junkers aircraft were brought to Moscow. Interest at the Moscow Higher Technical School was high; Tupolev longed to use the new metal for aircraft construction, and I. I. Sidorin, already a recognized metallurgist, committed himself to organizing the production of a similar alloy in Russia.

Although the relative weight of the dural was three times greater than wood, it was much stronger than that of a wooden biplane wing of the same lifting surface. While it was evident that duralumin was highly adaptable to aircraft construction, hurdles still existed. It was susceptible to corrosion. All the parts from the dirigible and aircraft that had been shot down were covered with a gray film, a layer of oxidized metal. Tupolev felt that since oxidation was no more a drawback than rotting wood, and since the Germans were not afraid to use duralumin in the Zeppelins and the Junkers, it was the right metal to use. He just had to find the right protective coating for that metal.

Tupolev compared his own conclusions and calculations, sketches of subassemblies, and how he imagined them for the future machines with the very sparse information received from abroad. He also looked at designs in related technical fields, especially in automobile construction, and then formulated his credo: "The future of airplane construction is in light metals!" Since Russia could not count on purchasing duralumin abroad, Tupolev and Sidorin proposed initiating its production at home. The protocol of the Supreme Council session states:

Comrade Tupolev, representative of the TsAGI, believes that metal airplane construction must rely on a strong metallurgical base, such as may be found

only at the Kolchuginskiy and Krasnyy Vyborzhets factories of Gosprom-
tsvetmet [State Committee for the Nonferrous Metals Industry], which have
long experience in producing alloys out of nonferrous metals.

Test meltings began in 1922 at Kolchugin, and in August bars and rolled
products appeared. It did not go smoothly. The German metal did not
want to yield to Russian hands, but Sidorin and his assistants Butalov,
Muzalevskiy, and Babadzhanyan finally succeeded in producing Russian
duralumin-type alloy.

It would seem that metal airplane construction was the direction to
take, but the proponents of wooden aircraft did not surrender easily.
Wooden airplane construction in Russia was widespread. The Duks
aviation factory in Moscow was producing sixty aircraft a month. Several other such factories existed in Petrograd, Kiev, Kharkov, and
Odessa, all with qualified personnel—workers, designers, engineers,
and wood technology scientists. Russian woodworking machine tools
had been developed, and tool production had been set up. These factories were well established, bringing both economic benefits and a sense
of tradition to Russia.

And what did these totally unknown young engineers propose to
replace all of this? Some light metal called "duralumin," known only to
a limited group of specialists. No one knew how to make it, there were
no factories for its smelting and rolling, nor specialists—nothing at all!
The new metal's only merits were that it would make accurate calculations possible, and would give designers the ability to create aircraft
lighter than those made of wood. But when hundreds of wooden airplanes were successfully flying, not only in Russia, but also abroad,
these factors did not seem significant enough. In the ensuing debates,
Tupolev was called "the Iceberg" for his uncompromising attitude. He
understood clearly what resistance the proposal to build the new machines out of light alloys would meet at all levels, and how much effort
it would take to defeat the majority's viewpoint.

"I was sent," recalled Tupolev, "when we were being condemned in
the institutions and at the aviation factories. Would it not be the greatest
mistake to be led by a young, 30-year-old engineer, who had not yet
distinguished himself in any way? And if he is mistaken, will we not

be left with nothing for our pains? Will we place our reliance on non-existent metal airplanes, and be left even without wooden ones? It was difficult, but we were convinced of the correctness of our conclusions, and turned out to be right."

By decision of the Supreme Council of the National Economy, a commission was established at the TsAGI for the construction of experimental metal airplanes. The government later ordered that aviation factories begin to prepare for the construction of metal airplanes.

Now, of course, it is clear that Tupolev then made a far-sighted decision, one for which nature had so generously endowed him. "Out of the new material we will build, for example, aerosleighs. We will overcome all of the difficulties involved in mastering the new work techniques, and we will study the behavior of duralumin, not on an airplane, where mistakes often cost people's lives, but on apparatuses that move along the ground. This is much safer. And the workers and engineers will also gain experience on them."

The TsAGI personnel (by this time KOMPAS had ceased to exist) built eight types of sleighs. In creating them, Tupolev passed in sequence through all stages of the change in design: from two-seat, completely wooden sleighs to five-seat sleighs made entirely of metal.

To determine their suitability, the Revolutionary Military Council and the Supreme Council of the National Economy organized a race to Tver. They also decided to make it a propaganda tool. The sleds were decorated with red calico bearing slogans, and they took newspapers and leaflets with them. Tupolev drove sled number 2. After a farewell meeting in the village of Vsekhsvyatskiy, raising a cloud of snow the column set off.

"At first everything went smoothly," Tupolev recounted, "then troubles began. Horses reared up and threw themselves from side to side from the roar of the motors and strange sight of the sleighs. The militia intervened and demanded that we move off the road. As we moved off the highway, some of the sleds tipped over into the ditches. There were breakages and, in short, plenty of difficulties. On the other hand, we developed speeds up to 70 kilometers per hour off-road. The race proved that the propeller sleighs were very maneuverable, and ran easily off-

road. We passed our first examination, and passed it well." The tests confirmed that aircraft construction from duralumin was viable.

In the Tver race the participants discovered in Tupolev's character a trait they had not before seen—recklessness. He wanted to run faster than anyone, and always be first. Engineer T. P. Saprykin, an automobile enthusiast who worked with him, recalled, "We had an ancient Benz automobile at the TsAGI. One day I was riding with Tupolev from Khodynskiy Airfield on the Leningrad Highway, alongside a lane along a riverbank. A rider on a trotting horse was trotting along the lane. 'Let's go,' he says, 'beat him!' I stepped on the gas but we still lagged behind. He leaped up and shouted 'Step on it!' and slapped me on the back. We were not able to overtake the trotter. Tupolev became outraged, called me a bungler, and the Benz a sloth, and stopped using it."

The ANT-1

With the success of the duralumin tests came the end of an era. The concept of the TsAGI was now fully realized and it was time to shift to solving more pressing tasks.

To those not favorably disposed toward the newly created institute, it may have seemed that building aerosleighs was too frivolous an occupation for a scientific center, and that, protected by the imposing figure of N. Ye. Zhukovskiy, the young people were simply amusing themselves by driving across the snow. This view was extremely short-sighted. In those long-ago 1920s, on Voznesenskaya Street, in two impractical huts, the TsAGI was scraped together, headed by N. Ye. Zhukovskiy and A. N. Tupolev. They were joined by S. A. Chaplygin, and the brothers A. and V. Arkhangel'skiy, Ivan Ivanovich and Yevgeniy Ivanovich Pogosskiy, Aleksandr Ivanovich Putilov, Nikolai S. Nekrasov, and a little later Vladimir Mikhaylovich Petlyakov. All gravitated toward airplane construction and aligned themselves permanently with the institute. They were aided by Boris Sergeyevich Stechkin, B. Yuryev, V. Vetchinkin, G. Musinyants, K. Ushakov, and G. Sabinin, who were drawn toward scientific research, and finally I. Sidorin, Aleksei Mikhaylovich Cheremukhin, and G. A. Ozerov, who were interested in metalwork and the strength of aviation designs. Around this nucleus there formed a strong group of general workers, such as patternmaker V. Romantsov, metalworkers P. Komolov, N. Lysenko, A. Novikov, and I. Borisevich, and mechanics I. Ivanov, M. Shcherbakov, and A. Vasin, the majority of whom soon occupied leading positions in the institute's experimental production, having become chiefs of factory shops and departments. The TsAGI's administration took shape as well, headed by D. Orlov and S. Losev, who remained in their positions for years.

Planning for the first airplane built at TsAGI began once the aviation department group became convinced of their chief designer's abilities and capabilities. Tupolev called A. Arkhangel'skiy, brothers I. and Ye. Pogosskiy, Aleksandr Putilov, and Nikolai Nekrasov together. After heated debate, it was decided that the first airplane would be a single-seat monoplane of mixed construction. To make it entirely out of metal still seemed risky, and to make it out of wood would mean that they acknowledged the weakness of their convictions.

The initiation of ideas about the first machine coincided with a major change in Tupolev's personal life. When he grew tired of living in dormitories, he moved to his sister Vera's home on Voznesenskaya Street. It was next to both the TsAGI and the Moscow Higher Technical Institute. The war was going on, and a large hospital for the wounded was opened in a Moscow Higher Technical Institute building. It was here that nurse Yulenka Zheltikova met medical orderly Andrei Tupolev.

Having completed the Rzhevskiy private *gymnasium,* where one of her friends was the future poetess Marina Tsvetayeva, Yulenka was attracted to art. It was a time of controversy, of changing ideas, and establishment of the new art. Such opposing talents as Serafimovich/Erenburg, Yesenin/Mayakovskiy, Stanislavskiy/Meyerhold, Moskvin/Tairov, Yuon/Tatlin, Zholtovskiy/Iofan, and Glazunov/Prokofiev, each in his own way affirmed the new era. Gifted with artistic inclinations, Yulenka was attracted to theater, concerts, and literary pursuits. She could in no way be called a beauty, but she was uncommonly sweet, possessed great charm, and easily fascinated those who met her. Young people gathered at her house in their few free evenings. Yulenka sang and played the piano, and the group argued fervently about art. Their acquaintance began in this environment and their mutual attraction gradually became more serious.

Yulenka's mother, Yenafa Dmitriyevna, was somewhat shocked at first by the outward appearance of her future son-in-law—the inevitable long-belted blouse and trousers, tucked into high boots. Taciturn, and glowering from under his brows, he seemed such a lone wolf. She did not yet know that when he was absorbed in aviation, he opened up completely; his circle of friends were just as obsessed as he was. In this group he became cheerful, witty, and even charming. It might seem that a man who was far removed from art would not attract Yulenka, but the strong and passionate characters understood one another, even as they were held captive by totally different callings. Yulenka also sensed another aspect of his soul. Andrei Tupolev was a one-woman man. Having chosen his companion, he remained faithful to her throughout his life. It was a happy marriage, a union of loving hearts that lasted through the joys and sorrows of life's long journey.

The wedding was modest. The country was still just getting on its feet and it was not possible to purchase everything for a wedding. But joy, happiness, and domestic bliss were abundant.

Now to his beloved work was added a no-less-beloved family. Unexpectedly for Yenafa Dmitriyevna, her son-in-law turned out to be a tender family man. After the wedding he changed greatly—none of the lone wolf remained.

In the spring Tupolev went in search of a dacha to rent. On Kalyaevskaya Street there was not a tree, it was dusty, and the sun hardly shone in the gloomy yard. "It is not far from the TsAGI to Sortirovochnaya Station on the Kazan Road. I will search along that road." He preferred Kraskovo. There were pine woods and sand around, and a stream nearby. The artist I. V. Kosmin rented the neighboring dacha and the two families became acquainted. Later Tupolev asked him to paint a half-length portrait of his revered N. Ye. Zhukovskiy. This outstanding canvas even now hangs in Tupolev's former office.

The area along the Kazan Road so pleased the family that after about two years Tupolev built a dacha in Ilinskoye. They spent the summer months there, and at times even came in the winter on the holidays with the children.

Family matters in no way interfered with Tupolev's design activity, however. The first airplane project would be a cantilever design monoplane, that is, one without braces or wires, a design Tupolev had nurtured in his mind for years. Designing took place on the second floor of the merchant's house, manufacture of parts on the first floor, and assembly under the awning of the inn.

Testing the new structure was a problem. An airplane is not a sled and can be destroyed in one trial flight; therefore, the theoretical calculations had to be checked by static tests. The main strength elements of the wing, its longitudinal beams (longerons), were fastened to the inn building and the aircraft builders sat on the longeron to test its strength with their own weight. Tupolev, Nekrasov, Putilov, and the Pogosskiy brothers took their places.

"You, sit on the very end," Tupolev asked Putilov. "You are the lightest, and if we break the longerons, you will be hurt less than the others." Moving gradually along the longeron, like a chicken on a roost, Putilov reached a point four times greater than that expected in flight.

But the excessive strength meant unnecessary weight. Putilov was exasperated. "Andrei Nikolayevich, as you wish, but in order to lighten it, it is necessary to do it over again. If we calculate so carelessly and make mistakes, we will have to pay for it with excess weight. Our machines will be the heaviest in the world!"

Tupolev dissuaded him. "Listen to me, friend. Doing it over will delay output greatly. I promise you that on the next machine I will watch strength and weight myself."

Putilov resigned himself eventually, but he muttered for a long time.

Building proceeded slowly, from summer 1922 through fall 1923. Lack of experience in designing parts had an impact, and it was necessary to redo a lot, at times more than once. However, in the end an assembled airplane appeared under the awning of the inn. It did not have a name, and since airplanes, like people, cannot get by without one, the workers decided to christen it with their chief's initials—"ANT-1."

When the name was presented to him, Tupolev smiled. "Quite recently we named our daughter, and now there are further christenings. I am a happy man." He affectionately stroked the little airplane. The dimensions of the ANT-1 were truly small: the wing span was seven meters, the total length five meters. The Anzani motor was rated at 35 horsepower. The machine weighed 205 kilograms, less than a present-day passenger car. They called her "Little Bird," and she was the forerunner of all the subsequent Tupolev designs.

When it came time to fly the airplane, Tupolev proposed, "Let us go to the other side of the Yauza, in the field where the Cadet Parade Ground is located."

"But that area is full of debris!"

"We will clean it up ourselves. I figure that she will take off after 80 to 100 meters, so it won't be all that much work."

After several days of labor with shovels and rakes, they had their "airfield," a clean strip 20 meters wide and 400 meters long. On 21 October 1923 everything was ready. If the first flight of each aircraft is an event, then the first flight of the first aircraft designed and built with one's own hands is like the birth of one's first child. Tupolev walked off to the side, away from the crowd. His friends understood and left him alone.

Pilot-engineer Yevgeniy Ivanovich Pogosskiy raised his arm, and the designers holding the machine by its wings leaped to the side. Raising a cloud of dust, ANT-1, after running less than 100 meters, took off into the air. The airplane's creators and the omnipresent young boys witnessing the flight froze. After making several circles over the scrap heap, Pogosskiy glided down and landed.

After the chief and his friends greeted the pilot with great enthusiasm, Pogosskiy said, "Andrei Nikolayevich, the airplane is very stable and is easy to handle. I did not even know that I was flying, it took to the air so quickly."

After towing ANT-1 to the inn, the group decided to mark the event. They got on tram no. 24, transferred onto a "bukashka"[7] at Pokrovskiy Gates, and went to Akvarium Garden on Sadovaya Triumfalnaya. Obliviously engaged in building their aircraft, they did not know that this was a gathering place for NEPmen.[8] Dressed in torn and soiled leather jackets, the group was greeted with a demand for payment in advance. But nothing could cast a shadow on their happiness. On this day they knew they had stepped into their future.

It would not be inappropriate for the Moscow Soviet to place a memorial placard in the beautiful square on Krasnokazarmennaya Street, in the exact place where the first Tupolev design flew. This was a historic event, indeed; after all, Tupolev's airplanes originated with the ANT-1.

It was necessary to halt test flights because the motor on the ANT-1 kept overheating. It was removed and given to Stechkin for repair and adjustment.

For the second airplane, their first all-metal construction, Tupolev envisioned a two-seat passenger airplane. There were still no domestically produced motors, so they settled on the British-made 100-horsepower Bristol-Lucifer. The plan was the same: a cantilevered monoplane with high-set wing. It was a test of Tupolev's general style and technical concept. The fuselage was given a triangular cross-section, outlined around a seated man. Above, the width corresponded to the width of his shoulders and narrowed downward to the width of his feet. The passengers sat facing one another. The pilot was located behind the motor, and his head stuck out, shielded by a celluloid windshield.

The dimensions of the airplane were a bit larger: the wing span was 10 meters, and the length of the fuselage 7.5 meters. The ANT-2 already weighed 836 kilograms.

It was significant that beginning with the ANT-2 the group had to work without referring to previously built aircraft. No one in Russia was building metal machines. Germany had the Junkers, but that information was not available to Russians. Although the Russian-German company Deruluft operated the Junkers Ju-13 airplanes, German personnel repaired them in closed hangars, where Russian mechanics were not allowed.

Pilot-engineer N. I. Petrov flew the maiden flight of the ANT-2 on 26 May 1924 at Khodynskiy Airfield. None of the eyewitnesses suspected that they were present at a historic event, that the ANT-2 was the forefather of all metal airplanes in our country. It happens that way in life. One may be present at a first flight, sail, or launch of one of the new creations of a designer's curious mind and not even suspect that one is an eyewitness to an event of the century!

By August, when Petrov had finished his flights on the ANT-2, the group was exhausted and decided to take a breather. Someone suggested going out of town, into the country, to sleep on the hay, drink milk, and walk along the stream with a drag-net. Tupolev grew excited: "Let us go to my home in Pustomazovo." They went to Zavidovo Station on the Nikolaievskiy Railroad, from where through Korchevo the road ran through Ilinskoye to Kimry. Nekrasov, Petrov, Putilov, and Tupolev took passing carts. By evening they crossed on a ferry near Korchevo across the Volga, and by night reached the Tupolev residence. For three days they helped the peasants gather the harvest, collected mushrooms, caught fish, cooked fish soup on the open fire, and rested, forgetting about spars, ribs, and other parts. Tupolev later said, "This was my last encounter with my beloved places, my home, my childhood. Later Natalya [his sister] went there and said that the graves of our ancestors were intact."

Reporting in fall 1929 to the TsAGI Collegium on his department's work, Tupolev said, "The date 26 May 1924 should, in fairness, be marked in the history of Soviet aviation. On that date, at Khodynskiy Airfield, took place the first flight of the first Soviet monoplane made entirely of metal." He failed to mention that he himself flew as a passenger on one of the ANT-2's flights, with some of his close friends. Tupolev also flew on the ANT-2, ANT-9, ANT-14, Tu-70, Tu-104 and many other airplanes, not out of bravado but because of a deep conviction in the reliability of the machines he created.

Having calculated, designed, and built their first two airplanes in the half-homemade shops with primitive machine tools, Tupolev's group was confident that much had been accomplished. They succeeded in developing design techniques and technology for light-metal construction on aerosleighs and two small airplanes. Production of duralumin-type pipes, sections, and corrugated sheets for the aircraft skin had been mastered. A competent group of production workers, foreman, technicians, and engineers had taken shape. "Our youth ended with the construction of the ANT-1 and ANT-2," Aleksandr Aleksandrovich Arkhangel'skiy recalled.

The ANT-3 Reconnaissance Airplane

It was time to attempt to solve more complex problems. At the request of the People's Commissar of the Directorate of the Air Force, in mid-1927 Tupolev's design bureau began to design and build the first combat reconnaissance aircraft, the ANT-3. For the first time the designers had conditions to meet. The first two airplanes were built according to the principle "Fly and confirm your ideas." But the ANT-3 had to follow the strict technical specifications of the Red Air Force. Its flight parameters were stipulated in advance: speed, altitude, lift capacity, range, number of crew members, and armament. Even the speed of a 360-degree turn (which defined the maneuverability of the airplane) was specified, because in an air battle the victor was the one who could reach the enemy airplane's tail most quickly, from where it was easiest to shoot it down. It was this requirement that made a sesquiplane the choice for the ANT-3. Tupolev absolutely did not want to give up on the monoplane, and he sat for days with the aerodynamics engineers, exhaustively delving into their calculations, and at times even doubting their methodology. Nevertheless, the theoretical calculations irrefutably showed that the sesquiplane was much more maneuverable than the monoplane. Moreover, the materials from which the machine was built, and the entire technology, did not yet impart confidence in the strength and reliability of the combat machine, which had to carry out rapid and complex maneuvers in the heat of battle.

They left the fuselage triangular, like the ANT-2, following one of Tupolev's rules: once found, do not throw out a successful design solution until a better one appears.

Choosing the engine caused much difficulty. Soviet motors were still not yet in production. The first consideration was the American "Liberty" engine, and then the British Napier. They finally settled on the French Lorraine-Dietrich. It must be emphasized that such apparent wavering was not at all due to a lack of decisiveness on the part of the industrial managers and designers, but because many elements needed to be considered: the international situation, cost of the engine, agreement of the firm not only to sell several motors but also to give consultation to the machine-building industry, and a number of other factors.

Working on the machine, the designers for the first time understood what it meant to design an airplane according to a precise formula. They studied the specifics of military aircraft, became familiar with weapons mounted on combat airplanes, and began to imagine realistically their operating conditions. Moreover, Tupolev came to the conclusion that aircraft engineers alone were not sufficient for creating combat aircraft, and he created a section of weapons specialists in the design bureau, putting in charge A. V. Nadashkevich, with whom he was already acquainted through the Imperial Moscow Technical Institute and the Aeronautics Circle.

Those who called Tupolev "the fist" or "the petty proprietor" when the TsAGI was being organized were to a certain extent correct. Without these gifts, it is difficult to become a chief or general designer. It is not enough to design an aircraft successfully. It is necessary to assemble a collective of people devoted to the project, to seek out an enterprise where the machine is built, and, finally, to find the pilot who tests it. For all this it is necessary to possess an imperious character. But wisdom is displayed in sacrificing one's imperiousness to the common cause when necessary.

So it was during the ANT-3 project. Convinced that the aircraft was reliable and promising, the government decided to send the machine on a flight to the Western European capitals. Everyone agreed the pilot would be M. M. Gromov, who had tested the airplane. But a flight mechanic was also needed. Tupolev wanted the marvelous flight engineer, Ye. I. Pogosskiy, who had put much effort and energy into the creation of the ANT-3, and who had designed the engine mount of the machine. But Gromov had more trust in Ye. V. Rodzevich, who had worked with him for two years.

When Tupolev tried to order Gromov to fly with Pogosskiy, Gromov replied, "I will not fly without Rodzevich. Harmony and mutual understanding between the pilot and the flight mechanic are the foundation of success. We understand one another in an instant, I might say at a half a glance. And with Pogosskiy I am merely acquainted, no more. I cannot guarantee success in that case!"

Tupolev conceded. "If the success of the mission depends on Rodzevich, then fly with him. What is important is not the last name, but the final goal."

Over four days in the fall of 1926, Gromov and Rodzevich, on a test ANT-3 named *Proletariy*, flew to Berlin, Paris, Vienna, Prague, and Warsaw, covering 7,150 kilometers in 34 hours. In August 1927, in another ANT-3, called *Nash Otvet* [*Our Answer*], pilot S. Shestakov, with mechanic D. Fufayev, flew to Tokyo in 10 days.

The young Soviet aviation was demonstrating to the entire world its initial, but in no way bashful, steps. Recalling these 1927 flights, Tupolev in his book, *Na Vostok i na Zapad* [*To East and West*], wrote,

"Overtake and surpass," that is how the Central Aero-Hydrodynamics Institute formulated its task, as in 1923 it approached the question of developing the work of experimental airplane construction. The flights of *Proletariy* were an examination, which we took before Western European technology. And I must say that we passed it, and passed it with excellence. This means that we are already on a par with them, and means that in the matter of design and construction we have become equal. What is the secret of our successes? Here we should note theoretical preparedness for the work, and participation in it of the complex of TsAGI Laboratories. Gradually, step by step, all of the questions that arose in the process of developing the experimental construction were worked out in these laboratories.

He stated further that experimental aerodynamics study, testing of materials, and questions of design strength were solved with the closest interaction between the experimenter and designer.

Tupolev's main idea is contained in this article. All of his successes were the product of the foresight of N. Ye. Zhukovskiy, who insisted on a symbiosis of science and practice. Tupolev never forgot Zhukovskiy's words: "The most important thing is the developing link of practice with theory."

Building the TsAGI

And so the Supreme Council of the National Economy decided to open the TsAGI, the Central Aero-Hydrodynamics Institute. There was space with the house and the inn. The founders were determined that the new scientific center would be not a "temple of basic science," where scientists occupied themselves in solving abstract, theoretical problems far removed from real life, but a center where scientific inquiry was conducted in order to solve the specific requests of industry, specifically aviation industry. For the 1920s, when there were no Soviet aircraft, this required developing designs that met the needs of the national defense.

The very name TsAGI defined the sphere of activity: the field of science associated with the movement of the air environment and air streams. The scientific studies were aimed at creating airplanes, hydroplanes, and speedboats, and finding ways to harness the energy of wind and water for the most modern wind engines, hydroturbines, and dams for hydroelectric plants. Moreover, it was planned that when solving these tasks meant touching upon related disciplines the TsAGI would not await assistance from others, but would create the necessary laboratories itself. Plans took shape to design equipment models that would be suitable for series production in industry.

As a result of this far-sighted policy, from the very outset, along with the scientific laboratories design groups grew with production shops at the TsAGI.

From the outset it became obvious to the institute's colleagues and leading specialists that locating the institute on the grounds of the Mikhaylov house and the Rayek Inn was only temporary. In fact, after only one year the fast-growing collective was overcrowded. In the summer, when a significant portion of the work was carried out under the awnings of the inn, it was tolerable, but in the winter, when bad weather forced everyone inside, it was unbearable. There were no areas adequate to develop scientific laboratories. And they could not even think about fire safety.

The number of assignments was growing rapidly, and the topics were becoming more complicated. The complex studies and wind-tunnel tests of future aircraft were impossible to accomplish in the existing laboratories and wind tunnels of Moscow State University and the Moscow Higher Technical School. The clients—the Red Air Force, the Civil Aviation Fleet, and the Navy—demanded increased speed, altitude, and range in the airplanes, as well as faster ships. The existing airplanes and ships were no longer sufficient to meet the clients' needs, and it was necessary to seek new designs. The designers wanted to know in advance what advantages one or another wing profile promised, whether a machine in flight or a ship afloat would be well balanced, what reaction would arise from one or another control surface deflection, what new materials could be used to reduce the weight of a design, and what lines to give to the hulls of fast ships.

More subtle and refined experiments and more accurate measuring instruments were needed. From the field of art, and inspiration or enlightenment, aircraft design shifted to the exact sciences and engineering. It was not possible to achieve progress while the dimensions and equipment of wind tunnels and, especially, the scale at which the lift force of the wing of a model being analyzed was measured were so primitive that great errors resulted. Experience dictated the need for more modern wind tunnels.

In exactly the same way, a water channel with laboratory was needed, which would make it possible to substantiate scientifically the most modern hulls for boats, seaplanes, hydroplanes, and torpedo boats and the shapes of the blades of hydroturbines, dams, and sluices of the already designed Dnieper Hydroelectric Power Plant, the ZA Hydroelectric Power Plant, and other hydroelectric power plants awaiting design.

The laboratories needed to test aviation designs for strength, and to test new materials and new engines. It was not possible to test wind engines without erecting towers. Finally, there was no completion unit capable of embodying the concepts of the scientists and engineers at the TsAGI. The ramshackle shops, which by miracle had been able to construct the aerosleighs and the first tiny ANT-1 and ANT-2 airplanes, were so overburdened that they literally could not handle the orders.

There were no longer any doubts that it was necessary to build, and quickly. Despite the Soviet Union's financial difficulties in recovering from the difficult years of the war and blockade, in the spring of 1924 the Soviet government found it possible to allocate significant sums to the construction of the TsAGI. It was also remarkable that a portion of these funds was provided in hard currency. This made it possible to acquire, especially in Germany and Austria, precise measuring instruments and apparatuses, which it was not possible to create in the RSFSR in those years.

The Moscow Soviet of Workers' and Peasants' Deputies set aside for construction an entire block, bordered by Nemetskaya and Voznesenskaya Streets, and Kirochnyy and Demidovskiy Alleys. Several dilapidated wooden houses were removed, and later a long-closed Lutheran church. S. A. Chaplygin was named chairman of the construction commission of the TsAGI, and engineer N. I. Vorogushin became his most active assistant.

It was necessary to find the appropriate architect to head up the project. In those years, architecture was an art filled with vast differences of opinion. Urbanist tendencies had replaced the eclectic approach of Tsarist Russia. The designs of many buildings were given most unusual shapes. For example, the theater in Rostov-on-Don was built in the shape of a tractor! In Moscow little had been built, and Chaplygin and Tupolev rightly assumed that conflict would arise surrounding such a major structure as the TsAGI.

"If those 'progressive' designers such as those in Rostov gain the upper hand," Tupolev told Chaplygin, "then you and I will turn red every time we pass by. What an obscenity they have erected!"

It was necessary to obtain the consent of one of the truly universally recognized architectural masters, A. V. Kuznetsov. Both Chaplygin and Tupolev knew Professor Kuznetsov from Moscow Higher Technical Institute. He was a talented artist and a demanding builder. They met one evening at Chaplygin's and it became clear that they agreed on the fundamental appearance of the future TsAGI. Hence, a creative cooperation was established.

According to the design, the institute complex included four major structures: an experimental factory with design bureau; a hydrochannel to be used to find the most modern contours for seaplanes, with a 200-meter-long water surface; a complex of wind tunnels, of which one was rather large; and a laboratory of wind engines.

No one had experience designing and building such specific structures. They could not expect assistance or consultation from foreign specialists, who knew that a scientific center for aviation was being constructed but were not eager to assist in its creation. The architects needed guidance, however, and it was decided that all department and laboratory supervisors would give input in developing the design. Kuznetsov later acknowledged that without this assistance they could hardly have managed the project.

Tupolev consulted on the construction of the hydrochannel and the experimental factory, and his intractability disturbed Kuznetsov. The chief architect said, "Tupolev chose an expanse of 70 meters for the assembly shop. I asked him, 'Why do you need such a large shop? After all the wing span of your ANT-1 and ANT-2 are only 7 and 10 meters.'

"He responded, 'They are already built. How do you know what I have here?' as he slapped himself on the head.

" 'I, of course, do not know what you are thinking, but on the other hand I can anticipate that there will be plenty of troubles with the experts. None of them will agree to approve such a bulky thing.'

" 'Do not trouble yourself, I will take this on myself. I will bet that they will support me in the government. They know that Sikorsky used a wing with a 30-meter span on the Ilya Muromets. Are you assuming that we will not be able to do this?' "

Numerous problems arose with the hydrochannel project. With a length of 200 meters, one million buckets of water were required to fill it, which was a significant portion of the daily productivity of the Mytishchinskiy water pipeline feeding Moscow. Unshlikht, Bubnov, and Baranov, members of the Revolutionary Military Council (which ordered the majority of the airplanes), arrived to familiarize themselves with the project. Having looked at the model of the channel and listened to the explanations, they approved its construction.

Many unanticipated troubles arose with the institute's instrumentation. At that time, Soviet industry could not yet develop and construct the institute. In 1928, Tupolev and a group of employees were sent to an aviation exhibition in Berlin. He was assigned to place an order for apparatuses and measuring devices. Believing that they were monopolists, the Germans requested extremely large amounts of money. Fortunately, at this same time academician A. N. Krylov, a famous Russian shipbuilder, was in Berlin. Tupolev went to him for advice. Krylov willingly gave his help. He said that the equipment for the test basin built in St. Petersburg before the revolution was ordered from Professor

Gebbers in Vienna, and recommended that Tupolev go see him. The Austrian equipment was much less costly and turned out to be of the highest quality. Thus, Academician Krylov helped to secure an excellent apparatus for the TsAGI's channel, and to avoid the unnecessary expenditure of a considerable sum of hard currency.

There were especially long discussions about the design of the buildings for the wind tunnels and about the dimensions of the large tunnel. The small tunnels at the university and the technical institute made it possible to study only the tiniest models. It was truly remarkable work to make them a precise likeness of the airplanes being designed, but it was very time-consuming and the little models continuously delayed the designing of the subsequent ANTs.

Meanwhile, the dimensions of the aircraft grew and grew; the wing span of the ANT-4 reached 29 meters. The existing tunnels were no longer adequate for the new designs. The new tunnel had to be built truly *large,* so that it could be used to test not only models, but also full-sized airplane parts. It would have been possible to design such a tunnel, but the city electric power network was so overloaded and dilapidated that the Moscow Regional Power Administration established a strict limit of 700 kilowatts. This restriction determined the dimensions of the new tunnels. First of all, it was necessary to determine what air-flow speed was required in the new tunnel. The airplanes of those years flew no faster than 120–150 km/h. If the air in the new tunnel was forced to move at a speed of 300–350 km/h, their needs would be met for the next five years. For this a 600-kW motor was required to turn the fan. The cross-section of the operating chamber was analyzed and it was determined that models with wing spans up to 1.5 meters could be studied with sufficient clearances. This was a great step forward. There was only one problem. Due to the limitations of the electric power network it was not possible to test entire parts of an actual aircraft in the tunnel. The TsAGI collegium, aerodynamics specialists, and aircraft designers looked hopefully at the wind-tunnel designers: Is this task really insoluble? Looking at their work one day, Tupolev said, "Will you really be able to pull it off?"

And they were able. Tunnels T-I and T-II, besides the fixed portions, also had a mobile portion. This construction made it possible to change the working section of the tunnel. The six-meter length was used to test models with a wing span up to 1.5 meters, and the three-meter length to test full-sized airplane assemblies.

The wind tunnels were intended to answer the question of how an airplane would behave in flight. When an airplane model was placed in the operating chamber and a stream of air was blown in, lift occurred on the wing of the model, as on an airplane in flight, attempting to change its attitude. The model was connected to sensitive scales through a system of rods. The operator could evaluate the amount and vector of the force from the deviation of their indicators. In the small tunnels at the university and school the scales were imprecise. For the new T-I and T-II tunnels it was necessary to design scales capable of registering the slightest deviation of the model under the effect of the airstream flowing around it. G. M. Musinyants, one of the pioneers of the TsAGI, solved this problem. The completed T-I/T-II tunnel, which made it possible to carry out both high-speed and full-scale tests, was a great success for the TsAGI.

Meanwhile, the layout of the institute itself gradually developed on the drawing paper in Kuznetsov's shop. It was not easy to do. The four-story residential house standing within the block interfered with a final decision. Kuznetsov and Vorogushin sought to have the Moscow Soviet authorize its destruction, but the value of the multistory house was considered too great under the grave housing crisis.

In despair, Kuznetsov suggested, "Won't you tear down the Mikhaylov home; you see for us it is like a fish bone in the throat, it interferes terribly."

"No, not for any reason!" Tupolev sharply protested. "Zhukovskiy worked in that house, he greatly loved it, and when we finish construction, we will create a memorial museum there."

A layout was finally approved, and the question of the external appearance of the buildings was raised. Tupolev insisted, "Everything must be simple, laconic, but worthy. No pilasters, Doric, or Ionic caps, or other bagatelle. Just let the passersby see that this is truly a world-class scientific center!"

Construction was difficult. In those years there were neither excavators nor tower cranes. Everything was done by hand. There were not even any skilled builders. Construction was neglected altogether in the four years of the imperialist war and three years of the civil war, and the skill was lost. So in the mornings Kuznetsov, with Chaplygin and Tupolev, had to walk through the construction site to see whether the workers had ruined something somewhere. When they demanded that some slapped-together partitions be dismantled and rebuilt, Tupolev threatened the foreman with a monetary fine so that he would be shown proper respect as the "owner."

The construction project was large, and the residents of Bauman District displayed some interest in it. Its name bothered them. No one could decipher the four letters of which it consisted. In general, in those years few could fully imagine the growing importance of aviation. The airplane flights, which were infrequently demonstrated at Khodynka, were in their opinion for foolhardy daredevils and akin to circus tricks. They could not even conceive of the fact that in the former merchant's home and the Rayek Inn young enthusiasts, united under the puzzling name of TsAGI, were designing airplanes on which these residents would some day fly as passengers.

In October 1925 the first construction line was put into operation, and on December 31 a stream of air was forced into the large tunnel designed by A. M. Cheremukhin. Tupolev had a very warm relationship with Cheremukhin. Later, in an article dedicated to his memory, Tupolev wrote,

All of us who knew Cheremukhin personally loved him honestly and simply. I do not remember when he came to us, but I also do not remember a time when he was not with us. This means that he immediately became a dear and needed member of our collective. He helped us in everything. We began to build the TsAGI and brought in Cheremukhin, and he built the original wind tunnel, out of assembled elements, which was the largest in the world in those days. The question arose of creating new experimental apparatuses at the TsAGI, and Cheremukhin became the main engineer for their design. It was necessary to develop a helicopter, and he himself conducted its flight tests and set a world record. The country needed airplanes, and he put his whole heart into that cause. Formally he was in charge of our durability department, but in reality he was in charge of a wider area—questions of durability, and of how to put together a design so that it is strong. His great gift—technical intuition permeating the work of the design itself—helped us in everything. When he said that it was so, it could be believed. And I do not remember a case when his conviction was erroneous. In questions of durability, he was my reliable, strong, talented right hand.

The building complex was excellent. Today it retains its charm, even with the addition of buildings of a later architectural type. Tupolev's colleagues at the TsAGI breathed a sigh of relief, as it seemed that the problem of finding space for the employees and laboratories had been removed from the agenda, if not forever, then undoubtedly for a long time. But the tempo taken by the country was such, and the stream of scientific assignments increased so, that soon it became crowded again. It was necessary to build again, first filling up the remaining free areas on the allotted parcel, and then on a neighboring parcel across the street. When it became clear that there was no more free land nearby, and it was necessary to build even more than had been built, construction was moved out of the city, where land was plentiful. After a long search for an appropriate site, a river valley, with nearby knolls overgrown with pines, was selected. At first Tupolev wanted to dam the river and test hydroplanes and boats on the reservoir that was formed. On one side of the reservoir plans were made to build a future airfield, and on the other side a city. According to calculations, several thousand people would live in the center. And here a snowballing process began. For them to live there, in addition to houses, it was necessary to have stores, baths, schools, kindergartens and nurseries, clinics and hospitals, guest houses, a culture park, bakery and laundries, stadium, club house and city soviet, telephone station, and intracity transport. But these were all secondary. The main things were the hangars, the experimental design factory, laboratories, research facilities, runways, and airfield buildings with radio stations, beacons, and position finders.

Tupolev recalled, "When on May 23, 1936, as the chief engineer of the Main Administration for Aviation, I approved the project, it became clear what a tremendous construction project it would be, and how many funds would be needed. I gave a report to People's Commissar G. K. Ordzhonikidze. To the question of the possible cost, he answered, 'We will give as much as you need!' Nevertheless, I was troubled and consulted with my colleagues. Some told me, 'Take 200 million.' Others said, '200 million is too little, request 300.' I went to Ordzhonikidze and asked for permission to inspect some other factory that cost 200 million to construct. He advised me: 'Recently a metallurgical complex in Nizhnyy Tagil was completed. It cost 180 million; go there.' I went there and saw what 180 million would buy. It was a huge factory. An entire city was adjacent, with its own educational institutions, its own steamships. I was amazed, and I told Ordzhonikidze that we did not need more than 150 million. These funds were allocated.

"I became acquainted with the chief of the Tagil construction project, M. M. Tsarevskiy, and when the question arose of who should head up construction of the new Central Aero-Hydrodynamics Institute, I recommended him. The proposal was approved."

Tupolev enjoyed his creative participation in building the TsAGI complex so much that for the rest of his life he remained not only an aircraft designer, but also an architect who delved devotedly and deeply into the design of aviation factories, institutes, residential houses, and houses of culture[9] for aviation industry workers. He was a well-versed specialist, deeply convinced of the need to make a structure beautiful and comfortable, and in a way that did not spoil the face of the land or the environment. Tupolev's keenness of observation regarding architecture was enviable. Once while I was working with him, we were driving along the Yauza, when at Sadovaya Street he became interested in seeing the stone lattices that had been laid on the upper floor of Soyuzazot.

"They were pretty good, don't you think? Let's go take a look, Kerber L'vovich," Tupolev stated.

"It is simply bad. The rooms are dark. Typical architectural rubbish."

Another time we were driving to Vnukovo along Leninskiy Prospekt. On the left was the oil institute building. He was interested in the arrangement of the entrance.

"There is something fresh about it, don't you think? We should go take a look."

This happened repeatedly. The vigilant and curious designer noticed everything unusual, and always wanted to incorporate something similar in one of the buildings being built for us.

After the revolution and civil war, several houses were built for the employees in the settlement outside Moscow. At the discretion of the suppliers, bricks—some white and some red—were hauled in from various factories. At that time the walls were not plastered on the outside. Seeing this structure, Tupolev became enraged.

"Look at that! What an abomination! Have you no conscience? You say you do, but I think not! I will make you inscribe on the pediment: 'Built by work superintendent so-and-so,' and your whole life you will be embarrassed to walk by." He insisted that the red and white bricks be laid not haphazardly, but in the form of an ornament, and it became truly beautiful.

For the construction of the test flight base, a parcel of land in a pine forest was selected. When the outlines of all the structures were determined on the ground, he called for his construction assistant and said, "Keep in mind that I do not authorize a single tree outside the perimeter of the buildings to be cut down. You will say they will be more difficult to build. Perhaps, but that is your work. For me another thing is important. The people who will live here, and there will be many, will always be grateful to you. Moreover, it is easier and more peaceful to work when you see branches out the window, and fewer mistakes will be made under such conditions. And you know what mistakes mean in our business."

And truly, the buildings of the base even today are buried in the verdure of the pine forest. Tupolev built his base as he did everything, with scale and perspective. Among the production buildings and hangars there was also an administrative building. According to the fashion popular at that time, each structure had a turret on the roof, and a pediment with balusters, in short, architectural extravagances. On the second floor was a suite of rooms for the general director: a reception room, office, conference hall, and restroom. Their trim was to be done according to Kondorskiy's sketches. I do not know whether this was carried out precisely, but when Tupolev invited Aviation Minister Petr Vasilyevich Dement'yev to take a look, Dement'yev remarked with his characteristic keenness, "A precise image of Petrovorets. All you forgot were the fountains!" In his customary low-key way of making a point, our minister summoned Tupolev to "Peterhof."[10]

Later, when jet airplanes began to take off from the base, their engine noise chased the birds from the pine forest. The biological equilibrium was destroyed, some kind of harmful insects developed, and the pines began to waste away. Tupolev was troubled. He invited scientists, an ornithologist, and a forester, and he consulted with them for a long time on how to correct the matter. According to their recommendations, poisonous chemical dust was scattered about, but it helped only a little. He said regretfully, "In my passion for technology, I hope you have no doubts. Nonetheless, we are not doing everything right. Perhaps, in the technical specifications for all types of production it is necessary to introduce requirements that protect nature: animals, fish, and birds. We must not adhere to the views of French King Louis XV: 'Après nous le déluge.' I not only have grandchildren, but great-grandchildren. Will we preserve for them the Russian nature celebrated by Pushkin, Lermontov, Tolstoi, Chekhov, Blok, and Prishvin that we delight in ourselves? One thing is clear. We must think about this."

A great number of such examples can be given about his concern about beauty, aesthetics, and comfort, things that ease people's work and lives.

The TB-1 (ANT-4) Heavy Bomber

It became obvious that new aircraft designs were needed as soon as possible for both civil and military aviation sectors. But where to get them? Should new airplanes be purchased abroad, or should Russia attempt to design and build them itself? Where should they be built, and which designer should handle the task? This problem was discussed for several days in the Military Industry Soviet, with the participation of military and industrial leaders, under the chairmanship of P. A. Bogdanov. Having thoroughly examined the problem, they decided to reject purchasing the bombers abroad, and to entrust their design to A. N. Tupolev, chief designer at AGOS (Aeronautical and Seaplane Experimental Construction, a division of TsAGI). Having Tupolev design the new equipment would free the country from foreign dependence, and they knew he had been successful with both the ANT-2, made entirely of metal, and the ANT-3 combat reconnaissance aircraft.

The final technical specifications for the creation of the first Soviet Air Force heavy bomber were formulated in fall 1924. Numerous debates preceded this decision, and Tupolev had reason to believe that the design would be awarded to the TsAGI. Moreover, little by little, he was carrying out the design calculations. Critically interrupting his previous work, and thinking about the design of his first bomber, he became convinced that only a monoplane, built entirely of metal, with a cantilever wing would work. His mind was constantly occupied with the ANT-4. At meetings, or in the evenings sitting at home, Tupolev sketched on rolled-up scraps of paper the external contours and individual assemblies of the future bomber.

On Tupolev's instructions, draftsman B. M. Kondorskiy drew variant after variant on sheets of Whatman paper. Tupolev refined each variation. At this stage of the project, Tupolev's work was like that of a sculptor who moves several steps away from his work to gaze at it critically, then grimaces involuntarily and starts anew. So it is here. Carefully assessing the sketch of the nose section of the aircraft, Tupolev sees that it can be made still smoother, thus reducing wind resistance and increasing speed. However, in such a cockpit the pilots will be cramped. Consequently, their seats will have to be moved back where it is a bit wider. Stop! This must not be done as it will complicate landing. The pitiless eraser wipes out the just-made contour. Tupolev sketches a new one with a soft pencil and looks at it: "God knows how to do it right, but honestly this is totally rotten! Let us erase it again and redo it." A mountain of rejected sketches grows on the floor. Tired, tormented Tupolev pushes away the chair and stands up to rest. But what is this? From here the drawing looks different, and he notices other flaws. A new round of exploration begins. It is not easy to find the compromise between lightness of construction and strength, between favorable external shapes of the aircraft and comfort of operations, between aerodynamic perfection and the protruding radiators, machine guns, antennas, and generators. In each such case, only the chief designer is able to find the best solution to the numerous such details, and only when he achieves these will the external shapes of the aircraft crystallize.

Great tenacity, strong nerves, and tremendous industry were needed, but at times the correct solution did not present itself for months. Gradually the precise forms of his first bomber, the TB-1 (ANT-4), began to take shape. The contours of the new bomber were not yet finalized, but day after day the severe straight lines of the fuselage and the mighty wing with two engines neared what had to appear in metal a few months later.

Satisfied, Tupolev took the drawing to the modelers on the first floor of the house. Their shop smelled like wood shavings, and a small stove was on, where a "bath" of joiner's glue bubbled, and a potato was baking. The elderly master modeler V. P. Romantsov understood Tupolev at once, and the work went well.

The dimensions of the models for the tests of the first ANT airplanes differed little from one another, for they were dictated by the restrictions of the wind tunnel at the Moscow Higher Technical Institute. It is understandable that, having only rarely seen a small wind-tunnel-model TB-1, Tupolev's workers could in no way assess completely the innovative design of a monoplane with a 30-meter wingspan!

Results of the wind-tunnel tests were encouraging. It turned out that with two engines of 560 to 620 horsepower each, the flight data of the TB-1 corresponded fully to the performance standards of the design. Although the concept worked, Tupolev was not yet satisfied. It was still necessary to develop an aero propulsion for the monoplane design, solve the questions related to structural integrity, and attach the weapons and instrumentation without which the TB-1 was no more than a huge air frame. Gradually, the simple and refined lines of the huge monoplane came closer and closer to the design perimeters, until the day when they fully satisfied Tupolev.

It was now necessary to involve his assistants with the TB-1 in more detail, so that they could work with him. When he gathered A. A. Arkhangel'skiy, N. S. Nekrasov, Yevgeniy Pogosskiy, and N. I. Petrov in his office and hung the large-scale design drawing of the TB-1 on the wall, Tupolev did not anticipate their reactions. The architectural shapes of the TB-1 were so unusual, and differed so greatly from the traditional, that his assistants were doubtful they could succeed in building such a huge machine with an absolutely unprecedented layout, and especially whether it would be able to pass the strict conditions of flight testing. His friends' skepticism deeply troubled Tupolev. He had to spend much time patiently (which he did not enjoy) proving that he was right.

"Understand, without breaking the established patterns and techniques we will not obtain a qualitative leap forward."

"That may be so," they objected, "but is this not too risky a leap? If we are wrong, won't the whole idea of metal monoplanes be ruined?"

"No, no, a thousand times no!"

And Tupolev spread out before those present a whole pile of rough drawings and sketches of aircraft design parts, and rough calculations, the physical proof of his activity during the previous months, over which he had sweated during all his free evenings and Sundays. He spoke so convincingly and temperamentally that gradually he inclined to his side even the most stubborn opponents. Recalling this episode, Tupolev acknowledged, "I was very upset. It was the first serious technical conflict in our group. Would it cause a split in our ranks? Deep within my mind the thought even crept in: Are you absolutely sure you are right? But I forced it away. For long years I had nurtured and tested in my mind the future airplane design, and had come to this conclusion. And when my friends believed me, it was as though a mountain had been lifted from my shoulders.

"It was already late at night when we parted. Hurrying to the street-car with Arkhangel'skiy, I suddenly began to whistle some operetta tune.

" 'Stop, for heaven's sake,' said Arkhangel'skiy. 'You are unbearably out of tune. You changed our minds, so be happy, but don't whistle, it will make us ill.' "

Gradually his closest associates—A. A. Arkhangel'skiy, V. M. Petlyakov, A. I. Putilov, N. S. Nekrasov, N. I. Petrov, and A. M. Cheremukhin—joined in the building of the TB-1. At first the work could not be called attractive. Tupolev himself selected the strength configuration of the aircraft, the number of spars (longitudinal beams) in the wing (there were customarily five in the TB-1) and in the fuselage, as well as the spacing of the transverse load-bearing elements (ribs in the wing, and frames in the fuselage). Now it was necessary to create the full-strength primary structure of the future airplane, to put the pieces together. In the era of the TB-1, the spars were made out of pipes, and the frames out of duralumin alloy sections. The attachment points where they came together were welded from steel. At times they turned out to have a rather complex and even intricate design. Tupolev considered it necessary to become personally involved in the design of each attachment point. Usually he attempted to sketch them in perspective. He "attempted," for he did not draw well. After modifying the initial sketch several times, he would call on Kondorskiy, the "hands of the ANT," as his fellow workers called him, to build a three-dimensional model of the attachment point out of scantlings, plywood sheets, and modeling clay.

A. M. Cheremukhin, the structures specialist, together with Tupolev carefully evaluated the structural design. At times one of the attachment point rods was not attached quite correctly; the forces running along the rod were applied not to the center of mass of the attachment point, but to a shoulder. If it was not corrected, it would need thicker metal, and therefore be excess weight. In such cases, Kondorskiy would be summoned and shown what needed to be corrected.

The designers became involved at this stage: V. M. Petlyakov if it was a wing point; A. A. Arkhangel'skiy if a fuselage point; and N. S. Nekrasov if it was from the tail unit assembly. With their assistance the attachment point gradually "came together," and was put down on the drawing paper in all three projections. Finally, the technologists came in. Was the attachment point compatible with manufacturing requirements for production? How many and what kinds of stocks for attachments were necessary for its manufacture and assembly? An ever-growing number of engineers, technicians, and draftsman were drawn into the work, until in November 1924 the entire AGOS group was fully employed with the TB-1.

When they worked on their own with small aircraft, the need for wooden full-scale mockups did not arise. As a matter of fact, on the ANT-1 the pilot sat at the height of a dinner table, and on the ANT-2 just a bit higher. The small sizes of the first airplanes made it possible to draw even the insignificant details in their layout and design. The TB-1 was an entirely different matter. Here the navigator and pilots were seated at the height of the second floor of a small house. The two-dimensional capabilities of a drawing did not make it possible to determine on paper whether the navigator would be able to sight and bomb conveniently, or whether the pilots would be able to taxi, take off, or land conveniently. New projection techniques were needed. Without a three-dimensional representation of the design it was not possible to determine where the sight should be located, what part of the cockpit should be made of glass so that the navigator could see the target ahead of time, how to seat the pilots so that they would have excellent visibility, and so that it would be convenient for them, both while in flight and while coming in for a landing, to look both at the ground, and at the instrument panel. Was it accessible for the gunners to place the machine-gun mounts? How many hatches should there be, and where was it necessary to cut into the casing and construction of the machine?

The group discussed how best to answer these questions for a long time, and concluded that it was necessary to build a full-scale wooden airplane, a mockup out of bars, scantlings and plywood. All of the elements of the machine's design (spars, frames, skin, crew seats, supports of all kinds) and all the equipment (motors, propellers, undercarriage legs with wheels, control wheels and pedals, compasses, instruments, sights, bombs, machine guns and their shells, radio sets, gasoline and oil tanks) were made out of wood. Only through the mockup did it turn out to be possible to evaluate the good and bad points of the designers' concept. Dressed in flight suits with parachutes, the pilots, navigators, radio operators, flight engineers, and gunners received the opportunity to imagine actual flight conditions while still in the shop.

It was difficult to make the first mockup. Tupolev had to resolve such conflicts as when two construction teams were working in the same space in the airplane at the same time. He spent his days with the designers at the drawing boards, and in the evenings walked to the empty shop, where, perched on the workbench, he thought about what could still be improved in the machine. At times he crawled into the mockup and carefully evaluated the comfort of the crew's work places. Sometimes the creaking of nails being removed was suddenly heard, and then the unsuccessfully installed device or apparatus, trailing an arc in the air, flew out of the mockup and crashed onto the floor.

Anyone still present could hear his high voice muttering, "Bunglers, nitwits, good-for-nothings, lazybones . . ."

In the morning the "bunglers" and "nitwits" were ordered to redo the error by a specific time. At precisely that designated time, Tupolev would arrive and examine the problem. Either everything was done once again, or he said, "Good lads, clever hands, that is the way it should be."

So the new trade of mockup carpenters was born, people with strongly developed spatial comprehension, capable of making parts and instruments, at times of uncommonly complex design, out of wood. The most capable formed a group that became an integral part of the design bureau. It was said they could "shoe a live flea with wooden horseshoes, attached by wooden nails."

A commission was convened to evaluate and approve the mockup. Supposedly several participants asked with disbelief, "And where is the upper wing?" That is how uncommon a monoplane with a 30-meter wingspan was!

Drawings began to arrive (several thousand were needed for such a large airplane as the TB-1), and it was necessary to carve out time to examine them. On their heels, parts flowed out of the shops. They could not be overlooked, for the airplane had to be simple to manufacture, in order to produce it in large series production.

In August 1925 construction of the TB-1 was completed. The largest component, the central part of the fuselage with a piece of the wing, was assembled on the second floor of an empty house on Voznesenskiy Street. It did not fit through the door and window openings, and it was necessary to dismantle the partitions and manually carry them out on scaffolding into the yard. At night the assemblies were hauled to the Central Airfield on carts by teams of three horses.

Two months were spent assembling and adjusting the airplane at the airfield. Only then, outside of the shop, was it possible to assess its appearance. Its smooth and flat fuselage, with pointed nose and tail, somehow reminded one of a destroyer. In the forward part of the cockpit, which was glassed on the bottom, sat the navigator. Behind him, covered by a transparent visor from the oncoming stream of air, sat the pilots, and slightly behind them the mechanic. The two gunners, one of whom was a radio operator, were located where the trailing edge of the wing intersected the fuselage. The upper wing, with its palisade of supports that interfered with vision and firing (a distinguishing feature of wooden airplanes), had disappeared, and the crew could see the air around them without any obstructions. The engine pods came out from the wing, which was thick at its root, and the chassis supports were located under these pods. Toward the ends of the wing the height of the airfoils tapered smoothly, and it seemed very flexible, and, when the airplane was taxiing, looked like the live wing of a fantastic bird. There was nothing excessive in this airplane; nothing other than what was absolutely necessary stuck out from its contours. Everything was sensible and functionally justified. The design of the TB-1 was unusual, and was elegant in its engineering. Of course today one might criticize the corrugated skin of the vehicle. Increased speeds have necessitated switching to more modern aerodynamic shapes.

In the midst of the work on the TB-1, Tupolev's son Aleksei was born. Tupolev was immeasurably happy. A son would continue the name, and perhaps even his father's path. When the joyful father brought Yulenka Nikolayevna and their son home, Yuliya awaited them. The parents were concerned about how she would feel about her little brother, whether she would be jealous. But no, little Lyalya, as she was called at home, immediately clung to her brother. Every evening, when the baby was put to bed, she cried bitterly. When she was allowed to be present during his bath, she was overjoyed.

Yulenka Nikolayevna was worried that Aleksei's crying would interfere with his already overtired father's rest, and offered to switch to the next room. With his characteristic decisiveness Tupolev refused her request.

It was not long before he referred to his son during the airplane construction work. Crawling into the pilot's cockpit, he scratched his leather jacket on the sharp edge of the frame corner plate. Foreman A. Novikov was nearby.

"What do you think, I have had a son, so you say that just any old way will do? It will not do. Direct that all edges be smoothed off immediately. I will bring him tomorrow [he was 17 days old], and what if he scratches himself? What then?"

On the morning of 26 November 1925, pilot A. Tomashevskiy took the airplane on its first flight.

"Of course I was worried," recalled Tupolev, "although I was also absolutely sure of the aircraft. The numerous skeptics who did not believe in the reliability of a monoplane with a 30-meter wingspan were also irritating. Before the first flight there were many. After it," he smiled, "few remained."

After thoroughly testing the machine in the air, Tomashevskiy assessed the TB-1 a "first-class airplane." This was a difficult, but complete victory for Tupolev and the airplane builders. Moreover, it signaled the birth of an entirely new class of large, metal monoplanes, the external shapes, design, and layout of which were far advanced from other airplanes of the era. Comparing the TB-1 with the French Farman-Goliath two-engine, wooden biplane bomber made it clear even for those with no experience in aviation.

In the assessment of aviation historian V. B. Shavrov, "The TB-1 was the world's first two-engine monoplane bomber built entirely of metal. In 1925 nowhere in the world were the attributes of a bomber so completely and successfully combined into one airplane as they were with the TB-1. This airplane became the decisive prototype of all subsequent bombers. All of the 'Flying Fortresses' and 'Superfortresses' were essentially developments of the TB-1 and TB-3 types."

By governmental decision, in the fall of 1929 one of the series TB-1 airplanes, *Strana Sovetov* [*Land of the Soviets*], was sent on a flight from Moscow across Siberia to the United States. The crew, S. Shestakov, F. Bolotov, B. Sterligov, and D. Fufayev, flew 21,242 kilometers in 137 flying hours to New York, where they were greeted triumphantly. After becoming acquainted with the TB-1, the Western press roused a noisy campaign. The Farman, British Vickers, and Italian Caproni biplane bombers, still built out of wood and steel, were greatly inferior to the TB-1. The press accused their designers of a lack of foresightedness. With the construction of the TB-1, the USSR seized primacy in airplane construction and 37-year-old Tupolev became one of the world's greatest aircraft designers.

With accolades, however, often come problems. The Junkers firm brought an action against Tupolev, accusing him of violating the patent for a metal wing. This action dragged on for a long time, but failed because, although like the Junkers the wing of the TB-1 was metal, it was of original design.

A Trip to America

The successes achieved in creating the TB-1 were somewhat obscured by the fact that its engines were imported. Russia was still not producing its own engines. This was an old story. Throughout World War I the "Allies" supplied Russia with information about airplanes, but not about engines. Apparently, they believed that we could manage to create our own airplanes, but not our own engines. When, in 1928, the Soviet aviation industry set up production of the Grigorovich's I-2, Polikarpov's I-3, and Tupolev's R-3 and TB-1, the engines still had to be purchased abroad. In order to eliminate this dependence, the Soviet government decided to build several large new engines. Moreover, experimental models of Soviet engines were already being bench-tested. These were the 500-horsepower M-17, on which V. Ya. Klimov was working, and the 600-horsepower M-34 design by A. A. Mikulin. Work was going well, and it was hoped that soon they would be in series production. These were liquid-cooled engines; work was going on less successfully with an air-cooled engine. At that time air-cooled engines had become widespread in the United States. It was necessary to purchase models of the best engines of this type, and to obtain an agreement on technical assistance for the design and construction of several large factories. For this purpose, at the end of 1929 Tupolev, engine specialist B. S. Stechkin, TsAGI chief N. M. Kharlamov, and engineer Ye. V. Urmin were sent to America. Petr Ionovich Baranov, chief of the Air Forces, assumed leadership of the delegation.

Baranov, back before the war, as a 20-year-old youth had entered the ranks of the Bolshevik faction of the RSDRP [Russian Social Democratic Worker's Party] St. Petersburg organization. In 1915 he was arrested for dissemination of illegal literature, and sentenced to eight years of penal servitude. Released after the February 1917 Revolution, by decision of the Central Committee Baranov was sent to do party work among the military units at the front. He spent the Civil War in the Red Army, first as a regimental commissar, and later as a special detachment commander. As a delegate to the 10th Party Congress, he took part in the assault against the rebellion at Kronstadt. Immediately after the fortress had been taken, by telegram under the signature of V. I. Lenin he was summoned to Moscow and named Deputy Chief of the Air Fleet for Political Affairs (Political Commissar). Displaying uncommon capabilities, he quickly mastered the new branch of arms, industriously studied the technology, and soon became a full military leader. In 1925, the All-Russian Central Executive Committee confirmed him as chief of the Air Forces.

It is quite natural that in his new position Baranov wanted to become immediately familiar with the TsAGI, the main center where combat aircraft were being designed and constructed. Tupolev gave him complete explanations. Their mutual interest in the creation of the air fleet, as well as their shared belief in finding the best way to achieve this goal, gave them much to discuss. Unfortunately, the burden of work did not leave much time for dialogue. This unexpected, lengthy trip together to the United States would give them that opportunity.

On 15 December 1929 the steamship *Columbus* departed Bremen for New York. *Columbus* swayed mercilessly in the Atlantic's winter storms. One after another the passengers became sick. Soon all the members of the delegation surrendered. Last of the group to leave his position, Kharlamov was troubled: Wouldn't *Columbus* suffer a shipwreck, and wouldn't the captain lose sight of the beacons? Tupolev somewhat darkly tried to calm him: "I do not think that there is any reason to seek beacons in the middle of the ocean, and besides, after a shipwreck they will hardly help us."

Baranov and Tupolev were not subject to seasickness, enabling them not only to enjoy the bounty of the ship's kitchen, but, most important, to discuss in detail their plans for becoming familiar with the U.S. aircraft engine industry. They compiled a list of purchases, as well as an agreement for assistance in the construction of engine factories, which they hoped to have signed.

On 22 December *Columbus* made fast to one of the piers of New York. After quickly concluding their business at the Soviet consulate and at Amtorg,[11] they headed for California, where the majority of the aircraft factories were located. They drove rather quickly, stopping only to eat at roadside restaurants. At one stop everyone laughed at the inscription from the days of the Wild West: "Please do not shoot the piano player. He is playing as best he can."

The Soviet delegation's trip was completely successful. Models of the best aircraft engines were acquired, and a contract was signed with the Wright Company, for technical assistance in building aircraft engine factories and setting up series production of engines.

Tupolev's impressions of America are interesting. They suggest his perspective on both American technology and culture:

One cannot help but admire the industry, organization and complete lack of bureaucracy in America. One's word is trusted more than we with our enumerable papers. To say means to do. Disdaining bureaucracy, I hope in my heart that soon we will overcome it just the same. I had a remarkable impression of their factories. You go from shop to shop, and nowhere is there a single idle person. Once I even poked Baranov: "You must be expecting, as I am, that over behind that door probably the workers are standing around and smoking." But no, we were not able to see anyone smoking and talking! On the other hand, outside of the gates there are plenty of people smoking and talking. They stand around waiting, and suddenly are satisfied, and get back to work!

I especially want to mention their aircraft design bureaus. The windows shine with cleanness; there is more than enough light. There are many adaptations that ease the work of the designer: vertical boards, adding machines, paper with which copies are made directly from under the pencil, electric pencil sharpeners. And what is very important, technical service personnel make all of this possible.

On the other hand, there are many things that for us are incomprehensible and unpleasant. Leave aside the unemployment and racial question, which are unworthy of a civilized state. I cannot comprehend a woman with children and belongings evicted for not paying her apartment rent. Still more terrible are the "shanty towns," and we saw them on the outskirts of every city. These are settlements of the unemployed living in improvised shelters made of sheets of tin, pieces of boards and rags. In winter they will truly be chilled to the bone.

Let us take an entirely different aspect—standards. Of course, they are irreplaceable in production, but in everyday life or art they are incomprehensible.

Let us take films. Either they are vulgar, narrow-minded melodramas, or cowboy westerns. Let us look at the people. There are millions, and are in hats, bow ties, and watch chains from their waistcoat pockets. When it is hot the millions take off their jackets, hang them over their backs, and roll up their sleeves. Damn! Are you alive or mannequins?

The dinners, breakfasts, and suppers are rather tasteless, but satisfying. However, they serve them lightning fast and hot. If it is tea, it is boiling; if lemonade, then icy. Tablecloths, forks, knives are blindingly clean. Their service is such that you would think that to feed us is the happiest dream of their lives. It is bad that everyone smokes everywhere: in the dining halls, theaters, and offices. This is repulsive, but my personal feelings enter in here. I cannot stand the stench of tobacco.

And in general, everything is not so simple, and there are things to learn from them, only to learn, and not to slavishly copy. Although, I might add, we could slavishly copy their industrial production practices.

The TB-3 (ANT-6) Heavy Bomber

Work on the TB-1 was complete. The first Soviet bomber was in series production. It now belonged to the military, and they would give it its final evaluation. But how could one know how objective that evaluation would be? What future directions were required for large bombers? Having read the literature of air power theory, Tupolev was a frequent guest at air tactics conferences, which included air theoreticians such as V. V. Khripin, V. K. Lavtov, F. I. Ingaunis, and A. I. Lapinchsky. At these conferences foreign specialists were also studied, including Generals Mitchell (United States), Douhet (Italy), and others, who believed that ultimate victor in a future war would be decided by air power alone, specifically by bomber aviation. Tupolev, however, did not agree. Always soberminded and keenly analytical, he believed that "single" theories were a good application of the dictum of Kozma Prutkov: "Every specialist is like an abscess." Tupolev said, "Aviation cannot become the decisive branch of arms. An enemy territory cannot be seized by airplanes alone. It seems to me that this will continue to be the job of the ground forces."

Time took its course. Soon after work was completed on the TB-1, a landmark was noted in the Tupolev family: Aleksei entered school. It was necessary to look for a school a bit closer to home, as the boy was independent and did not want to be led by the hand. Now in the mornings, as soon as breakfast was finished, all parted, Papa to work, sister with brother to study. The house was empty. Mama nervously awaited the children's return home. On Sundays Papa had to check the children's homework. Lyalya was already factoring polynomials and beginning to solve equations.

Tupolev's mind, however, soon became focused on his next major project, building the ANT-6, designated Heavy Bomber 3 (TB-3). This project was a logical development of TB-1, but a much more ambitious design with four engines. Whereas the appearance of the TB-1 on the airfields summoned admiration and delight in the Red Army Air Force pilots, the TB-3 caused a real sensation. And no wonder! A length of 25 meters, with wingspan of 42 meters, a forward cockpit that rose almost five meters above the ground, with four A. A. Mikulin– designed engines and a double-bogie undercarriage—the airplane's dimensions were large and striking. When the prototype was moved from the factory to the airfield it was necessary to recruit the help of the streetcar directorate of the Moscow City Soviet. The overhead streetcar wires had to be removed in places so that the test aircraft could pass.

Always remembering instances when short-sighted people complicated the accomplishment of his "too progressive" projects, Tupolev recalled, "When we insisted on having the assembly shop in our factory made 70 meters wide, these 'stewards of the people's money' literally groaned, 'Well, why do you need this? After all, your airplanes have a wingspan of 15–20 meters, and here you are talking about 70. You will probably not make such machines even in 20 years.' And how many years passed? A total of five, and the TB-3 had a wingspan of 42 meters. Had we listened to these advisors we would have been unable to make such progress."

The technical specifications of the TB-3 were ambitious for 1930. The bomber carried three tons of bombs at a speed of 200 km/h, for a distance of 3,000 km. The crew consisted of seven men; a special radio operator was added, with a separate cockpit to protect him from the roar of the four engines. For defense, the TB-3 had six machine guns. With refinements, the TB-3 could cruise at 300 km/h. The armament was strengthened, a transparent cover over the pilots was introduced, wheels with brakes were mounted, a radio direction finder was attached, and many other innovations were applied.

These truly grand bombers were a significant achievement for our country. The government decided to exhibit them in the European capitals. In 1933–34 three TB-3 bombers visited Warsaw, Prague, Rome, Vienna, and Paris, generating great excitement among military and civilian specialists. At that time there was nothing similar in the Western countries. The USSR's long-range bomber capabilities were now formidable.[12]

By 1934, militaristic Japan possessed considerable control over Korea and Manchuria. Japan's penetration into these areas, bordering the USSR, alarmed the Soviet government, which then decided to transfer three TB-3 aviation brigades (150 airplanes) to the Far East. The transfer was brilliantly organized. Every two days, 17 TB-3 bombers took off from one of the airfields near Moscow. They arrived at the coast precisely on schedule, without a single late airplane. From that point, the TB-3 bombers represented a potential threat to the very center of the Land of the Rising Sun. This shrewd rebasing of our strategic bombers cooled the ardor of the Japanese army and navy. Conflict with Japan was averted.

In 1936–37, there were massive maneuvers with our airborne troops, who were dropped from TB-3 aircraft during training operations by the Kiev and Belorussian military districts. After these parachute troops seized the airfield, the same TB-3s delivered armored cars, tanks, and artillery—in short, everything necessary for an effective air assault operation. The Western military attachés who were invited to the maneuvers favorably evaluated the success of this operation.

Subsequently, the Air Force experimented with so-called parasitic aircraft, which were tested on TB-3s. Several fighters with pilots were attached to the wings of the large four-engine bomber, and were fed gasoline from the TB-3's fuel tanks. It was anticipated that the mother airplane would deliver them to the area of combat, where they would disengage from the bomber and defend the formation of their own bombers. They were to return using their own fuel.

The idea of the parasitic aircraft ultimately found application in the Great Patriotic War of 1941–1945. For a long time our Il-4 bombers hit the railroad bridge across the Danube near the city of Galati without results. The target was too small to hit from high altitude, and the enemy antiaircraft artillery was too dense to allow our bombers to approach at low altitude. This bridge was especially important because across it funneled supplies for the entire southern sector of the front of attacking Fascists. In the Crimea, two I-16 fighters, each with two 250-kg bombs, were attached to a TB-3. After delivering them across the Black Sea at low altitude almost to the target, the TB-3 detached both I-16s. They gained altitude, then dove on the bridge and destroyed it.

There are many glorious pages in the history of the TB-3. The Arctic explorer Papanin flew to the North Pole in a TB-3 under conditions of the polar night in search of Levanevskiy's lost aircraft.[13] TB-3s carried on their external stores especially large and important equipment to remote construction sites in Siberia and the Far East. They also made their contribution in World War II. Designed 12 years before that war began, they, of course, were obsolete. Nevertheless, they were used at first as night bombers, and then to drop combat equipment into partisan areas, and finally as transport aircraft in air assault operations throughout the entire war.

In considering the influence of the TB-3 with reference to any number of military-technical categories, one cannot help but conclude that this aircraft made a profound impact on bomber aviation worldwide. Had it not been for Tupolev's TB-3, airborne forces would not have appeared in those years, and the idea of the parasitic aircraft would not have arisen.

The TB-3 was one of the most simple and reliable airplanes to operate. Once these bombers ended their operational life in the Air Force, they were still used as cargo aircraft by Aeroflot—supplying polar stations on the coast of the Arctic Ocean, delivering furs from Yakutiya and Northern Siberia, and hauling water in the Central Asian republics.

TB-3s were built in a massive series production at several factories. To assist in managing their production, Tupolev did not just send specialists but frequently visited the factories himself. Tupolev was well known in the oblast' [administrative district] party committees of those oblasts where the factories were located, as he demanded factory expansion, completion of the construction of new shops and housing, receipt of equipment, and the laying of streetcar tracks to the enterprises.

When he visited the aircraft factories, he was interested not only in the manufacture of aircraft engines but in all phases of production. Believing that people determine productivity and quality, he did not forget about the dining halls or canteens, and was interested in whether the food was tasty and the locations clean. When he found a bone one day in his fish stew, he gave the dining-hall director such a dressing down that those nearby were afraid that he might damage his own health. Another time he surprised a factory director by suggesting they take a stroll into the shop bathroom: "Let us see how it is there!" When he saw that the floors were covered in thick mud, he stated somberly, "Now it is clear why there is such a high percentage of waste in the factory." The director understood and took measures. Knowing Tupolev, the director had no doubt that on his next visit he would take another look. And so it happened. Finding everything in order and being suspicious that it was a "Potemkin village,"[14] Tupolev was placated only after he had visited several such places.

Tupolev was just as well known in the Air Force units that operated other Tupolev-designed aircraft. He was interested in how the pilots evaluated his aircraft, what defects they had detected, and what sort of modifications they wanted him to make. Tupolev demanded that his design team take into account the fliers' comments. He was particularly annoyed by any slovenliness in design or production. Gazing through his strong glasses at some unsuccessful part, he would remark with amazement, "Who managed to design this garbage? Talent is necessary for this!" He would then summon the creator of the part, and very eloquently and dryly characterize both the part and its creator.

Sometimes such visits included side trips for his beloved pastime of fishing. On Sundays he led a group beyond the noise of the city. While water in buckets boiled over the fire on the river bank, he and his fellow fishermen set up a drag-net. Large, heavy, with untanned white skin, wearing only pants, "Himself" would become excited and start to shout as the fish were caught. After hauling the catch to shore, the fish soup was made in three steps. It was boiled three times with the small fish. The broth was then strained. Pieces of sterlet, purchased from local fishermen, were dropped into the broth. Tupolev himself always took charge of the finale. He tasted the concoction excitedly, and cursed if there was not enough pepper or laurel leaf. The group ate from bowls, lying on the grass, occasionally drinking a shot glass full of vodka cooled in the river—even as they constantly brushed off the fierce Volga mosquitoes that descended on the party.

On one fishing trip, Tupolev suddenly swam away from the bank with powerful overhand strokes. He was an excellent swimmer, but we became worried, and I set out with I. F. Nezval to overtake him. We managed to do this with difficulty, somewhere in the very deepest part of the stream. Laughing and spitting out water, he suggested, "Let's go to the other side, to Uslon." Although this was before the hydro-electric stations were built, the width of the river at this point was some 800 meters, and we talked him out of it. After we returned, he stretched out on the sand, snorted, and said, "Now that is a sport, it is not like jumping rope."

In general, it can be said that during the work on the ANT-4 and ANT-6 the creative thinking of the collective, if not always smooth, was nonetheless clearly focused. Only external circumstances plagued them, for example, the frequent reorganizations. In August 1931, for example, while Tupolev was out of the country, it was decided to combine all experimental airplane construction into one mighty organization. The second large design bureau, which existed at the Menzhinskiy Factory, was merged with the Aeronautical and Seaplane Experimental Construction Department of the TsAGI. S. V. Il'yushin was named chief of the combined center, and A. N. Tupolev appointed his first deputy.

Upon his return Tupolev protested the change. He believed that any monopoly contributes to decline, and that progress is possible only in competition. By this time, the accuracy of his objection was being confirmed. The huge experimental multi-element bureau became too difficult to manage. Work bogged down. Dozens of types of airplanes were being designed and built simultaneously. Sometimes small parts needed for one nearly completed airplane were not available, yet the production facility was loaded down with parts for a different airplane still in design. Sometimes an assembly for one airplane was mistakenly fastened to another. Naturally, this engendered friction between the various design teams and their leaders. Finally, by January 1933, the absurdity of a single design center was recognized at all levels, and the authorities ordered a return to two bureaus. As before, one became the Aeronautical and Seaplane Experimental Construction Department of the TsAGI, and the other the Menzhinskiy Central Design Bureau. Later, recalling this episode, Tupolev said, "This rather muddle-headed reorganization took place because the people involved had too many ideas about which they had too little knowledge." It must be said that this conclusion was a fair one.

During the years in which the TB-1 and TB-3 were being created, Tupolev formulated an idea that he subsequently adhered to—the Soviet Union needed a passenger aircraft. "I think that this can be done by borrowing the wing, chassis, tail section, and sometimes even the cockpit and the engine-propeller unit from a previous military airplane that has undergone severe operational testing, and by manufacturing a new fuselage," he asserted.

In precisely this way, based on a TB-1 after it had been in operation for several years, the nine-seat passenger ANT-9 *Wings of the Soviets* was produced. On May Day 1929, the airplane was exhibited at Red Square, next to the cathedral of Basil the Blessed. That same summer, pilot M. M. Gromov, together with a group of aviation specialists headed by A. A. Arkhangel'skiy, flew this airplane to Berlin, Paris, Rome, London and Warsaw.

Later, the 36–seat ANT-14 *Pravda,* a design based on the TB-3, flagship of the Gorkiy Political Agitation Squadron, took to the air. In ten years of operation, approximately 40,000 passengers received their "aerial christening" in this aircraft.

Tupolev Becomes Chief Designer

Georgiy Konstantinovich Ordzhonikidze, People's Commissar for Heavy Industry, spent a long time selecting the candidate for chief engineer of the main administration of the aviation industry. He decided on Tupolev, not only because he was the chief designer of the heavy aircraft being developed, but also because of his organizational capabilities, sobermindedness, decisiveness, and patriotism. And this choice proved to be a wise one. Whereas N. Ye. Zhukovskiy had nurtured the young student Tupolev as he grew into a great engineer, G. K. Ordzhonikidze gave Tupolev the scope of a great state figure. Tupolev related with great warmth and humor the first lessons that Ordzhonikidze had given him on how to manage Glavkoavia [the main administration of the aircraft industry]:

Initiative must never be stifled, even if it is incorrect. Imagine [he used the familiar form in addressing the majority of his close associates, but this was not familiarity, but, to the contrary, his Georgian disposition] that someone comes to you with the idea of making some unusual airplane, and you have doubts. Say that you looked at it and didn't see anything faulty. But you continue to have doubts. Here there may be two decisions. First is that there is little money in the country, and we will not build it. But would this be right? But wait, my friend, is it a talented design? Let us give some thought to how it should be. And who told you that namely for this project there is no money? There probably isn't, but just the same, one must not shut off an idea. After all, if the country needs this airplane, we will not niggle, we will give as much as necessary! Now let us look from a different viewpoint. And suddenly, is this bashful designer a gift from God? You do understand, don't you, my friend, that he might be wasted? It is true that we do not have very much money, not much to waste, but to create a new and needed airplane, can we really not find any? You and I, my friend, would be criminals if we could not. And this is the science of leadership! Is that not so, Comrade Tupolev? What is the difference between the first and the second solution? None at all. The most important thing is not to waste talent, but to help him. That is the way I ask you to work. And if you feel a need for assistance, do not worry, I am always with you, and will always support you.

Tupolev appreciated Ordzhonikidze as a leader in 1930, serving as People's Commissar of Heavy Industry. The majority of his predecessors were improvisers, who concentrated on the problems that arose instead of moving ahead.

By the 1930s Tupolev enjoyed popular acclaim and official recognition. The Supreme Soviet awarded him the orders of the Red Star and the Red Banner, and then the highest award of the homeland, the Order of Lenin. In February 1933, the general meeting of scientists of the USSR Academy of Sciences elected him an associate member. That same year TsAGI celebrated its fifteenth anniversary, a festive occasion at the Bolshoi Theater. In attendance were members of the Politburo and Central Committee of the party, the various people's commissars, and many prominent scientists and military men.

Opening the session, Ordzhonikidze remarked:

Our aviation industry is experiencing indisputable and tremendous success. The Central Aero-Hydrodynamics Institute has played a decisive role in these achievements. Several years ago our aviation industry depended entirely on foreign equipment. This has now fundamentally changed. A special merit of the TsAGI is the fact that it never walled itself off from the factories, from industry. The holiday of the TsAGI is a holiday for all those engaged in technological work. They said that the Bolsheviks would not be able to handle the task of building a socialist economy because they did not have their own engineers. Andrei Nikolayevich Tupolev is a representative of this best and numerically large part of the new Soviet technical intelligentsia. Saluting the TsAGI, I salute its founders and leaders, comrades Tupolev and Chaplygin, and in their person all the Soviet intelligentsia.

Then K. Ye. Voroshilov spoke.[15] He stated, in part,

The Central Aero-Hydrodynamics Institute is dear to us all, not only because it has achieved tremendous results in its creative work, but also because in it are raised new cadres of scientific research workers. The TsAGI has shown what people inspired by the ideas of socialism may attain. Talented companions-in-arms N. Ye. Zhukovskiy, A. N. Tupolev, S. A. Chaplygin, A. I. Nekrasov, and others have led the TsAGI from victory to victory. Our military combat aviation is very indebted to the successes and growth of the TsAGI, and, in particular, to Andrei Nikolayevich Tupolev.

Tupolev responded:

Today I have been given the great honor of expressing for the entire TsAGI collective gratitude to the party and government for the exceptional attention and high awards given to us. The firm leadership of the party and government ensured the successes of the TsAGI, and its becoming one of the world's leading, largest aero- and hydrodynamics institutes. In the future as well, the TsAGI collective will fight to introduce the latest achievements of science and technology into the aviation industry of the Union, and to create a powerful Red Air Fleet to protect the borders of the Land of the Soviets.

Working for the Navy

When recounting Andrei Tupolev's career one must not forget the Navy. At the end of the 1920s, the TsAGI received an assignment to build a torpedo boat for the Navy. After the aerosleighs were completed, Tupolev built two hydroplanes, the GANT-1 and GANT-2, in association with Zhukovskiy.

Both Tupolev and Zhukovskiy tested and made many calculations on the behavior of hydroplanes. It was necessary for them to engage in theories and experiments involving hydrodynamics, stability, and a number of other new problems and questions. At TsAGI, it was discovered that in a number of fields related to hydrodynamics they had been simply ill-informed. Moreover, the classics of Russian shipbuilding, and in particular the works of A. N. Krylov, contained detailed information about how to design and build a battleship, but nothing useful for the design of test boats. To fashion their own test data, they obtained the calculations from D. P. Grigorovich's M-9 flying boat and studied them in detail. Grigorovich's pioneer designs had been developed in the prerevolutionary period.

In mid-1921, the first test boat, the GANT 1, was pulled into the yard for engine tests. The craft's propeller was replaced with a "dummy" that threshed the air when the motor was turned up, without creating thrust. Then they took it to the Yauza River, which ran near the school. At first the trip went well, but near Semenovskaya Street, opposite the Provodnik Factory, GANT-1 ran headlong into the shoals. They could not pull it off themselves, but a crowd of bystanders helped. Someone brought ropes, and to shouts of "Ready, heave," they pulled the GANT into deep water. The Yauza turned out clearly to be unsuitable—it was narrow and windy. In order to continue the tests, they decided to rebase to the Moscow River. Near the Khamovnicheskiy

Barracks, opposite Neskuchnyy Garden, they fenced off a section along the bank and organized a port for test boats.

Tupolev was not always serious—he could be mischievous as well. Josef Fomich Nezval, one of the engineers who began working with Tupolev in the 1920s, recalled an incident from the time they were testing the test boats. "One day we arrived, turned on the motor, and began to go toward the Vorobyevyye Mountains. Nearby were two lovebirds in a pleasure boat, who were kissing so sweetly that they didn't oblivious to everything. Tupolev turned the wheel sharply, made a loop around them, and laughed. The boat rocked, the frightened youth jumped up, balancing himself, threatened with his fists and shouted. Having made sure that the boat was not taking on water, Tupolev raced the boat farther, and, smiling into his beard, said: 'I got them a good one, eh?' "

After many long disputes, and then experiments, with GANT-1, the collective decided to build the next test boat out of corrugated duralumin sheeting. It was necessary to develop the technology for airtight riveting of the hull in order to prevent corrosion of the light metals in the water. Needless to say, the collective also had to create the apparatuses, tools, and building slips for assembly. They figured it all out themselves, and did everything with their own hands. Sometimes things did not work out and assembled and prepared connections turned out to be so bad that there remained nothing to do but simply throw them away and laugh. And Tupolev laughed louder and more expressively then anyone. Most important was that, while confronting an entirely unknown field, the young people did not throw up their hands in desperation. To the contrary, the more difficult it was, the more furiously they worked to overcome the difficulties.

No one working on GANT-2 thought that it would become the precursor of the torpedo boats—the "mosquito" fleet of our country. Having assessed the results of the tests of the two Tupolev boats, the Navy command decided to continue the construction of such boats. OSOAVIAKHIM, a popular volunteer youth movement, also showed an interest.[16]

Based on the Navy's report, the government ordered Tupolev and TsAGI to look into the possibilities for constructing combat torpedo boats, which became the G-3. In order to achieve the Navy's required speed of 40 knots, the designers mounted two 600-horsepower aircraft engines on the G-3. The boat was built out of a duralumin-type alloy. Tupolev wondered how well the light metal would perform in seawater. "Bring a couple of railroad tank cars of seawater from the Black Sea," he instructed D. O. Orlov, chief of supply. "We will test it out in the same water in which we will be operating!"

The TsAGI's Aviation Materials Department was filled with tubs of seawater in which samples of sheets, profiles, and entire parts were kept for months. In the saltwater the duralumin alloy corroded quickly, a problem overcome only after long and difficult tests.

The boat was built and underwent testing. Not everything went smoothly. From the beginning the G-3 simply failed to reach its design speed. The collective toiled over this day and night. They changed the screws. In search of the optimal shape they made design changes, adjusted the carburetors, and tested various brands of gasoline and lubricants in the laboratory. The boat continued to be stubborn. Tupolev was summoned. Having evaluated the situation, he proposed several steps. It was then that the legend was born of how Tupolev looked at the screw, took a hammer, struck it once or twice on the blades, and the speed immediately increased to design speed. Of course, this is a convincing story for children, but adults understand that the shape of the blades, which are calculated theoretically and checked experimentally, cannot be corrected in this way. Tupolev, as a great designer, in no way needed the dubious laurels of a conjurer. In actuality he ordered the entire chain inspected: engine, shaft, screw, hull outlines, and so on. It turned out that the piston rings needed to be changed; because of an irregularity in the engine block, one of the screw blades was nicked, apparently from striking a rock; and most important, the boat, which had been sailing for six months, had not been hauled out and barnacles and marine growth had formed on its bottom. It was these "tricks" at which Tupolev was the master. He was always using Prutkov's expression, "The evils are at the root." The "first-born," as the sailors called the G-3, reached its design speed and entered the rolls of the Black Sea Fleet.

When V. Zof, the Commander of Soviet Naval Forces, was told the results, he asked the Central Committee and state authorities to charge Tupolev with the design of the next torpedo boat, the G-4, to be armed with two 450mm torpedoes, which reached a speed of approximately 50 knots (almost 90 km/h!). And a year later the G-5 was built, with new, larger 533mm torpedoes, and with two 1,000-horsepower motors. Tests of the new boat concluded on 17 June 1929. The official document stated: "The commission of the Naval Forces considers that the GANT-5 torpedo boat is the best torpedo boat of all in existence." It was recommended that its large-scale production be organized at the shipbuilding factories. The shipbuilding factories had no experience in working with duralumin. This was when Tupolev performed his next "trick." He formed a team of workers, foremen, technicians, and engineers at his factory, and sent them to the Mari Shipbuilding Factory with the words, "Go there not to criticize, but to teach. Assemble the first boat with your own hands, from nose to stern. Build the next two with their workers on equal terms. They will make the next two themselves and you will help, not so much with your hands as with advice. When the Navy accepts all five your mission will be complete, and you may return."

Meanwhile, their appetite whetted, the Navy authorities decided to order the TsAGI to build a boat "a little bit larger." The number of torpedoes was increased to six, and a gun and three machine guns were required. According to calculations, it turned out that to reach design speed 800 horsepower were needed, that is, eight aircraft engines. The dimensions of the boat had grown to a true ship—35 meters in length, 6 meters wide, a displacement up to 70 tons, and a crew of fifteen.

Doubts arose as to whether it should be built. Those who prevailed pointed out that a new and possibly very valuable weapon could be assessed only after it had been tested under real conditions at sea, during fleet maneuvers. Its model was approved and construction began. Immediately the question arose as to how to deliver it from Moscow to the Black Sea. This was a challenge because the Volga-Don Canal did not yet exist. It was decided to move it on a barge to Stalingrad, from there on a trailer specially built at the TsAGI, along the Steppe roads to Kalach on the Don, then on another Don barge down the Don and Sea of Azov. What kind of road was there from Stalingrad to Kalach? Tupolev sent a truck with specialists to investigate and found that the road was passable in dry weather, but that the bridges across the streams and ravines would have to be strengthened.

When everything had been specified, Tupolev's collective designed and built the trailer themselves. At night, the gigantic G-6, wrapped up in canvas, was transported to Nagatino and loaded on a barge. Tupolev and N. M. Kharlamov supervised the loading. Why such senior supervisors? No lift cranes existed, so success depended on rigging ability. The 37-meter-long vessel had thin dural plating, which would require months to restore if it was bent or broken. After some six weeks the G-6 arrived at the test site, and a few days later the Black Sea waves tenderly rocked her.

The G-6 turned out to be a "tsar-boat." The main idea of the mosquito fleet was the small size of the boats. With the combination of tremendous speed and good maneuverability it was becoming invulnerable. And here was a regular steamship! Speed was down, maneuverability worsened, the boat was detectable as a target much sooner than the G-5. The maneuvers confirmed that three G-5 boats, totaling the same six torpedoes as one G-6, were much more effective. In short, the naval command rejected the idea of gigantic "mosquitoes," and the G-6 was not built in series production.

The Great Patriotic War, 1941–45, became the crucible to test this design of Tupolev. Hundreds of G-4 and G-5 boats, in all of the fleets from the Baltic to the Black Sea and the Pacific Ocean, did their work. The enemy felt the full force of the strikes by the Soviet torpedo boats. Tupolev had every reason to be proud of his contribution to the naval might of the country.

Hydroplanes

Naval aviation underwent rapid advances. The flying boats built by aviation designer D. P. Grigorovich were considered the best (he began to build them back before the war of 1914–18). The British and French requests in 1916 for the drawings of the Grigorovich's M-9 flying boat (the ninth variant of the naval reconnaissance aircraft) for series construction in their own factories confirmed Grigorovich's reputation as a designer. Naval aircraft were built using the same materials as those used on land—wood, percale, and plywood. Unfortunately, all those materials were quite vulnerable to saltwater and by the end of the Civil

War, the young Soviet aviation found itself without flying boats. By the 1920s the majority of the land-based aircraft were obsolete, and the naval aircraft had simply rotted away. Since the hydroplanes were the "eyes" of the squadron commanders, the naval department insisted that flying boats be included in the Central Institute agenda of experimental work.

In 1925, Tupolev began to design his first hydroplane, the ANT-8. Considering that the topic of his dissertation had been *Design of a Hydro Airplane,* it is evident that Tupolev had been working toward this for a long time. He assigned one of Zhukovskiy's students, engineer-pilot I. I. Pogosskiy, the role of lead designer. Henceforth, Pogosskiy became his permanent assistant in naval aircraft construction. A group of designers organized to assist him became the embryo of the future team of designers of naval aircraft at the Central Institute's Aeronautical and Seaplane Experimental Construction Department. By this time the metal airplanes, the ANT-3 *Proletariy* and ANT-4 *Strana Sovetov,* had performed a satisfactory number of test flights, so it was decided to build the naval ANT-8 out of the same duralumin-type alloy.

It was possible to begin the hydroplane design phase, but there was a priority for land-based airplanes, which the Army needed even more acutely. The ANT-8's design was held back, and the prototype entered testing only five years later, at the end of 1930. Economizing on resources, Tupolev decided to use the wing and tail assembly from the passenger ANT-9, which had already been flight tested, on the MDR-2. Only the hull of the boat and the motor mounts were newly designed. The water channel had not yet been built, and the boat's lines were chosen mainly by intuition. Of course, Pogosskiy's analysis of several foreign boat designs was of considerable use. Their first seaplane floated, took off, and landed on the water well.

Noting the lack of an adequate testing area for the new hydroplanes, the TsAGI leaders gave authorization for a new base. Tupolev went to Sevastopol, and fell in love with a piece of land not far from the city on the bank of one of the numerous bays. There he built a hangar with a slipway for launching assembled airplanes onto the water, and a shop building. He also sought approval to build a settlement of several houses nearby. Thus, the "naval" affiliate of the TsAGI arose on the Black Sea shore.

By the end of 1930, the disassembled ANT-8 was hauled by rail to the south. Flights began in the spring. Pilots M. Gromov and N. Kastanayev tested the hydroplane. Although the design specifications were met, the forced delay in construction caused the MDR-2 to be obsolescent, and it was not placed in series production.

Of course, this distressed the naval team. Tupolev calmed them: "As the saying goes, the first step is always the hardest, but there is no need to be upset. Just think, we have put together a group of seaplane designers. Second, you were able to choose good lines for your first seaplane although the water channel had not yet begun to operate. I consider this an indubitable success. Third, and this was a great achievement, our materials specialists found a means to combat the corrosion of dural in seawater. Finally, the airplane turned out not bad at all. So you don't have to worry about unemployment."

As a matter of fact, orders for seaplanes came one after another. At the end of 1932 an order came for a very interesting aircraft, a "sea cruiser," designated the ANT-22. The naval leadership's requirements were set down in laconic lines: "Sea Cruiser 1 (MK-1) is intended to strike task forces of enemy ships, and must have a bomb and torpedo load of 7–9 tons, a speed of 220–230 km/h, and a range up to 2,500–3,000 kilometers." Seagoing qualities were specially stipulated: The Navy demanded that the airplane take off and land in waves up to 1.5 meters high, be capable of a long and autonomous stay in the open sea, and be able to be towed by ship.

The most difficult task in designing such a mighty seaplane was to achieve excellent seaworthiness. If they were to proceed in the usual way, the wingspan of the MK-1 would have been 50 meters. With such a span, if a single-hull design were used, the heel could become so great that the ends of the wings would begin to sink into the water. The only means of preventing this from happening was to place support floats on their ends. But in order to be adequate for such a huge machine as the MK-1, these floats would be so large as to lead to a substantial loss of speed, and probably would make it more difficult to keep the machine on course. Even more troubling was the question of wing strength on a 25-meter wing. The force of a 1.5-meter wave would create a considerable dynamic load. Day after day Tupolev and Ivan I. Pogosskiy sketched variants of the MK-1, but could not find a solution. They sat in the office gloomy and somber.

"Which variant shall we examine, Ivan Ivanovich?"

"We have probably already examined dozens, Andrei Nikolayevich."

"You know, Ivan Ivanovich, it sometimes happens like this: You push literally against an unseen wall and spin your wheels, and just can't find the right solution!"

As often happens, it came suddenly—a catamaran! It is impossible to say now which of them first came up with this thought, but the arrangement was much easier and more productive.

The catamaran had two hulls, each covered by a wing, above which six AM-3R engines were mounted in three gondolas, placed in tandem behind one another in three sets of two. The wing, resting on the two hulls, did not require floats. The solution they had found freed the designers from the majority of their previous difficulties. The glassed-in cockpit protruded forward from the center part of the wing. Gunners with their weapons were located in both hulls.

Pogosskiy's team began designing the machine in January 1933 and the finished MK-1 reached the fleet in July 1934. The sea cruiser turned out to be truly huge. On 8 August pilots T. Ryabenko and D. Ilinskiy lifted the MK-1 into the air for the first time. During the tests the aircraft's speed turned out to be somewhat slower than planned. The propellers on the rear engines, which rotated in the air stream from the forward engines, turned out to be less efficient than expected and did not provide the required thrust.

But what turned out to be outstanding was the MK-1's seaworthiness. The water channel, which had been built and entered service, now made it possible to evaluate the seaplane while it was afloat, and during takeoff and landing. They hung the two-hulled model MK-1, made of wood and paraffin, under a self-propelled trolley that rolled on rails along a channel above the water. The trolley dragged the suspended model at various speeds along the water that filled the channel. It was uncommonly apparent in its smooth surface how the airplane, picking up speed, split the wave, how it moved up on plane,[17] and then lifted from the water. By slowly lowering the model while gradually slowing the movement of the trolley, it was possible to imitate the landing itself.

Here the sea-cruiser model touched the channel water surface. First the keels of the two hulls made two slender lines on the smooth surface of the channel, and then both hulls came to a stop and raised fountains of spray. The model slowed. This was a very graphic and convincing exercise.

Along with measurements on the trolley, a camera took pictures of the simulated takeoff and landing. Analyzing the test materials and looking at the films, Tupolev and Pogosskiy tried to find better and better hull contours, until they decided on the most "ideal," in their opinion. It must be noted that the final shapes of the MK-1 hulls differed considerably from the initial, theoretical shapes. The model, reworked based on the test results, and already with a full-sized wing, was tested for stability. An apparatus for wave-formation was lowered, and waves corresponding to 1.5 meters in height ran along the surface of the channel. The model MK-1 rocked, but the catamaran design was so stable that no matter what the angles at which they met the waves, the ends of the wing did not touch the water.

Satisfied and pleased, Tupolev summarized: "Now, having in addition to three aerodynamic pipes, a water channel as well, we are capable of fulfilling any assignments."

How useful the channel was! During tests the MK-1 confidently took off and landed in the open sea in waves of 1.5 meters, floated stably in a strong wind, moved easily in tow behind a ship, and taxied. The MK-1 concluded state tests in fall 1935, at a time when many countries were still reexamining their seaplanes. Large, slow-moving machines were giving way to smaller, fast ones. The USSR anticipated this. The ANT-37 Rodina was already flying, and the ANT-40 (SB) had been moved to the airfield. At the same time, people were discussing the possibility of using ordinary land-based airplanes for overseas operations. The reliability of aircraft had increased, and forced landings had become rare. It also should not be forgotten that through standardization it was possible to sharply increase output, so that more airplanes were available.

No matter how burdened Tupolev was with his innumerable jobs (as chief of AGOS and chief designer, he was also a member of the Central Institute Collegium, handled the construction of the institute, and was constantly brought in to decide questions in the People's Commissariat, and the Supreme Council of the National Economy), he stole time to travel to the seaplane base. This was yet another characteristic trait. He always kept track of the course of events himself. From his youth to his old age, he remained light on his feet and without any hesitation flew to distant places, if only to see personally how things were going, and to intervene and speed them up.

Nearly simultaneously with the MK-1, Tupolev worked on the MDR-4 long-range reconnaissance seaplane. The history of this airplane is very unusual. In 1932 designer I. V. Chetverikov transferred his hydroplane, the MDR-3, to the Navy for testing. The airplane did not achieve its stated specifications, and it was discarded. Finishing the vehicle required substantial design and production modifications, which the small Chetverikov group could not handle. Considering the Navy's very acute need for reconnaissance aircraft, the People's Commissar for the Navy and the People's Commissar for Heavy Industry decided to transfer the aircraft to a stronger organization for completion, and they assigned Tupolev to get it into shape in the shortest possible time.

Tupolev and Pogosskiy discussed the situation and decided that the MDR-3's shortcomings could not be fixed without substantial redesign. The wing, with numerous developed struts, needed to be replaced with a cantilevered wing, of the type that had been well tested on the TB-1 and ANT-9. Instead of four BMV-VI engines in two tandems, it was more reasonable to mount three more powerful Mikulin M-34R engines. It was necessary to remake the complicated strutted tail assembly into a simpler one such as that of the MK-1. Having left only the hull virtually unchanged, by March 1934 Tupolev's bureau was able to begin flight tests of the fundamentally rebuilt machine.

"Although the flights went well on the whole, the pilots did not like the takeoff of the aircraft, and Pogosskiy asked me to come to the base," recalled Tupolev. "In the morning in Sevastopol the weather was good, sunny, and a light breeze was blowing. The flight was set. Pogosskiy decided to conduct it himself and demonstrate to us the peculiarities of takeoff. A small, rolling wave was coming in from the sea, but it did not reach us in the bay. We took a boat to the liftoff. A little later a whine could be heard. This was Pogosskiy starting the engines. The flying boat began to accelerate. I must say that, at that time, many people were flying on the heavy Dornye Val machines. They had a peculiarity in that they did not want to lift off from water. In order to make the takeoff easier, seaplane pilots thought up the idea of rolling the machine, saying that this made it easier to take off. I watched, and Pogosskiy and Volynskiy, already at the very outlet from the bay, began to roll the MDR-4. I jumped up, shouted, and waved my arms—stop! But how could they hear me in the thunder of three engines in takeoff? When the machine, already at considerable speed, left the bay and entered the sea, it began to beat against the waves. When it struck the first wave it bounced, but not yet having reached a speed at which the wing was capable of carrying it into the air, it sat down and cut into the second wave. The struts of the little engine could not withstand the blow and broke off, and the engine crashed into the cockpit and killed both pilots."

Tupolev grasped his head, covered his face with his hands, moaned, and literally crumbled to the seat of the boat: "Hurry there!" The boat scurried about the site of the catastrophe in vain. The sea did not give up the bodies of pilots I. I. Pogosskiy and A. A. Volynskiy, or engineers G. S. Noskov and K. K. Sinelnikov.

One must understand what grief he felt. Close collaborators had been killed, including Pogosskiy, one of those with whom he had worked shoulder to shoulder, arm in arm, since the very birth of the TsAGI. How could he, returning to Moscow, withstand the gaze of the relatives of those killed? How could he explain to his coworkers the absurdity of this tragedy? This first catastrophe in the group revealed to his fellows the profoundly personal aspects of Tupolev's character. He always treated the employees of the Experimental Design Bureau and the factory, regardless of their rank, as he did his own family. He knew the majority of them by their first names, and was interested in their lives, successes, and schooling. The accident changed him: he became haggard, gloomy, introspective, and reproached himself constantly for not having foreseen it.

It was necessary to take care of the families of the deceased. Tupolev managed to obtain allowances and pensions for them, and helped them in every way possible. Of course, accidents in aviation are unavoidable, especially in testing, but each time they happened Tupolev took them hard.

It may be as a result of his fatherly attitude toward his coworkers that they called him "Papa." When, in the factory or the design bureau they said, "Papa asked, Papa said, or Papa commanded," everyone understood who they were talking about. He had many nicknames: "Papa," "Ded" [Grandfather], and "Starik" [Old Man]. These names show how warmly his coworkers regarded him, despite his demanding and at times even autocratic nature. He knew the majority of these nicknames, and he accepted them without offense. One day they had gone to the Volga for fishing. It was shallow along the bank, and the motorboat was not able to reach it. They decided to cross on some inflatable rubber boats that were on board. But Tupolev did not want to risk sitting in one of them. He said, "I weigh too much," and he remained on the boat. Two of his assistants floating in the boat were talking quietly, "It is cool and damp, and better that Andryupolev not catch cold."

In the still of the night the sound carried far, and, sitting on the deck of the boat, he heard them. "How is that?" Increasing laughter rang out from the boat. "Andryupolev? That is a new one!"

After the catastrophe with the experimental MDR-4 and Pogosskiy's death, the deputy of one of the deceased engineers, A. P. Golubkov, headed up the seaplane team. Careful investigation revealed no defects in the aircraft, so they decided to build a duplicate. It was completed rather quickly and tested in the summer of 1935. They delivered to the Navy an aircraft that could fly approximately 2,000 kilometers with two torpedoes or a bomb load.

Tupolev also developed a passenger airplane, based on the MTB-1 naval torpedo-carrying bomber, to transport resort visitors and tourists along the Crimean-Caucasus coast of the Black Sea. In its civilian form, the MTB was intended to have sixteen comfortable passenger seats and a separate compartment for baggage. Possessing a speed of approximately 240 km/h, the machine could fly from Yalta to Sochi in two hours, and from Odessa to Batumi in five hours. At the time this was an extremely rapid means of transport over the sea. To give passengers the opportunity to appreciate the beauty of the Black Sea coast, Tupolev proposed a window cut in the aircraft hull to the side of each seat. When the speed of the MTB bomber no longer satisfied the Navy, production was halted and the passenger ANT-29 airplane was never pursued.

The ANT-20 *Maksim Gorkiy*

Nothing can give such a vivid impression about the chief designer's creative capabilities and his design department's dedication as can the production rate of the test aircraft of the Experimental Design Bureau between 1929 and 1934. Over the course of these six years, thirteen test aircraft were created in the design bureau, that is, more than two aircraft per year. It should be recalled that among them were such machines as the four-engine TB-3, the five-engine *Pravda,* the eight-engine *Maksim Gorkiy,* and the six-engine MK-1 cruiser and ANT-25 (*Rekord dalnosti* [*Distance Record*]).

The ability to introduce thirteen airplanes in such a short period of time is indicative not only of Tupolev's extraordinary skill, creativity, and organizational talent, but also of the significance of his closest assistants and the entire makeup of the design bureau. Undoubtedly, the enthusiasm that gripped the workers, the technical staff, and all the people of our country in the years of the first five-year plans played a role as well. In any event, no other design organization at that time was capable of accomplishing such a gigantic amount of labor in such a short period of time.

By the beginning of the 1930s, the Air Force and Civil Aviation had obtained all the types of series-produced Soviet-made airplanes they required. These airplanes included I-2 fighters (designed by D. P. Grigorovich), I-3 fighters (N. N. Polikarpov), I-4 fighters (Tupolev), R-3 and R-6 reconnaissance airplanes (Tupolev), and R-5 reconnaissance airplanes (Polikarpov), TB-1 and TB-3 bombers (Tupolev), the ANT-9 and ANT-14 (Tupolev), K-5 (K. A. Kalinin), and Stal-3 passenger airplanes (A. I. Putilov).

The attention of leading aviation designers then turned to creating competition-class airplanes capable of setting world records. Proven by their successes in range of continuous flight and the ability to lift maximum loads, they knew they had enough experience and talent within the Russian aviation industry to be competitive. France, England, America, and Germany were attempting to design record-breaking aircraft as well.

The need to increase load-lift capacity stemmed from the ideas about increasing bomb load. It was believed that once the airplane was at the target, it had to destroy it as thoroughly as possible. The competition raged. The press reported that the Italian designer Caproni was building a wooden bomber with a nine-ton load capacity; the British Blackborn Company was threatening to lift ten tons of bombs on its wooden biplane; and in Germany, Dornier AT was designing a ten-engine flying boat, the Do-X.

Soviet military doctrine adhered to these same views. Our airplanes' bomb load increased every year. In 1925 the TB-1 was required to carry 800 kilograms of bombs, in 1930 the TB-3 had 2,000, and in 1933 the TB-4 had 4,000 kilograms.

Neither the load-carrying capacity nor the range problem could be solved without large-scale preliminary tests and experiments. Naturally the government assigned these problems to the TsAGI, naming Tupolev as leader of both projects.

The end of 1932 coincided with the fortieth anniversary of Maksim Gorkiy's literary career. M. Koltsov, editor of *Ogonek,* proposed that a gigantic airplane be built in Gorkiy's honor. The aircraft would be dedicated to political propaganda. The idea was met with enthusiasm. Soon more than 6 million rubles were collected in voluntary contributions. Even before construction of the gigantic aircraft was begun, on 17 March 1933 a special agitational squadron named after Gorkiy began to operate as part of the Civil Aviation Fleet. The ANT-14 *Pravda* became its flagship.

A construction committee was formed under M. Ye. Koltsov. At the time Tupolev had just completed the design of the six-engine ANT-20 passenger airplane. When Koltsov consulted with Tupolev, they decided that the ANT-20 would be adapted for the new Gorkiy flying machine. The committee had unique requirements for the aircraft, named the *Maksim Gorkiy.* Necessities included a salon on board for press conferences, cabins for 70 people, a small press, a powerful radio center for broadcasting to the ground, photographic and film laboratories, two film projectors (it was proposed that the screens be raised up 6–8 meters with the aid of a pair of small balloons), several light projectors, an autonomous power and telephone station, a buffet, kitchen, restaurant, and sleeping berths.

All of the equipment and passengers exceeded 14.5 tons in weight, more than a six-engine aircraft could lift, but Tupolev had already decided on eight engines for the *Maksim Gorkiy*. Its dimensions became impressive: length 33 meters, wingspan 63 meters, height of the vertical stabilizer 11.3 meters. Each undercarriage leg carried two wheels two meters in diameter. The wing root was so thick that it could easily accommodate the sleeping cabins. The eight engines, totaling 7,200 horsepower, raised the 42-ton machine into the air and carried it at a speed of more than 220 km/h. Up to that time no one had built such a huge monoplane.

The main problem on the *Maksim Gorkiy* was its unprecedented design configuration. The ANT-20's design did not fulfill the new requirements even though it was an upgrade from the large TB-3, ANT-14, and TB-4, all of which had been well tested in operation. Tupolev did not want to introduce the design innovations into the ANT-20: "It is too responsible a task, creation of an agitation airplane. We have too little time, and we do not have enough personnel to work on two experimental designs simultaneously."

Only the entire bureau could handle the design of such a "goliath," especially in such a short period of time. Tupolev, retaining overall supervision for himself, entrusted the team of V. M. Petlyakov to design the wing, the team of A. A. Arkhangel'skiy the fuselage, Ye. I. Pogosskiy and A. A. Yengivaryan instrumentation, and N. S. Nekrasov the tail unit. B. A. Saukke was placed in charge of the airplane. He had begun work at the Central Institute back in 1924. A number of scientific research institutes, such as TsAGI, TsIAM [Central Scientific Research Institute of Aircraft Engines], VEI [All-Union Electrotechnical Institute], TsRL [Central Radio Laboratory], and several others, were involved in the creation of the ANT-20. The factories also became involved, including the Gelets printing equipment factory, Kinar film apparatus factory, Kauchuk projector factory, Popov and Morse radio factories, Moscow Electrical Factory, Lepse Factory, and a number of others. In essence, all of Soviet industry cooperated to build the flagship of the agitation squadron.

Once the Experimental Design Bureau had built the full-scale wooden mock-up of a fully equipped ANT-20, with all of its equipment, Tupolev and his wife, Yulenka, invited the committee members to examine the airplane. It must be emphasized that Yulenka Nikolayevna was not only the wife of the well-known designer and mother of his children, but also his creative assistant. Without being on the staff of the Experimental Design Bureau, she was an active coworker. She invested much work and energy in creating the internal decorations of the passenger salon of the aircraft to give people a feeling of at-home comfort in flight. She also assisted in furnishing the salons and their interiors of all subsequent ANT and Tu passenger airplanes.

Committee chairman M. Koltsov, committee members Ye. Zozulya, A. Khalatov, and S. Uralov, as well as Air Force Commander Ya. Alksnis, Chief of Civil Aviation and Deputy People's Commissar for Defense I. Unshlikht,[18] Army Commander M. N. Tukhachevskiy, academician S. A. Chaplygin, writer A. N. Tolstoi, people's artist I. M. Moskvin, and a number of others came to look at the mock-up. The guests examined the huge airplane, which seemed even more grandiose inside the shop, saw the print shop, radio center, film projectors and laboratories, salon for meetings, and sleeping cabins.

Tupolev was a cordial host. He loved this role, and it suited him. His high voice and expressive laughter could be heard everywhere as he called a group of guests to the chassis of the aircraft: "Come take a look. Two wheels each two meters in diameter. Quite something, eh?" He banged his palm on the wheel and turned to Tolstoi. "Look here, Aleksei Nikolayevich, how would you like those on an automobile!" Both men laughed.

"You know, one inventor proposed that the tires of the *Maksim Gorkiy* be blown up with hydrogen, to obtain additional lift. And true, if we made them out of thin rubber, they would fly like children's balloons. But you cannot do that! Each wheel will have a burden of 30 tons upon landing," he said seriously. Combining humor with laymen's explanations, Tupolev charmed everyone. While he acquainted the guests with the technical side of things, Yulenka Nikolayevna demonstrated the comforts of the cabin and salons. The show ended, and the guests departed, amazed by what they had seen.

A few days later the official document of the mock-up commission arrived at the factory, approving the arrangement of the agitation airplane. This is probably the only approval document in the history of world aircraft construction in which along with the signatures of the leaders of the Civil Air Fleet and the Air Force are the autographs of Soviet writers.

It was fascinating to observe how Tupolev, confronted for the first time with the need to accommodate a diversity of equipment on an aircraft, boldly solved all the problems. For example, the weight limits adopted in the preliminary design had been greatly exceeded. Tupolev discovered that the film, photo, radio and print equipment placed on the *Maksim Gorkiy* was the same as those widely used in clubs, studios, and print shops, all powered from alternating current. However, at the time only low-voltage direct current was used on airplanes, so the assistants placed transformers on the *Maksim Gorkiy* to supply the necessary power.

Tupolev boiled over. "Something is wrong here! Your transformers' efficiency is 50 percent. This means that a good half of their weight will be useless. Yet all of this equipment has been working well for many years on alternating current in the cities and villages. Why can't we reproduce an ordinary alternating current grid on the airplane?"

When his assistants replied that alternating current was not used on airplanes, Tupolev responded characteristically, "Well, what of it? Have you not thought about why? Probably nothing more than pure conservatism, nothing more. If it is safe on land, then apparently it will also be safe on an airplane. Use it, and thereby take off several hundred excess kilograms from the airplane. And concerning world practice, I don't think it is worth talking about here. You know that I am notoriously skeptical about it."

"But alternating current may cause errors in the compass readings, and interfere with radio reception."

Tupolev spoke with the flight station. "Stoman, is that you? Yevgeniy Karlovich, I am sending our electrical engineers over to see you. Let them experiment with alternating current on one of the ANT-6s." The engineers were to put some source of alternating current on the airplanes, and lay a temporary electrical power grid within the airplane. They would then observe whether the current affected the accuracy of the compasses, and whether the radio operator noticed interference with his reception of radiograms. Tupolev was confident that there would be no interference.

Not waiting for the end of the experiment, Tupolev contacted the All-Union Electrotechnical Institute, and asked academicians V. S. Kulebakin and A. N. Larionov, well-known specialists on electrical equipment for automobiles and aircraft, to share their experience. Neither found any insurmountable obstacles, and they outlined how to provide the *Maksim Gorkiy* airplane with alternating current. Larionov agreed to design and build an on-board power plant for the gigantic airplane. It worked beautifully. Historically this was the first alternating current electric power network on an aircraft. After the war this type of current began to be used on large airplanes, and now it is also used on many others.

Inspecting the airplane's printing machine, Tupolev asked what material was used to make its case. When he was told it was cast iron, he did not even get angry, but was simply at a loss. "Allow me, if you gave yourself the goal of making it as heavy as possible, then you would have cast it out of platinum! Send your engineers to us, and we will tell them how to make it no less strong, but half as heavy, out of duralumin."

"Aleksei Mikhaylovich," he asked Cheremukhin, "you take care of this along with Golovinyy [our chief metallurgist]. If they are not able to cast duralumin in their factories, help them and cast it here. Take everything under your control."

The cases were cast out of duralumin. One was made for the airplane, and Tupolev ordered the others sent to the Max Gelets Factory, where the rotor was being prepared for extended tests. The factory designers objected, claiming that after operating for a few days it would go out of commission. This did not happen; in fact, the rotor worked for a long time. Then they decided to shift part of their production to duralumin, because so little of this metal was being produced that it was difficult to satisfy the aviation industry's needs.

Tupolev found a solution just as simply when he was told that, due to the great aerodynamic loads, the pilot would not have enough strength to control the *Maksim Gorkiy*'s huge rudder. Tupolev ordered the engineers to install a remote-controlled electrical auxiliary motor, a servo, ignoring their protests that no one had done it before and that it would not work.

Let us recall that this was in the early 1930s, when designers were very unwilling to incorporate such innovations, and did so warily, for they had no experience. Distrust in mechanisms could be overcome only by personal courage and conviction that your abilities were sufficient to the task at hand. And Tupolev possessed that supreme self-confidence.

Understanding clearly that they could hardly expect to create new, complex, and at times even unique equipment for the *Maksim Gorkiy* using standard practices in a limited period of time, Tupolev, along with N. V. Babushkin, secretary of the Central Institute party organization, brought in not only the leaders of the supplying factories, but even the public at large, to fulfill these tasks. As soon as it appeared that an apparatus would not be delivered on time, they appealed to the party organizations, trade union committees, and communist youth league organizations. Having seen how effective this was, Tupolev subsequently made use of their support, believing quite correctly that public opinion more than ever before would contribute to accelerating technical progress.

In addition to the external questions there were significant internal ones. Noticing that an ordinary entrance had been put on the mockup, where the passengers went up the stairs, he chided his workers: "This will never do. The entrance must be grand. What kind of agitation is it to clamor up into the machine on these little perch-like stairs, like sparrows?"

He fell to thinking, scratched his head, chewed his nails (signs of active thought), and, finding the required solution, approached the fuselage. "It is necessary to make an opening here. The lower part of the fuselage, with the grand entrance stairs and hand rails, drops to the ground and, when the guests enter, it mechanically raises them up and closes the hatch."

Thus, on the *Maksim Gorkiy* the lower surface of the fuselage opened in the form of a ramp. The French later used the same technique on their Caravelle aircraft. Yakovlev on the Yak-40 and Yak-42, and many other designers, used this solution as well.

Tupolev, in inspecting the machine, struck the toe of his shoe on the threshold and his forehead on a spar, and commanded that they be illuminated with lamps. "For shame! We built an eight-engine agitation airplane, and we will be agitating with bumps on our foreheads!" Subsequently such lamps were installed on all passenger aircraft.

"We need a normal laboratory and container. When the airplane arrives, it will be cleaned and then on its way again." This solution was also introduced in all subsequent aircraft. Thus, by being present every day at the airplane as it was being built, our chief designer found all kinds of design solutions as the construction went along.

The *Maksim Gorkiy* was built in a very short time. It was begun on 4 July 1933, and hauled to the airfield on 4 April 1934. On 17 June M. M. Gromov made the first flight. On 19 June 1934, while Moscow was greeting the *Chelyuskin* survivors,[19] the *Maksim Gorkiy* flew regally over Red Square, demonstrating to thousands of Moscow residents yet another victory of Soviet aviation. On 18 May 1935 the final two flights were scheduled, after which the airplane was transferred to the agitation squadron. And on the eve of these flights, on 17 May, French pilot Antoine de Saint-Exupéry, author of the charming fable, *The Little Prince,* and special correspondent to the *Paris-Soire,* made a test flight on the *Maksim Gorkiy.* He was amazed at how easy it was to pilot such a gigantic machine.

Employees of the Experimental Design Bureau were to fly on one of the flights, and "Stakhanovite"[20] workers at the factory, those who had most distinguished themselves during the design and construction of the airplane, were to fly on the other. Before the flight, as often happens in such instances, a dispute arose concerning who should fly first. The underlying reason for the conflict was the age-old dispute:

"If we had not designed the airplane you would not have been able to build it."

"And if we had not built it, you would have nothing on which to fly."

They decided the dispute by chance. They threw a 20-kopeck coin into the air, and the factory's thirty-six best workers ascended up the boarding ramp.

The flight was photographed from an R-5 airplane for a film chronicle. For purposes of comparison, pilot Nikolai Pavlovich Blagin flew next to the *Maksim Gorkiy* in an I-5 fighter. Blagin began to make acrobatic maneuvers in the immediate proximity of the *Maksim Gorkiy*. Miscalculating, Blagin, possibly coming out of a loop (it could not be established from the film taken from the R-5), struck the wing of the ANT-20 near the outside engine. The gigantic machine shuddered and tilted. The spars of its right wing lost stability and the torn-off engine nacelle and large piece of the wing's skin snapped. Broken into several pieces, the airplane fell into the forest. All of the passengers and eleven crew members perished, among them pilots N. Zhurov and I. Mikheyev. Fighter pilot N. Blagin was also killed.

The residents of Moscow, and the entire country, suffered greatly over the catastrophe. At the great concourse, urns with the remains of its victims were laid in the niches of the wall of Novodevichi Monastery, and were literally drowned in flowers and wreaths. On May 20 Saint-Exupéry placed a note in *Izvestiya:*

Deeply shaken, I feel the mourning into which Moscow has been plunged today. I have also lost my friends. . . . The gigantic airplane was not to be. But the country, and the people who created it, are able to call to life even more amazing craft, marvels of technology.

With dispatch, the hand of People's Commissar K. Ye. Voroshilov declared Blagin an "air hooligan." In those harsh years "air hooligans" were not buried at Novodevichi Cemetery, and officials did not come to their orphaned families to notify them that they had nothing to worry about, that they would not be touched. It is unlikely that the well-known TsAGI test pilot on his own initiative began to do unexpected tricks near an airplane with passengers. We note a number of violations committed by the TsAGI administration while carrying out this flight:

1. Combining a turnover-acceptance flight with an aerial jaunt by the best employees of the plant was prohibited.
2. A joint flight by three different types of airplanes required the necessary preflight briefing. Of the three crews, pilots V. V.

Rybushkin and N. P. Blagin received their flight mission only on the evening prior to the flight.

3. Instead of eight crew members on the flight there were eleven.

A statement by I. N. Kvitko,[21] who in those years worked at the Scientific Research Institute of the Air Force, merits attention: "The turnover flight was being carried out. Design Bureau pilot Zhurov was transferring the airplane to agitation squadron pilot Mikheyev. For effect it was decided to have pilot Blagin join the flight in a fighter. His task was to include doing acrobatics at some distance from the *Maksim Gorkiy*." We will probably never know the true guilty party in the deaths of dozens of people and the destruction of this unique airplane.

The *Maksim Gorkiy* was the last in a number of increasingly large, grandiose, but slow-moving airplanes. But one more large aircraft of this type was destined to appear. By decision of the government, the collection of funds among the population was begun to build sixteen airplanes similar to the ANT-20. Sixteen million rubles were collected. Tupolev assigned supervision over the construction of these airplanes to Boris A. Saukke. By this time A. A. Mikulin had manufactured more powerful engines, making it possible to remove the two upper engines and have only the six located in the wings. The first airplane left the factory in 1939. It operated under the designation PS-124 on the Moscow-Mineralnyye Vody passenger line until the start of the war. In 1942 it was destroyed through the fault of the crew. With the start of the Great Patriotic War production of the PS-124 was halted.

Despite the tragic loss of the *Maksim Gorkiy,* the design and construction of such a grandiose machine was an unquestionable success for the chief designer and the collective of the experimental factory. A tremendous number of questions were addressed and solved during its creation. In addition to the dimensions of the airplane, they added a chassis with an oil-pneumatic shock-absorber and braking wheels, an alternating current electrical network, an aircraft control system with autopilot and servos, and much new and original equipment. All of these innovations not only enriched aviation science, but also contributed to increasing the assortment of aviation instruments and equipment, especially in the field of on-board radio systems.

Aiming for a World Record

While they were creating the *Maksim Gorkiy,* Tupolev and his assistants were also working on the design of the ANT-25, a vehicle capable of earning the world record for continuous flight. In the 1920s the Americans owned the record for absolute continuous flight range. In 1931 the French gained the record. Pilots (Bossoutrot and Rossi) flew 10,611 kilometers on a Bleriot-110 airplane without landing.

Tupolev understood clearly that the creation of the ANT-25 would require a significant amount of research. Without the assistance of scientific circles, and possibly even the Academy of Sciences, it was clear that some problems could not be solved. To break the flight record many flight and engine experiments would be required, using more than just one or two airplanes. Consequently several airplanes and no fewer than 20 engines were needed immediately. It was just as clear that the overloaded airplane would not take off from the usual dirt airfield, and that a concrete runway was necessary. These considerations were set forth to the government by chief designer Tupolev and Ya. I. Alksnis, chief of the Red Army Air Force, in a joint report. It was evident that carrying out such a large amount of work required substantial capital investments. Attached to the note was a sketch of a single-engine monoplane, with a Soviet-made M-34 engine designed by A. A. Mikulin as visualized by Tupolev, who had been thinking about it thus far only on his own.

Initially there was skepticism, but the government approved it, and created a commission at the Soviet of the People's Commissariat for long-range flights. The importance that they placed on it is indicated by the appointment of K. Ye. Voroshilov as commission chairman. The commission members approved included G. K. Ordzhonikidze, People's Commissar for Heavy Industry; M. A. Rukhimovich, Chief of the Main Commission for Aviation and People's Commissariat for the Fuel Industry; academician and Arctic explorer Otto Yulyevich Shmidt; Ya. I. Alksnis, Commander-in-Chief of the Air Force; and Chief Designer A. N. Tupolev.

All elements of the TsAGI Aeronautical and Seaplane Experimental Construction Department took part in designing the long-distance aircraft, under Tupolev's direct supervision. Pavel Osipovich Sukhoi was designated chief designer. An associate of Tupolev, he was a talented engineer and later chief of the design bureau that built numerous "S"-type aircraft. The wing of the ANT-25 was designed by the V. M. Petlyakov group, the fuselage was made by V. A. Chizhevskiy, and the tail assembly by N. S. Nekrasov.

Flight tests were headed up by engineer-pilot Ye. K. Stoman, who had earned the St. George's Cross in the Tsarist Army, and after the revolution became one of the first Red military pilots. For his feats in the Civil War, Stoman received the Order of the Red Banner soon after it was established. With the staff in place, it was time to begin developing the ANT-25.

The Bleriot airplane on which the French pilots had established the record was a classic biplane with thin wings. It was not possible to accommodate the gasoline tanks in those wings, and they took up almost all of the fuselage. It was difficult to steal room for the pilots' seats. When they returned from their flight, pilots Aime Rossi and Lucien Bossoutrot told reporters, "Most of all we were depressed by the unendurable living conditions in the aircraft. In the open cabin the wind was cold, penetrating to our bones. There was no place to sleep, and it was not even possible to stretch one's legs. The food was cold, there was no toilet. It was awful!"

Taking into account the French designers' results, Tupolev considered the scenario for the proposed Soviet flight over the North Pole. They knew they would be flying 12,000 kilometers in three days, although they did not yet know the precise route. Success would be measured in the crew's ability to carry out normal activity. Consequently, the cockpit had to be closed and heated, with thermoses for food, a place to sleep, and a toilet, the total of which would take up basically the entire fuselage. This meant that the fuel tanks would have to be placed in the wings. "We have designed many monoplanes, their wings are sufficiently thick, and thus the fuel tanks on the ANT-25 will fit there perfectly. But keep in mind that a super-long-range monoplane must have qualities that approach those of sailplanes of the highest quality, with very large wingspans. . . . Our biggest difficulties will arise in designing and calculating the extremely long wing. Let us work on this first of all."

Tupolev met with P. O. Sukhoi and draftsman B. M. Kondorskiy in his office, and they began drawing a sketch of the airplane. "Let us begin with the wing. The area of the biplane cellule of the wings on the Bleriot was 65 square meters. If we take for our airplane, to start with, the same relative load, and for the Mikulin engine the same fuel expenditure as for the French engine, the wing area of the ANT-25 will increase to 84–85 square meters. But on a biplane two wings have a span of 14 meters each. We have one wing. Consequently, the span of the wing on the ANT-25 should be increased to 30 meters. To reduce harmful inductive resistance, we must use a wing as long as possible. Thus, to start with, let us take a wingspan of 34 meters for the ANT-25. Now let us begin aerodynamic studies of the models."

Combining various wing shapes and outlines, the designers drew up several variants. One after another was tested in the wind tunnels at TsAGI. Some were eliminated immediately, others required additional modifications. Analyzing the results, the researchers settled on an aircraft model with a wing that had an aspect ratio of 13.5. No one in the world had yet designed such wings. No theories or methods of making strength calculations yet existed.[22] Consequently, the designers began basic scientific and theoretical tests to confirm the accuracy of their estimates. The innumerable design solutions of the individual wing elements and components, made at full scale, were tested for strength and vibration resistance. They built such a model for dynamic tests in wind tunnels. They investigated propeller shapes for the M-34 engine for the highest efficiency. Hundreds of scientists, designers, and Central Institute employees were involved in the research work on the ANT-25.

The desire to overcome the unknown and create a record-breaking aircraft gripped all the participants, from the chief designer to the metal fitter. They worked with uncommon smoothness and tremendous enthusiasm. "A thought, expressed in the morning, became a draft during the day and a part by evening," recalled G. O. Bertosh, lead AGOS engineer for the airplane. "We were captivated by our goal. A country, recently quite primitive and half illiterate, intends to achieve the world record held in an airplane created by the elite Western European designers and engineers. We agreed that it was a bold dream."

First it was necessary to achieve a high-quality wing surface, to make it as smooth as possible. But they had to decide where to put the undercarriage legs, each with two wheels and the landing lights,[23] which protruded into the air stream, and, finally, how to reduce the surface friction of the corrugated skin. Having thought about all of these questions, at a session with Tupolev the designers decided to make the undercarriage legs like the claws of a bird, which while in flight lay back against the surface of its body, that is, the aircraft. The landing lights would fold. But the problem of corrugation was still to be solved.

Perhaps Tupolev himself best characterized the problems of creating the ANT-25's wing when he reported to his colleagues at TsAGI after a number of test flights of the ANT prototype:

For the ANT-25 we used a wing aspect ratio of 13.5—previously never seen in the world. The highest aspect ratio before the ANT-25 was on the order of 10. We were very afraid that the wings would vibrate, but this did not happen. Our scientists studied the behavior of numerous models of the ANT-25 in wind tunnels. Professors Ye. P. Grosman and V. P. Vetchinkin and engineer V. N. Belyayev worked out for the designers a sophisticated and correct method of theoretically calculating the wings for vibration. The wings of the ANT-25 differed in yet another fundamental respect. The gigantic fuel tanks, up to seven meters long and with a capacity of 6,000 liters, were an organic part of the design. Placing the tanks along the entire wing also provided a great gain in strength. The wing of a heavily loaded airplane experiences in flight great tensions from the aerodynamic forces supporting it, which are directed upward. The force of gravity from the fuel, directed downward, in contrast relieved the burden on the wing. Having chosen this decision we greatly reduced the weight of the design of the wings on this machine.

Somewhat later he said:

Gradually increasing the amount of fuel in the tanks, we brought the load to its maximum—6,230 kilograms—and forced the airplane to fly under difficult weather conditions, in the clouds under strong wind conditions, in short at maximum burdens. We were helped greatly by Grigoriy Filippovich

Baidukov, a master of flying blind, who took up the difficult and severe tests with the greatest energy.

Major and lengthy theoretical scientific and experimental work confirmed our calculations. The wings of the ANT-25 behaved magnificently. No vibrations were noted during the experimental flights. This was a great scientific victory!

The second problem that arose during the design of the ANT-25 was finding a way to navigate and maintain radio communications at the high latitudes of the Arctic. Under these conditions, magnetic compasses malfunction with significant deviations, rendering them useless for navigation. For example, in the vicinity of Ushakov Island, which lies between Franz Josef Land and Severnaya Zemlya, the compass needle deviates 42 degrees from true north. It was clear that ordinary magnetic compasses were unsuitable for the ANT-25. L. V. Sergeyev proposed creating a solar course indicator. A solar sight rotating from a chronometer gave the ANT-25 navigator the true direction to the Earth's geographic pole. M. M. Kachkachyan and D. A. Broslavskiy developed a gyroscopic compass with DGMK-2 magnetic correction, in which errors even at the Pole itself were reduced to a minimum. In those years no similar instruments existed in the world, and our scientists literally laid a path across the wilderness.

How would radio communications work? At the polar stations then operated by our Northern Sea Route organization, radio communications were continuously being lost. Sometimes, while communications could not be established with the required party, other stations far afield were accessible. It was necessary to study how radio waves propagate in the Arctic and northern Siberia. Much depended on the directional properties of the aircraft antennas. Professor V. I. Bazhenov from TsAGI, Professor Ye. S. Antseliovich from the Scientific Research Institute of Communications, and Engineer N. P. Shelimov from the Scientific Research Institute of the Air Force studied this matter. An ANT-4 airplane equipped with the radio station from the ANT-25 and antennas of various types, piloted by V. Gorodilov, flew from Moscow to Chita and back, and many things were clarified.

While the ANT-25 was being developed, it became clear that the existing airborne radio stations could not provide reliable radio communications with the ground throughout the entire flight. They ordered from the Morse Radio Factory a radio station especially for this airplane, to operate at shorter wavelengths with quartz stabilization. This was called the RD. Engineers S. A. Smirnov, Ye. Ye. Galperin, and S. A. Arshinov designed the radio station. It weighed only 50 kilograms and worked excellently. In the summer of 1934, while working at the Soldiers' and Peasants' Red Army Communications Institute, I was assigned to test it on a TB-1 airplane during a flight from Moscow to Chita in the Far East. Several types of antennas were mounted on the bomber. I maintained communications throughout the flight with the Moscow radio center. The results were reported not only at the Communications Institute, but also at the Scientific Research Institute of the Air Force, and then also to Chief Designer Tupolev. That is how I became acquainted with this remarkable man.

The next year I was sent to the Far East to equip TB-3 heavy bombers with radio direction finders. The aviation navigators were not familiar with them, so I found instructions at the Pacific Fleet Headquarters for operating radio direction finders on ships, put together a small leaflet for the aviation navigators, and acquainted them with the principles of aerial radio navigation. I felt that this was not sufficient, however, and took my summaries and developed them into the book *Aircraft Radio Stations and Their Operation*. At that time there was no literature on this equipment in the USSR, and TsAGI quickly published my work. The Air Force purchased the entire publication as a training aid. Tupolev became aware of this publication and invited me to visit, having decided also to place a radio compass on the ANT-25. Bypassing bureaucratic channels, he made it possible for me temporarily to be sent to work at the commission for long-range flights. At the flight headquarters I was assigned to familiarize the crews with the new apparatus.

The third problem for the ANT-25 was the crew's life support. The flight was to last three days across the top of the globe and would take place under conditions of severe cold. Below, for thousands of kilometers, there would be no cities or human settlements. The possibility of a forced landing on the ice could not be ruled out. That meant that light, warm, waterproof clothing, a supply of concentrated food, and an emergency radio were needed. Part of the route was above the ocean.

How could the ANT-25 be made unsinkable? Bags made of the thinnest rubberized fabric were placed in all the free compartments of the aircraft and connected by pipes with a foot pump. Using this pump it would be possible to pump up the sacks quickly to prevent the ANT-25 from sinking. It must be noted that in solving such problems the engineers did not yet suspect that they were taking the first shy steps in these fields for the future cosmonauts. Tupolev, along with P. O. Sukhoi and Ye. K. Stoman, became deeply involved in all these problems, and they must be given their due. Their experience, persistence, and firm leadership determined everything.

Gradually, involving an ever-increasing number of scientists and various scientific research institutes, the problems of supporting the flight were slowly reduced.

It must be said that Academy of Sciences President Aleksandr Petrovich Karpinskiy was very cooperative about Tupolev's request for assistance from the scientists in organizing flights. But Karpinskiy was sick and, as it happened, all ties with the Academy went through one of his deputies—O. Yu. Shmidt. At that time, Shmidt was preparing his own air expedition to the North Pole, again on a Tupolev ANT-6 aircraft. Their interests conflicted, and both parties began to move forward quickly. N. I. Bazenkov, one of the leading officials at the N. P. Gorbunov Factory, quickly re-equipped the airplanes going on the polar expedition. The Academy just as effectively solved problems associated with the ANT-25's flight to the United States.

Out of all the test models of the ANT-25, the three most promising were selected for the transpolar flights. Examining them even more critically, the final design of the airplane was outlined, and a model was built. It looked rather unusual. The wingspan of 35 meters was two and a half times the length of the fuselage. Everyone who saw the model ANT-25 took it to be a sail airplane. No one imagined that an airplane of such contours would cross the pole to America.

Unfortunately, the first flights showed that the ANT-25 could not reach its designated range. Despite the fact that it included many innovations, even fundamental innovations such as the smooth fuselage skin, the retracted undercarriage legs, and a number of others, it simply could not fly more than 7,000 kilometers, due to the engine. The A. A.

Mikulin engine with reducing gear was delayed in manufacturing; without it fuel expenditure was too high. Although they confirmed that the range would increase sharply with the new engine, Tupolev was not satisfied. He sought ways to reduce known "harmful" losses from the friction of the aircraft skin against the atmosphere.

During the winter of 1933–1934, studies were conducted at TsAGI, the Flight Research Institute, and Central Institute of Aircraft Engines. To reduce the harmful resistance of a wing with corrugated skin, M. A. Tayts proposed that percale be sewn on top of the corrugated surface, which Tupolev approved after examining the results of Tayts's tests. The leading edges of the wing were lacquered. The propeller blades were polished. To improve the fuselage's internal aerodynamics, all holes were sealed. Parts and elements that protruded out of the aircraft were provided with fairings. Inspecting the airplane personally every day, Tupolev found more and more streamlining possibilities for its external "refinement." The fault-finding, at times seemingly trivial, bore fruit. When the Mikulin engine with reducing gear was mounted, and test flights were conducted, the success surpassed all expectations— a flight range of 13,000 kilometers could be anticipated.

By this time a combat engineer battalion had completed construction of the airfield runway. One of the most difficult tests could begin: takeoffs with gradually increasing load. To fly to maximum range, however, it was necessary to pour into an empty aircraft weighing 4,200 kilograms some 6,500 kilograms of gasoline. The undercarriage could not support landing with such a large fuel weight, means to discharge fuel in the air had not been developed, and it was not possible to fly for hours and days until all the fuel was used.

Tupolev and Stoman, who had pondered the solution to this problem for a long time, found a brilliant way out. They purchased several tons of rejected bearings at a ball-bearing factory, which were poured into sacks and loaded onto the airplane, imitating the weight of the fuel. The ANT-25 would take off and Baidukov would fly the machine, weighing progressively more from flight to flight, into the clouds, where in blind flight it was subjected to the tests for loads that it would encounter in real distance flights. Having flown for an hour or two, Baidukov would turn toward a swamp located near the airfield, where the ball bearings would be poured out through the on-board urinal. That way they accomplished the tests for takeoffs with the weight that the ANT-25 would carry to achieve the world-record range for continuous flight. Everything was prepared for the long-range flight, but sending the airplane on the long flight route without more testing was an unjustified risk, and the flight commission decided to have Tupolev fly a closed triangle not too far from Moscow. K. Ye. Voroshilov reported the commission's decision to the Politburo, where it was approved.

On 12–13 September 1934 the weather forecast was good. At 10 A.M. pilots Mikhail M. Gromov and A. Filin, with navigator I. Spirin, rose off the concrete runway of Shchelkovskiy Airfield and flew a Moscow-Ryazan-Kharkov triangle in 75 hours, covering a distance of 12,411 kilometers. By the end of the flight the weather in Moscow had turned bad, forcing the airplane to land in Kharkov, where the commission members had flown ahead of time.

Evaluating the ANT-25 and what needed to be done in order to facilitate the three-day flight, the crew complained most of all about the lack of controls for the second pilot. "Just look at us. We changed positions every two or three hours, which meant that during the entire flight we crawled across the confined cabin some 25 or 30 times. The sharp edges everywhere ripped our leather jackets to shreds.

"It is bad that we cannot stretch out on the sleeping cot. But the fact that it is located on the warm fuel tank is outstanding. We were not cold at all. It is unfortunate that the food in the thermoses cooled down after one day."

These were considered minor details. The chief of the Air Force issued new leather jackets for the pilots.

Tupolev considered the ANT-25 a successful machine. But every man makes mistakes. One hot summer day during the testing phase, the aircraft was standing on the runway. Routine finishing touches, the last before the record flight, were under way. The wind above the airfield was blowing in such a direction that airplanes had to land across our stand. It was nearing lunch time, the flights were ending, and several men, tired out by the sun, were sitting in the shade under the airplane. The whine of the engine of one of the returning fighters was heard, possibly the last on this day. The whine increased in volume, until it became painful to hear. Apparently the pilot had increased his rpms. There was an unexpected crash against our airplane, a cloud of dust nearby, and a sudden silence.

Stunned, we ran to the smashed fighter. Pilot Bazhenov had miscalculated the height of the landing approach and sliced into the wing of our machine with the wheels of his I-16. Fortunately, the pilot survived the mishap. The wing of the ANT-25 was crumpled, but apparently without too serious damages. Stoman was in despair. Soon Tupolev and Sukhoi arrived. Careful examination showed that the required repairs were not so difficult, and two weeks later our airplane continued tests.

"If he had struck our wing a half meter lower it would have been the 'end of the game.' There could have been no flights," Stoman somberly summed up the results. "These flight tests are not all that peaceful an activity."

The international aeronautics club confirmed flight records for distance only if they were carried out in a straight line. Considering the result achieved by Gromov, Filin, and Spirin, the governmental commission thought it possible to prepare for such a flight. Tupolev assigned Gromov, Stoman, and A. Yumashev to research possible routes for a flight of 10,000 kilometers. After lengthy consultations with diplomats, meteorologists, geographers, and radio operators, two routes were prepared. The southern route, began in Moscow, ran across all of Europe to Gibraltar, continued over North Africa to Dakar, and then ran across the Atlantic Ocean along the east coast of South America to Rio de Janeiro or Buenos Aires. The northern route lay from Moscow across the North Pole, Canada, and the United States. Despite a number of difficulties of the second route (unknown areas of the Arctic, and a lack of permanent polar meteorological and radio stations), this route was chosen. The reasons were purely diplomatic. On the southern route the Soviet aircraft would fly across numerous countries, some of which had no diplomatic relations with the USSR. A flight in their air space could cause political complications. The northern route crossed only Canada, which immediately gave consent.

The flight began at dawn on 3 August 1935. The crew was led by S. Levanevskiy. With him flew second pilot G. Baidukov and navigator V. Levchenko.

Why did the government commission entrust Levanevskiy to fly the ANT-25 over the pole? On 13 February 1934, near Chukotka, the steamship *Chelyuskin,* destroyed by the ice, went down. The hundred members of the crew and polar workers sat on the ice 100 kilometers from shore. The whole world waited to see what would happen to them. The Soviet government, having collected the best polar pilots, sent rescue aircraft from Krasnoyarsk and Khabarovsk to the "ice camp."

In another flight, G. A. Ushakov and pilots M. Slepnev and S. Levanevskiy were sent to the United States to purchase airplanes. They were then to hurry to the camp across the Bering Strait.

On 27 March Levanevskiy flew out from Nome, Alaska. He had on board Ushakov, leader of the rescue efforts, and an American on-board technician. Flying in the fog, near Vankarem, the airplane had an accident. No one was killed, but slightly wounded Levanevskiy was unable to continue. Slepnev continued the flight successfully and took part in the evacuation.

On 15 April, through the efforts of pilots A. Lyapidevskiy, V. Molkov, N. Kamanin, M. Vodopyanov, I. Doronin, and M. Slepnev, all the people at the ice camp were removed to the continent. To mark the polar aviators' courage, the government established a new mark of distinction, Hero of the Soviet Union. The six participants in the heroic epic, and somewhat unexpectedly, S. A. Levanevskiy, received the award.

Upon their return to Moscow, Josef Stalin received the heroes. At this reception, Levanevskiy somehow charmed the ruler. In spite of Stalin's suspicious nature, Levanevskiy made a positive impression. Of all the pilots who participated in the Chelyuskin epic, Levanevskiy was the "foreigner." His parents and brothers lived in Poland and were not workers, but members of the petty gentry class. Stalin could change his opinion of people abruptly, which was well known and confused many. Thus, many people did not approve of his affection for Valery Chkalov, the daredevil who flew on a dare under the center span of the Trinity Bridge in Petrograd, and was discharged from the Army for insubordination. Some were at a loss about the appointment of the young designer A. S. Yakovlev as reviewer for aviation, as he had also come from a bourgeois family, and in addition was married to the daughter of a member of the "opposition," former Council of People's Commissar Chairman A. Rykov. All of these cases did not fit the script for achievement in Stalin's Russia.

This-preference toward Levanevskiy was again evident when crew commander M. M. Gromov had to have surgery and would not be able to pilot the record flight. The whole world had been informed about the date of the start, and it simply could not be postponed. But who would replace Gromov? Stalin chose Levanevskiy.

Those involved in the long flight preparations were distressed. Levanevskiy was superficially familiar with the ANT-25, but did not have sufficient experience in flights on the airplane. All hopes were placed on second pilot G. F. Baidukov, who was knowledgeable of the ANT-25 and a such a master of blind flights that he was called "god of blind flights" in the Scientific Research Institute of the Air Force.

At dawn on 3 August 1935 the ANT-25 set out for America. The press was given instructions to cover the crossing as extensively as possible. The morning newspapers were filled with articles, forecasts, photographs of the takeoff, and American press coverage of the successes of our aviation.

Unfortunately, our hopes were not borne out. When the airplane had flown 2,000 kilometers and was over the Bering Sea, the navigator noticed through his optical sight that oil was oozing out from under the engine cowling. A thin film of oil was creeping along the bottom of the fuselage. Restrained and calm, Baidukov believed that since the oil pressure and temperature were normal it was a temporary phenomenon, caused by excess oil frothing up in the engine crankcase, which was then being discharged through a breather designed especially for this purpose. Levanevskiy was concerned whether there would be sufficient oil to last through the flight, and decided to return. A radiogram was sent to Moscow. The specialists assembled in the capital believed there was excess oil in the engine crankcase. Frothing, it was being discharged through the breather in the form of an air-oil emulation. Recommendations were sent to the airplane about how to reduce, and perhaps even halt the discharge. The airplane was silent for some time, and then reported, "I am returning, and will land in the area of Novgorod. Levanevskiy."

That evening the machine landed at Krechevitsy Airfield, not far from Novgorod. In the morning Tupolev, Chkalov, and other specialists, myself included, arrived. Seeing the ANT-25 at the airfield with a charred outboard wing cellule, we became concerned. It turned out that the navigator, Levchenko, in crawling out of the cramped cockpit, inadvertently struck the buttons that turned on the underwing landing lights. In flight, the heat from the burning magnesium lights was drawn off by a stream of air, but here on the ground the skin of the aircraft started to burn. Some Red Army soldiers who were nearby quickly beat out the fire with their overcoats, but approximately a square meter of the skin had burned. Having determined the causes for the cessation of the flight, the accident commission began to pour off residual oil. The soldiers carried the full canisters of oil away. As the quantity of the oil being weighed approached the weight of that poured in the day before in Shchelkov, the alarm and tension that had overwhelmed Tupolev fell away. On the other hand, the commander grew somber. The commission established that there had been no leak in the engine oil system. As Baidukov had supposed, the oil was thrown out through the breather. The weight of the oil that had been thrown out was so little that it was not even possible to determine using the gross airfield scales. Later Tupolev would say, "Just the same, I do not know what required more courage—to continue the flight or to turn back."

The preparations for the crossing, takeoff, and flight over the USSR and the Bering Sea had been covered extensively by the press. The papers on the morning of 4 August were filled with articles on the crossing and, naturally, that the unsuccessful end had evoked many critical remarks abroad. Skeptics raised their heads, and proposals appeared recommending the end of research and no further flights.

Furious with the failure and with the blow to the country's reputation, Stalin convened all of those who had taken part in organizing the flight. Levanevskiy spoke first and explained what had happened: "The ANT-25 airplane is not suitable for such flights. It has been built in a deliberately negligent way. I personally do not intend to fly again on Tupolev aircraft." Those present, in particular G. F. Baidukov, recalled, "Tupolev sat with his face ashen. By the end of the session he became ill, and a doctor was called." Levanevskiy went on to say that a single-engine airplane was not suitable for a transpolar crossing due to low reliability. A multiengine airplane was needed. The participants in creating and readying the vehicle for the flight were, of course, disheartened, and perplexed as to why the pilot had been silent up to this time about his views and why he had agreed to fly on an unreliable airplane. The governmental commission's conclusion placed the many years of work of TsAGI scientists in doubt. After all, it was they who, after long reflection and studies, came to the unanimous conclusion that an airplane capable of flying without refueling 12,000 kilometers must be an ideal sailplane with a very great wingspan and relatively small engine. And here, due to a vexing incident and hasty conclusions, the many years of work of hundreds of people had been for naught. The machine was covered up and placed in a hangar. It became unseemly to talk about it, as about a foolish occurrence.

Perhaps, had G. F. Baidukov not been so insistent, our country may not have become the holder of the most prestigious aviation record. Profoundly convinced that the ANT-25 was capable of the record flight, he brought his friend Valery P. Chkalov to the airfield, flew with him on the beloved vehicle, gave him the opportunity himself to sit behind the wheel, set up a meeting for Chkalov with Tupolev, acquainted him with the designers, and when he was convinced that Chkalov had been won over, proposed, "Let us write Stalin. He trusts you no less than he does Levanevskiy. Let us say that the airplane is reliable and that it we will fly to the United States. The long-range record will be ours." It did not take long to convince Chkalov.

The newly organized crew, consisting of Chkalov, Baidukov, and navigator A. V. Belyakov, was received by the General Secretary of the Central Committee of the Communist Party of the Soviet Union. Stalin asked, "Is it sufficiently reliable to fly on one engine?"

"It is sufficient, Comrade Stalin," answered Chkalov. "One engine carries a 100% risk, and four engines a 400% risk!"

After this extraordinary meeting the situation changed quickly. The flights were to continue. For the first flight Stalin recommended that the airplane not leave the borders of the country, but fly under conditions close to the route of the future flight across the Pole, that is, over the

Soviet Arctic. Moreover, he wanted no mention in the press until the end of the flight.

On the eve of the flight Chkalov was asked, "But you have never flown in the Arctic, have you?"

"That is true, but so what? I believe in the ANT-25. This machine will be able to cross the Arctic."

On 10–12 July 1936, the crew, consisting of Chkalov, Baidukov, and Belyakov, flew the ANT-25 some 9,374 kilometers in 56 hours following a Moscow-Franz Josef Land-Petropavlovsk Kamchatskiy-Lake Udd (near the mouth of the Amur). It would have been possible to fly even farther, to Khabarovsk, but the weather had become very bad. One wheel broke during the landing on swampy, sandy soil. Stoman and a group of workers who arrived on a TB-3 from Moscow quickly repaired the wheel. It was possible to take off, but how could this be done on only sand? It was proposed that combat engineers from the Special "Red Banner" Far Eastern Army make a landing strip out of wooden parallel bars. However, this was so out of the ordinary that the arriving specialists protested. Stoman queried Moscow.

"Make the wooden runway, and as quickly as possible" came the terse answer from Tupolev.

Soon the ANT-25, which had taken off from the exotic strip, returned to Moscow, where a ceremonial meeting was set up for the crew.

Following this successful test flight, the Central Committee and the Council of People's Commissars sanctioned the ANT-25's flight across the North Pole to the United States. The Shmidt and Papanin expedition's landing at the North Pole helped this decision. On 21 May 1937, after pilot P. Golovin on an ANT-7 aircraft had flown to the Pole, returned to Franz Josef Land, and reported the situation, four specially re-equipped Tupolev ANT-6 airplanes, piloted by M. Vodopyanov, I. Mazuruk, V. Molokovyy, and A. Alekseyev, landed on an ice field at the North Pole and brought a courageous foursome of employees of the Academy of Sciences and Main Administration of the Northern Sea Route, Ye. K. Fedorov, P. P. Shirshov, I. D. Papanin, and E. T. Krenkel. The polar scientific research station, Severnyy Polyus [North Pole], was in operation! Now from the very center of the Arctic basin, Krenkel's

Morse code of dots and dashes was carried over the airwaves, transmitting to the crew of the ANT-25 timely weather reports.

An accident occurred right before the flight itself. Working near the undercarriage, one of the electricians lay a spare drill between the teeth of the wheel assembly mechanism. When he finished work and reported to his chief that the airplane was ready for flight, he forgot about the drill and went to lunch. Chkalov arrived and immediately set off in flight. He took off normally, and started the scheduled mission. Upon returning he transmitted, "The right wheel will not extend, I will land on the left alone." Stoman turned white. Only a few days remained until the start of the flight!

Chkalov, with a barely noticeable left bank, slowly descended and, as the pilots say, edged the machine toward the ground. The left leg touched the ground, the flying ANT-25 rolled on one leg and the tail wheel. Losing speed, it slowly banked to the right, the end of the wing struck the ground, and the ANT-25 turned and came to a stop. The landing was magnificent.

We raced toward the ANT-25. The end fairing of the wing was crumpled, but the rest was whole. Chkalov was rightly indignant: "Devil knows what! Look at you, soaked to the bones, expecting an accident!"

A piece of the broken drill had wedged into the gear box, and the leg unit could not be lowered. It was a difficult time, and they feared for the electrician, but it worked out. When they reported the situation to Tupolev he reached two conclusions. "How correct we were to choose a design with a monoplane wingspan of almost 40 meters. You see how she flies! And had the French landed in the biplane with a wingspan of 14 meters, the airplane absolutely would have crashed." And "There is slovenliness at the field. The electrician should be fired immediately. But, considering that he immediately confessed, and knowing him as an excellent worker, and seeing how shaken he is by what happened, limit yourself to a strict reprimand."

On 18 June 1937, at 3 A.M., red-winged ANT-25 No. 1, *Stalinskiy marshrut* [*Stalin Route*] with Chkalov's crew, moved from the hillock of Shchelkovskiy Airfield on its historic flight. Accelerating down the concrete runway, the airplane drew even with the chief designer, standing alone at its end. Rushing past him, Chkalov stuck his arm out the open window and waved to the aircraft's creator.

On the eve of the Chkalov flight, Tupolev reported that he personally was assigning me ("To the devil with all these considerations of who is subordinate to whom") responsibility for the ANT-25's radio communications with the ground, and for use of the radio compass for navigation.

The long flight went well, passing over the North Pole. Not until the ANT-25 reached Banks Island in northern Canada did the crew experience a breakdown in radio communications. Although it was late at night, the commission members assembled at the Central Airfield. Nervous and angry, Tupolev demanded a clear answer: "Tell me, what is the weather there?"

There were 6,000 kilometers between the aircraft to Tupolev, so the answer could not be specific. Chief meteorologist of the Air Force V. V. Altovskiy nervously began to report on warm and cold fronts, weather inversions and occlusions. Tupolev interrupted angrily, "To hell with inversions and occlusions, understand this: We are interested in only one thing. Can they see the ground, and will they be able to land if they have to?"

Becoming even more irritated, he lashed out at me, "Kerber, answer me, why are there no communications with the airplane?" I answered that there could be several reasons, and that I had to think about it. Tupolev, agitated, demanded in a squeaky voice, "Sit on the sofa and give me an answer in 10 minutes. And consider that unless communications are restored I may cause you a mass of unpleasantries." I already knew from friends that at times the "highest measure" of unpleasantry was a reproachful gaze, accompanied by the words, "And I thought better of you, and counted on more." But there was also another variety of unpleasantness. "Go away and write an order on your dismissal." The order was to be placed on the writing table with one's passport. Frequently it ended in the order being torn up, and the guilty party being slapped on the shoulder and let go in peace. S. P. Korolyov, the spaceship designer, often behaved in a similar manner. Probably he experienced them himself while he was still a student under Tupolev. But if there were a serious error in calculations, a gross mistake in design, nonfulfillment of his personal instruction, or, most intolerable, that someone was attempting to deceive him, Tupolev's anger caused everyone to tremble, from his closest assistant Arkhangel'skiy to the most ordinary engineer.

Fortunately communications with the ANT-25 were restored, everything was in order on board, and everyone calmed down. The ANT-25 crew dealt with several cyclones and icing over in the clouds successfully. One more sleepless night had passed.

On 20 June, three days after takeoff, the ANT-25, having flown 8,582 kilometers in a straight line in 63 hours and 16 minutes, landed triumphantly at Pearson Army Air Field at Vancouver, Washington, just across the Columbia River from Portland, Oregon. As they crawled out of the machine, the three gallant men discovered that their dress uniforms, sewn in the shop at the People's Commissariat of Foreign Affairs, had been left hanging in Shchelkova, and they had nothing to wear except for their flight clothing, consisting of high fur boots, and fur trousers and sweaters. The commandant of the Vancouver army garrison, Gen. George C. Marshall (during World War II in 1939–45 he was chief of staff of the U.S. Army), and A. A. Troyanovskiy, Soviet ambassador to the United States, who had flown to the landing site, came to the aid of the heroes. They were taken to a store in Portland, where the tailors helped them select respectable suits and fitted them properly.

Many people came to believe that the long-range flight record was gained with this first flight. While the bad weather on the route forced the crew to go around several cyclones, significantly increasing their route, they flew 11,340 kilometers, but not in a straight line. We were consoled by the fact that they had established the record nonetheless, for it was the first transpolar flight, linking the Soviet Union and the United States.

The ovations that greeted the first crew had not yet subsided when, on 12 July, the second ANT-25 took off, with Gromov, Yumashev, and S. Danilin on board. After flying a straight-line distance of 10,148 kilometers in 62 hours and 17 minutes, Gromov masterfully landed his red-winged machine right on a farm field in San Jacinto, not far from the U.S. border with Mexico. There remained fuel for another 1,400 kilometers in the airplane's tanks, but the lack of a diplomatic agreement with Mexico interfered with continuing the flight. Nevertheless, the world distance record that had belonged to the French was exceeded by more than 1,000 kilometers.

The American people greeted the flights of the two Soviet crews, that of Chkalov and especially that of Gromov, with delight. They recognized the talent of the Russian designers, praised our aviation industry, and admired our scientific achievements. They paid tribute to the crews, who had overcome the harsh Arctic conditions, with all sorts of honors and recognition. The people carried them on their shoulders, tossed them in the air, arrayed them in wreaths of flowers, and maniacally ripped all of the buttons off their clothing as souvenirs. It was a remarkable victory for Soviet aviation, and the world press gave their due to the pilots, the airplanes, and, of course, their designer, A. N. Tupolev.

But there was a negative side as well. The English journal *Aeroplane* spread a false report alleging that Papanin's drift-ice research unit was not a scientific research station at all, but a fuel supply base organized by the Bolsheviks, where both Chkalov and Gromov had refueled. A classic Stalinist reaction followed—Stalin ordered that delivery of the newspaper to our country be halted.

The ANT-37 *Rodina*

The long-range ANT-37 bomber, *Rodina,* followed the development of the ANT-25. It had two M-86 engines, and was designed by P. O. Sukhoi's team, under Tupolev's general supervision. A prototype of this machine crashed on 16 June 1935, due to a structural defect in the tail assembly. Although pilot K. Popov and engineer M. Yegorov, who had parachuted to safety, reported in detail on the causes of the accident, it held up further tests for a long time. When flights resumed two years later, the machine's flight data turned out to be somewhat worse than that of similar foreign airplanes.

It was considered advisable to curtail further tests of the ANT-37. However, Sukhoi's calculations showed that if the weapons and military equipment were removed from the aircraft, and additional fuel tanks were installed, the machine would be capable of flying more than 7,000 kilometers without landing.

In the Fédération Aéronautique Internationale (FAI) table of world records, in addition to the absolute distance record, the distance record for airplanes with female crews belonged to a French woman, Elizabet Lyon, who had flown from Ivry, France, to Abadan, Iran, a distance of 4,063 kilometers. It was obvious that the *Rodina* could exceed this.

When this was called to the government's attention, it was decided we would pursue the women's flight distance record. Preparations began. One day Tupolev urgently needed to see the chief engineer for the ANT-37, M. Yegorov. It was raining and when Tupolev walked from the factory to the Experimental Design Bureau building, both he and the drawings he was carrying were soaked. For a time he could be seen standing under the rain between the two buildings, staring first into one and then the other. When he returned, he ordered that a covered crossing between the buildings be designed without delay, at the second-floor level. The height of the floors differed, and on the sketch the crossing was inclined. "What are we to do here, roll down hill like during the holidays?" Tupolev asked B. A. Novoselskiy, the chief engineer of the factory. "Did you really not think about making the crossing horizontal and placing a ramp at the Experimental Design Bureau building?"

From that time on all of our buildings were connected by crosswalks. It was no longer necessary to wear coats. Later we began to transfer drawings, documentation, and parts on these crosswalks, and everything became simplified. It was so characteristic of Tupolev's mind that such a small problem as the soaked drawings could lead to such a sensible solution.

On 24–25 September 1938, the crew of Valentina Grizodubova, Paulina Osipenko, and Marina Raskova, having flown 5,900 kilometers in 26.5 hours, on a route from Moscow to Kerbi Settlement in the Far East, broke the existing women's long-distance record. When they reached the Tugursk Strait of the Sea of Okhotsk, they set course for Komsomolsk-on-the-Amur. Suddenly the fuel gauge lit up, indicating there was only enough fuel left for 30 minutes of flight. Then the weather closed in and they found themselves in the clouds. When they made their way through the clouds they saw below a swampy section with bushes. Fearing that the glass navigator's compartment might be crushed upon landing, Grizodubova ordered Raskova to parachute.

When communications with the *Rodina* had been completely cut off, Moscow sounded the alarm. Aerial searches were organized. On the ninth day they found the Far Eastern aviators. Unfortunately, there were human casualties. The day after the *Rodina* was found, against orders, an Li-2 (DC-3) and a TB-3 flew to the area of the landing. As a result of their collision almost all those on board perished. A parachute landing was conducted at the *Rodina*'s landing site. The heroines and their ANT-37 soon arrived in Moscow.

Tupolev had reason to be satisfied. In 1937–38 all distance records gained by Soviet pilots were on aircraft designed by his collective.

Supplying the Military

In 1933, when Adolf Hitler became Reichschancellor of Germany, far-sighted people understood that the storm clouds of war were thickening. Following the death of elderly Field Marshal-President Hindenburg, Hitler assumed his duties, and became the uncrowned king of Germany. From this time on, Hitler's party, the National Socialist, began a serious campaign to abolish all of the restrictions of the Versailles Peace Treaty, introduce compulsory military service, and create a new million-man army, and rearm it with the most modern weapons.

No one doubted that military solutions to Germany's territorial and economic claims were on the horizon. Improving the USSR's defense capability became paramount. It was necessary to plan how best to supply the Red Army with the best weaponry, and especially aircraft, as quickly as possible. For this purpose, in addition to other measures the structure of the TsAGI had to change. One of the institute's leading spheres of activity became testing aircraft construction. The number of theoretical and practical scientific questions arising during the parallel design of several diverse machines increased swiftly day after day. In order to answer them most effectively, the number of specialized departments and laboratories in the institute grew. The administrative control apparatus of TsAGI became more and more complex and multifaceted; the flexibility we once enjoyed now seemed lost. The time had come to reorganize the institute.

An Institute of Aircraft Engines (TsIAM), Institute of Aviation Materials (VIAM), and the A. N. Tupolev Experimental Design Bureau were created out of the departments of TsAGI. The reorganization was satisfactory. Now the institute's activities were regulated and simplified.

By the new year, almost all of the assignments for military aircraft designs that had come from TsAGI to the newly created Experimental Design Bureau were approaching completion. It became possible to slow down a little and formulate our views on the next generation of bombers. The majority of countries was replacing their large and slow-flying bombers with faster, smaller, more maneuverable machines. Evaluating our technical capabilities for making this change, Tupolev stated at a meeting at the Experimental Design Bureau:

Much has been done by the institutes of the People's Commissariat. Scientists at TsAGI have developed a series of new and highly promising wing designs. The theory of calculating monocoque fuselages[24] with stressed skin, which markedly reduces the weight of construction, was also created there. The Central Institute of Aircraft Engines, along with the V. Ya. Klimov Bureau, has set up production of M-100 engines and developed a system for cooling them on aircraft. Having the same capacity as the Mikulin M-34, they are lighter and have a smaller frontal cross-section. VIAM has developed and introduced into production a number of interesting alloys, materials, and plastics with higher strength and weight characteristics. We have also achieved a few things. The retractable undercarriages for the ANT-21 and ANT-25 are working reliably. We have mastered the technology of smooth skin on wings and fuselages, and this was rather painful and not at all simple. I think that all of this creates the objective possibilities for a qualitative leap to fast-moving machines. I believe that we should do this. Let us develop proposals for the new bombers.

After a number of meetings with the military, it was established that the Army needed two types of new bombers: a long-range bomber for 3,500–4,000 kilometers, and a medium-range frontal bomber for 1,000–1,500 kilometers. The long-range bomber was assigned to the S. V. Il'yushin Bureau, and the medium-range bomber to the A. N.

Tupolev Experimental Design Bureau. Since the time had come to replace the slow-moving, nonmaneuverable TB-3 strategic bombers with faster and higher-flying bombers, the People's Commissariat assigned Tupolev to carry out preliminary investigations into a new heavy machine. The medium bomber was identified as the ANT-40, and the long-range, high-altitude, fast machine was called the ANT-42.

The Red Army Air Force insisted that the new medium bomber carry a half-ton of bombs a distance of 1,000 kilometers at a speed of no less than 400 km/h. By 1934 fighter speed had approached approximately this figure. When preliminary calculations showed that with the new M-100 engines the ANT-40 medium bomber would develop a speed even somewhat greater, people became interested in the machine. The possibility of having a bomber capable of operating without fighter escort was irresistible.

In contrast to the traditional monoplane with low-set wing, Tupolev accepted a mid-wing monoplane for the medium bomber. Although this made a higher undercarriage necessary, it also made it possible to hang bombs beneath the fuselage. The M-100 engines, covered by well-streamlined cowlings, were shifted to gondolas, where the undercarriage was retracted. The crew of the ANT-40 consisted of three people. The navigator was located in front, and behind him sat the pilot. The gunner-radio operator was located near the rear edge of the wing, and was armed with machine guns to fire against attacking fighters.

Compared with the previous Tupolev aircraft, the new medium bomber was more sophisticated. The machine carried the same bomb load as did the TB-1, but was one-third smaller. Its smooth, polished aluminum skin (used instead of corrugated sheets and no longer painted brownish-green) contributed to its elegant external appearance. The unpainted aluminum was a much better means of camouflage against air defense observers. Use of this skin also caused the restructuring of the entire aviation industry to a higher level of precision. Instead of simply hammering out the rivet heads, it was necessary to countersink each hole in the sheets of skin, that is, to make a recess under the head of the rivets. It was necessary to rivet in such a way that the head did not come to the surface. At first rejects from the new technology were common. Production schedules were missed and the factory directors began to grumble. It was necessary not only to hold skill-improvement courses for the workers but also, unfortunately, to restrict the use of the countersunk rivet to the most necessary places, where it significantly affected the loss of speed. In addition, it was necessary to develop and introduce new accessories and tools. In the end, we managed to deal with this sudden flareup of feverish activity.

The aerodynamic design specialists believed that operation of the medium bomber would hold no surprises. After all, tests of a number of models of Soviet fighters at speeds close to those of the ANT-40 were going successfully. From abroad reports came in about puzzling disasters of some airplanes at high speeds, but their causes were usually related not to some new as-yet-unstudied phenomenon, but to ordinary design or production defects.

At the beginning of 1935 the ANT-40 was delivered to the airfield. Test flights began. Pilot K. Popov, who flew the prototype, gave the new design a high evaluation in the air, but noted that it shook at certain flight regimes. He did not notice any direct relationship between the shaking and speed. Nevertheless, the prototype was very carefully examined, but no discrepancies were detected. After consulting with the crew and engineers, and taking a number of precautions, Stoman risked sending Popov on a second flight. Unfortunately, the shaking was also repeated. Upon his return the pilot reported, "When I approached a speed of approximately 380 kilometers, the airplane again began to jerk, then increased, and in addition a rattle began to increase, as though pneumatic hammers were pounding on the aircraft."

Sad though it was, no doubts remained. A flutter had started, as a spontaneously increasing shaking was called at the time. The flights were halted. Stoman phoned Tupolev, who immediately went to the Central Airfield. They began to look into the problem. They examined the wings, tail assembly, ailerons, and wheels carefully but found no cracks. However, there were places on the skin of the aircraft where the skin was slightly swollen.

"Popov, it is good that you immediately reduced speed, otherwise it could have been worse. You saved the aircraft, and that is most important," Tupolev remarked. "Theoretically the reasons for the phenomenon are more or less understood. Unfortunately, we cannot say right now what should be done with the medium bomber. One thing is clear. There must be no more experiments in the air. This flight fortunately ended well, the next may not end so well. I think that we must bring in the Central Institute scientists to study the problem. And we will not delay. Tomorrow morning we will assemble a conference of specialists. I will phone you from Moscow about whom to invite."

While the "sick" machine was covered up and put in the hangar, storm clouds began to gather over Tupolev's Experimental Design Bureau. People began to talk: "The fast bomber was not successful, the ANT-40 is not suitable, and it cannot be built in series production." Inevitably the question of what caused the failure arose. Analysis of the calculation and results of the tests in the wind tunnel showed that the medium bomber had aileron flutter. First of all it was necessary to strengthen the rigidity of the airplane's wing. This type of flutter is successfully treated by introducing a weight compensation for the ailerons. After it was installed, the tests continued. Upon returning from a flight following the modifications, a smiling Popov reported to Tupolev, "You would not know the machine. The shake is completely gone, and at speeds up to 400 km/h the bomber goes as if through oil!"

Through the efforts of TsAGI and the Experimental Design Bureau designers, the insidious and threatening flutter of the high-speed bomber was reduced and overcome. From that time on, mounting weight compensation on the controls became mandatory for all high-speed machines. Having tested the medium bomber at maximum speed, they established that it could fly at more than 430 km/h. The speed had doubled compared to the old TB-1 and TB-3 airplanes.

By government decision, the airplane was placed in series production. Owing to the selfless labor of the workers, by the May aerial parade in 1936 over Red Square, several tens of airplanes were prepared. The silvery medium bombers, passing by rapidly at low altitude, delighted the Moscow residents.

Soon the factory developed mass production of the airplanes, and they began to enter the military units. As is usual in such cases, world records began to be achieved on the airplane. Pilot M. Alekseyev carried a load of 1,000 kilograms to a height of 12,695 meters and established a load-capacity record.

In parallel they tested and refined the medium bomber in operation. Shortcomings and defects began to be revealed. After some time the Air Force demanded complete elimination of all flaws. Tupolev agreed to eliminate only some of them, asserting that some of the "shortcomings" did not affect the quality of the airplane or flight safety. When they could not come to agreement, the representatives of the Air Force took a somewhat unusual approach to the problem. They wrote down the defects on sheets of paper and hung them on one of the airplanes. One of the medium bombers was turned into a kind of Christmas tree.

Tupolev was indignant. Having expressed his opinion rather sharply, Tupolev went to the director's office to phone S. Ordzhonikidze. Meanwhile, the representative of the Air Force phoned his chief, Alksnis, who in turn reported to Voroshilov. In an hour the People's Commissars arrived at the factory and a dispute broke out. It became obvious that it would not be possible to reconcile the parties and Ordzhonikidze decided to satisfy all of the military's claims.

By decision of the People's Commissars, two of the airplanes in series production were taken out, and a large group of specialists from the Experimental Design Bureau, headed by A. A. Arkhangel'skiy (the lead designer of this airplane), was required to eliminate all shortcomings. The group drove back to the factory, and over the course of a month the majority of the shortcomings were eliminated. Acceptance of the airplanes resumed, and normal life at the factory returned. Over the course of several years, 6,656 of the medium bombers were produced. A large number of these airplanes took part in military operations.

Remembering this incident for a long time, Tupolev believed that Ordzhonikidze was incorrect, and that too many "purely gastronomic" demands, as he called them, were accepted. When he was reproached and reminded that he did not want to eliminate all of the shortcomings, but that he had to just the same, he noted, "I am even today convinced that it is much easier to find shortcomings in someone else's construction than in one's own. Try to do something that everyone likes. It will not work!

"I am not entirely objective in this regard. It is always more useful to look from the side of others' eyes at the fruits of the embodiment of one's thinking, at the chosen design decision. The number of a man's successful and original decisions is not endless. And far from always do they come off without a hitch. And there are always plenty of deadlines hanging over us. How easy it is in such instances to make some trivial decision and think that you are creating a road in the taiga, and it turns out that you are just riding a bicycle along a well-worn highway. And how bitter it is to see later that at times things have been done poorly. And you deprived yourself of true creative joy. No, do not back away from retribution!''

On the Eve of World War II

The year 1936 came, when General Franco led an uprising against the lawful democratic government in Spain. Hitler and Mussolini deployed squadrons of combat aircraft to assist the rebels. For the first time bombs rained down on peaceful cities. Guernica was wiped from the face of the earth. This tragedy became the subject of a famous painting by Pablo Picasso.

To assist the republicans, the Soviet government sent volunteer pilots with I-15 and I-16 fighters designed by N. N. Polikarpov, as well as bomber crews on Tupolev's medium bombers, the SB-2. The bombers operated successfully, attacking Franco's ground forces, and easily repulsed in air combat the Fiat and Heinkel fighters. However, when the Fascists began to receive the new high-speed Messerschmitt Bf-109 fighters, the picture changed, and we began to lose medium bombers.

Having become familiar in Spain with a captured medium bomber that fell into the hands of the Falangists, the German designer V. Messerschmitt characterized A. N. Tupolev in one of the technical journals:

Tupolev is undoubtedly the most prominent Russian aviation designer. First, he has supplied the Red Army with aircraft of the most varied types. Second, he never rests on what he has achieved, but is continuously seeking new paths. Third, he, and this is very important, in designing his airplanes does not forget about the level of the Russian aviation industry, and makes them simple and easy to manufacture by workers of average and low skill.

As a mark of their admiration for the medium bomber, the Spanish people, singing a song noted by our volunteers, "The Katyusha Came to the Shore," gave the aircraft the tender nickname "Katyusha."

In addition to the flutter, one more unhappy episode is associated with this machine. In 1938 designer A. S. Yakovlev built airplane No. 22, an original machine with a wooden wing and the same engines as on the medium bomber. Squeezed as small as possible, very light, but without any weapons, it flew at a speed 130 km/h greater than the medium bomber. At the annual aviation day at Tushino two such airplanes, both painted bright red, flew swiftly over the airfield at low altitude, delighting the crowd. Stalin became engrossed. As a result, he believed that Tupolev was not concerned about improving his series-produced airplanes. He also wondered why Tupolev's aircraft—identical to these airplanes, with the same motors—were much slower. Stalin ordered series production of the new Yakovlev design immediately. The military protested because they did not need an airplane without weapons. When weapons were mounted, in a much smaller assortment than on the medium bomber, the speed advantage disappeared, and series production of the new machines was halted. Nevertheless, a certain stigma had been attached to the Tupolev name.

In 1939, in the operation along the Khalkhin-Gol River, approximately 100 medium bomber airplanes destroyed the combat formations of the Japanese invaders, who had invaded the Republic of Mongolia, which was friendly to us. Later, in the war against the White Finns in 1939–40, these machines were used to destroy the fortifications of the Mannerheim Line. By the time of the Great Patriotic War they, naturally, were obsolete, but they constituted more than 94 percent of frontal bomber aviation. In addition, the medium bombers were used to link the headquarters of the large formations with the military units.

As early as 1935, Tupolev had suggested creating a high-speed passenger airplane, the ten-seat ANT-35, based on the medium bomber. It was successful, and flew from Moscow to Simferopol' in three hours, an unheard-of speed for that day. The ANT-35 successfully operated on the Aeroflot routes. At the Paris aircraft exhibition in 1937 a number of Western countries, especially Czechoslovakia and Serbia, displayed interest in purchasing them. However, the worsening international situation at that time hampered the Soviet aviation industry from moving into the world market.

Tupolev's last prewar designs were two four-engine low-altitude bombers: the ANT-42 for the ground forces, and the naval ANT-44. In essence, the ANT-42 was a five-engine bomber, and the history of the appearance of this fifth engine warrants attention. In those years air tactical theory presumed that in a future war bombers would fly at high altitudes. Only there would they be out of range of both antiaircraft artillery fire and fighters. At that time there were no airtight cockpits for fighters, and it was too difficult to reach bombers at high altitudes and wage battle with them in oxygen masks from open cockpits. Based on these considerations, the Air Force demanded a new high-speed bomber, a high-altitude airplane, to replace the TB-3. In principle, such an airplane could be created by supplying the engines with additional air at high altitudes, where, as we know, they lose power due to the rarity of the atmosphere.

A classic system existed for solving this task through the use of turbocompressors, which forced the air. Very intense work was already under way to create them, but a number of difficulties had been encountered. Turbocompressors bring into rotation engine exhaust gases with temperatures near 1,000 degrees, while ambient air in flight is at a temperature of minus 60 degrees Celsius. Thus, the first problem arises—heat-resistant materials. The second difficulty pertains to the bearings. An aircraft turbocompressor must be small. Consequently, to deliver the required amount of air it must revolve rapidly. Heat-resistant bearings good for 25,000–30,000 revolutions were needed. Finally, the third difficulty was in creating a coating against corrosion, for the exhaust gases of an engine are a very aggressive medium. All of this together showed that creating reliable turbocompressors was very difficult.

Tupolev set about to solve this problem himself. In the morning, so that telephones and visitors would not interfere with him, he locked himself in someone else's office. As usual in periods of intense activity, he was short-tempered and unsociable. His closest assistants, dropping by for a decision on some urgent question, at times were treated unfairly. All of the ongoing work of the huge design bureau, with the endless number of organizational and technical questions arising each day in the series-production factories and in experimental production, fell on Arkhangel'skiy's shoulders.

Arkhangel'skiy was one of the most colorful figures around our chief designer, and was his friend and comrade. They began their joint activity during the early days with Zhukovskiy, and continued to work together for a half-century. Chief of the group of engineers designing the structures of the airplanes in the Central Institute, lead designer of a number of the ANT airplanes, chief factory designer, and one of the closest assistants of Tupolev, Arkhangel'skiy in 1941 became his first deputy.

Outwardly, and perhaps inwardly as well, they were almost polar opposites. Tupolev was of average height, and inclined toward heaviness. Arkhangel'skiy was tall and very thin. While in the prewar years Tupolev in his long-belted blouse and wide soft trousers could be taken for a Bohemian, Arkhangel'skiy in his invariable leather jacket and Navy cap hanging over the face of an ascetic looked like an automobile racer. The chief designer was severe and unyielding; his assistant was mild and delicate, incapable of harsh words. Tupolev personified sole command and lightning decisions; Arkhangel'skiy was inclined to make decisions unhurriedly and only after first consulting with his colleagues. A bystander might think that it would be difficult for such opposite types to get along. In reality their many years of work together indicated the opposite. Both of these men not only did not oppose one another, but to the contrary they complemented one another, and in so doing cemented the leadership of first the Central Institute Aeronautics and Seaplane Experimental Construction Department, and later also Tupolev's Experimental Design Bureau. When the general designer was on a trip or on leave, Arkhangel'skiy moved into his office. Accustomed to the demanding and even harsh instructions of the general designer, the workers now became somewhat flustered at hearing, "My old friend, try. We must solve the problem with this ill-fated autopilot. Again it has started to oscillate the airplane. Only I ask you in earnest, please take it easier on the people associated with this matter."

Although he was extremely soft-spoken, Arkhangel'skiy loved to make sarcastic remarks. After the war, when Tupolev was assigned a new "Pobeda" automobile we all fell in love with the first postwar automobile. But there were only a few available. Arkhangel'skiy had been selected to receive one, for of all people he undoubtedly deserved one. Unable to present our Experimental Design Bureau a second Pobeda, Petr Vasilyevich Dement'yev awarded us a "Moskvich" car instead. Arkhangel'skiy thanked him, but could not help but say sarcastically, "Dement'yev, why don't you order them to remove the front seat. At my height I can drive it from the rear seat." The minister laughed for a long time, but soon found it possible to give a Pobeda to Arkhangel'skiy.

In Arkhangel'skiy's presence it was much easier to solve problems at meetings. His charm could be heard even in his voice. During the war it was very difficult at times to telephone Moscow from Siberia. Arkhangel'skiy could ask, "What is troubling you, my friend? Give me the phone. Sweetie, please connect me with Moscow, it is very important. Well, thank you, my dear." Then, with a sly look on his face, he would say, "You have to be able to make an impression on the women even by telephone!"

In solving turbocompressor problems, Tupolev considered how time-consuming it was to search for new materials. The epic struggle to introduce light metals into aircraft construction was not forgotten, and there was no doubt that solving the problems of turbocompressors would be just as difficult. The Air Force needed high-altitude airplanes as soon as possible. People in other countries were working on turbo-compressors, and there was no doubt that in a few years they would appear everywhere, with high-altitude airplanes directly behind them.

Thus it was necessary to find a solution that would make high altitudes possible without turbocompressors. Forcing additional air into the main engines was required. Consequently, they needed to create a supercharging system. So that it would be efficient, they needed to use a tested scheme, known materials, and an autonomous energy source. Tupolev considered what assemblies already existed in industry. The air blowers used for mines or blast-furnaces were very high-capacity machines that required powerful electric motors for rotation. Much less air was required for the four Mikulin M-34 engines. He decided to adapt the industrial air blower to the needs of the airplane, using dimensions similar to a circular dinner table for four. To turn such a supercharger, some 700–800 horsepower would be required, equivalent to the capacity of a fighter engine. Assuming they could achieve this, where would they place it on the airplane?

Now Vladimir Mikhaylovich Petlyakov had his say. At the time he was the chief of a brigade of heavy airplanes. "In the fuselage, of course. There is nowhere else. But we have already designed it, and there is no free space."

"Yes, that is true, here you cannot dig up any free space. But what if . . . " They examined the drawing of the profile of the machine and set to thinking. Forgetting himself, Petlyakov took out his cigarette case and began to smoke. Tupolev started to cough.

"What an abomination. How can you breathe in that rubbish?"

Embarrassed, Petlyakov put out the cigarette, but distracted by the search for a solution he soon lit a second one. Finally Tupolev could not stand it and called his secretary. "Vera Petrovna, take his cigarette case and hide it." Petlyakov turned red and, without saying a word, threw the pack into the trash can.

With a template, rulers, pencils, and erasers, the two designers, both of them large, well-built, strong men, leaned over the drawing table. One could hear them talking. "Let's cut it off a little here." "Now let's tighten the outline of the pilot's canopy." "Excellent." "A little bit more." "Now it's not so bad, don't you think?"

"You are right, this is the only possible solution, but . . . but now some questions arise: fire safety, air intakes, fuel lines, and endless other questions! But these are the usual design problems, and we must assume that we will be able to solve them."

Tupolev said, "Now about the engine. To add to the four M-34 engines a fifth engine of the same kind is most sensible. But it is much larger and heavier than the M-100 designed by V. Ya. Klimov. Have them draw it as it should be."

When they examined the arrangement of both engines, it became clear that the Klimov engine fit easily, while the Mikulin engine hung out of the fuselage in several places.

"Whether we want to or not, we must use the M-100," Tupolev summarized. "And so let us settle on it. Let us persuade Vladimir

Yakovlevich to design a reduction gear box for the revolution of the central supercharger. The central supercharger itself can be built only in such a qualified institute as the Central Institute of Aircraft Engines [TsIAM]. Let us ask the people's commissar and the leaders of TsIAM to assign this task to K. V. Minkner, their chief specialist on engines, along with chief designer A. A. Mikulin. In my opinion, only they, along with the personnel of the engine institute, are capable of handling such a difficult task. Everything else amounts to the normal activity of our designers."

Everyone understood that Tupolev and Petlyakov had found the only decision which, even though it did not follow a well-trodden path, could lead to success. It is true that at first it took aback the designers of both the engine and the airplane. Their first reaction was not favorable. But the seed had been sown, and the chief designer and Petlyakov were sure that it would grow. Business meetings with the engine designers took place in their offices from morning until night. The design bureau worked at full capacity. It became clear that by the method proposed by Tupolev and Petlyakov it was theoretically possible to achieve excellent results; there was every reason to believe that an airplane with a central supercharger of air into the engines could fly at altitudes of 11,000 meters at a speed of 400 km/h, which was precisely what our Air Force needed.

Work on the ANT-42 was progressing. Its outward shape to some extent mirrored the TB-3. Keen eyes would note that the lines were smoother and much more noble, but the family likeness was clear. In the wing, which had a span of 39 meters, were four well-cowled engines. Most often, radiators were placed under each engine in fairings. But here it was necessary to retract the huge wheels in flight, which meant making nacelles for two engines. Since the forward part of these nacelles was not serving any purpose, why not place the radiators for both engines in them? The proposal was implemented and had a significant effect.

Turrets with guns and machine guns could be seen in the nose and tail section of the elliptical fuselage. The machine had very strong defensive armament. Guns and heavy-caliber machine guns were located at seven points. Under the forward turret was the navigator's cockpit, and behind it, in the upper part of the fuselage, the pilots sat one behind the other. The glassed-in canopy of their cockpit led to the fairing, under which the fifth M-100 engine, with central supercharger, was mounted. Bombs of the largest calibers, all the way up to three-ton bombs, could be hung in the large bomb hatch, located under the pilots' cockpit and the compartment with the central supercharger.

The importance of the ANT-42 for the armed forces could be judged by the fact that M. N. Tukhachevskiy, the People's Commissar for Defense, who was in charge of technical equipment for the Workers' and Peasants' Red Army, was named chairman of the mockup commission for the TB-7. The famous theoretician of military science and commander in no way limited himself to official representation at meetings. Inviting Tupolev and Petlyakov with him, he crawled over the entire mockup, examined the stations of all crew members, rotated the gun mockups, evaluated the angles of fire and vision from the cabins and the convenience of the work of the navigator, radio operator, pilots, and gunners. A number of his practical suggestions were implemented in the airplane. Subsequently he followed the construction of the airplane, and contributed a great deal to its successful completion.

The commission chairman had reason to be pleased. The most advanced tactical concepts were implemented in the ANT-42. Inaccessible at its maximum altitude to either antiaircraft artillery or fighters, it was at the same time the most powerful bomber in the world.

Pilots Gromov and Zhurov conducted the flight tests. Everything went successfully, with one exception. No doubt there has never been an experimental airplane that has not had such impediments arise. In this case it concerned the fifth engine. As soon as the engine was started at the park, the invited representatives of the weapons and armament suppliers who surrounded the machine became agitated.

"Comrades, what is this? The engine roars, the airplane shakes, and not a single propeller turns. Something is not right here! We must call Stoman."

After some time Stoman's emaciated figure appeared. With his entire appearance he demonstrated how they all bored him. Waving his right arm nervously, "Karlych" (that is what he was called at all the airfields from Moscow to Vladivostok, for there were few among the testers of

Soviet aircraft who were as popular as the chief of Tupolev's flight station) began to think up explanations for why the engine would operate and yet the propeller did not turn. He included everything—both easy improvisation and unabashed lies—all determined by inspiration. It depended on whom he had to work with, whom it was necessary to "fib" to a little, and whom he could intimidate with his enthusiasm. Is it necessary to say that the existence of this engine was considered a secret, and that requests to explain why "the engine operates but the propeller does not turn" could ruin the lives of both Stoman and all of the test personnel?

Soon the Central Airfield was closed for reconstruction, and it was necessary to deploy the ANT-42 to one of the secondary airfields around Moscow. Pilot Rybushkin ferried the airplane and, upon landing, wrecked the undercarriage. Repair under poor conditions occupied much time, and tests of the badly needed machine were delayed. Who knows, had it not been for these vexing little problems, by the end of the war we might have had many more of these formidable bombers on line.

Brigade Commander P. M. Stefanovskiy, in those years in charge of military aircraft testing, characterized the times as follows:

SB medium bombers and DB-3 long-range bombers were being mass produced. . . . These were quite modern combat machines, which significantly surpassed in their flight specifications the best foreign airplanes of their classes. The shortcoming of these machines was that they could not carry large-caliber aerial bombs weighing one, three, and five tons. To make bombing strikes against large defensive targets in the enemy rear area, the design bureau of A. N. Tupolev produced the TB-7 fast, heavy bomber. This airplane was badly needed. A great war was looming. Therefore, Air Force Chief Ya. I. Alksnis took the organization of testing on himself. This machine had many innovations. Four boosted engines retained their power up to an altitude of 3,500–4,000 meters. A fifth M-100 engine, to rotate a central supercharger, was mounted in the fuselage fairing, behind the pilots' cockpit. It increased the altitude capability of the main engines to 10,000 meters, and was switched on in flight as required. Owing to this, the multiton airship, with its maximum flight specifications up to 10,000 meters in altitude, surpassed all the best European fighters of that time. The height, speed, and ceiling turned out to be unexcelled. The machine, in our view, fully met its intended purpose. The designers reacted quickly to comments and proposals of the crew, and eliminated shortcomings.

When the front line approached Moscow, the majority of our airplanes lost the ability to reach Berlin. Only long-range aviation pilots on ANT-42 airplanes continued to make bold strikes against the capital of Fascist Germany. Owing to the fact that they could lift one-ton and even three-ton bombs, the effectiveness of their strikes was very great.

In April 1942, I. V. Markov, the lead Air Force engineer for testing this bomber, flew in to see Tupolev. It was necessary to decide whether it was possible to send one of these airplanes with a government delegation headed by V. M. Molotov to England and the United States. President Roosevelt offered to send an American airplane, but Stalin wanted to make the flight on a Russian airplane. The necessary calculations confirmed that this was possible on an ANT-42. Several seats and additional oxygen tanks were mounted on one of the bombers. It was necessary to fly across the front line at maximum altitude. First the airplane flew to London, and then across Iceland, Greenland, and Newfoundland to Washington, D.C. The Americans assessed Tupolev's machine very highly, giving their due to its unusually powerful weapons. The president, who had invited the airplane's crew, headed by pilot E. K. Pusep and navigator A. P. Shtepenko to the White House, noted this.

In producing the high-altitude ANT-42 (TB-7), the Soviet Union had surpassed the United States and the other European countries. In 1939 a series-production factory, with I. F. Nezval in charge of construction, began to produce TB-7 airplanes.

It has been said that Tupolev divided airplanes into "fortunate" and "unfortunate." The ANT-42 turned out to be "unfortunate." At the start of the war, when it was clear to everyone that it was necessary to produce as many Il-2 Shturmovik ground-attack aircraft as possible, airplanes that were sowing panic in Hitler's troops, it was decided to switch all of the factories producing Mikulin engines to the AM-38. Production of the AM-35, used for the ANT-42, was curtailed. I. F. Nezval, in charge of construction of the TB-7, mounted air-cooled ASh-82 engines on them, and mounted A. D. Charomskiy M-40 diesels on some airplanes. All of these modifications, of course, slowed the production of the large, difficult-to-manufacture airplanes, and by the end of the war they had been able to turn over only seventy-nine to the military.

The ANT-44 flying boat (it was called *Chaika* [*Sea Gull*] for its curved wing) entered testing soon after the ANT-42. A. P. Golubkov was its chief designer. He had been named chief of the naval team after I. I. Pogosskiy's death. Maj. I. M. Sukhomlin carried out the flight tests. Several records for speed and cargo-lift altitude were set on the ANT-44.

Although it had good flight and seaworthiness specifications, the large flying boat did not enter series production. The war interfered. It was necessary to build machines that were more needed by the front. The Il-4, Yer-2, and Pe-2 wheeled airplanes being produced by the aviation industry were successfully handling missions on the confined Baltic and Black Seas.

The ANT-44 was the last flying boat designed and built by Tupolev's design bureau. After the ANT-44 the seaplane team ended its existence and the designers were transferred to other teams.

Imprisonment

One would think the country would be proud of the creator of such remarkable machines, but, as N. Arkhangel'skaya, wife of Alexander A. Arkhangel'skiy, the closest friend and associate of A. N. Tupolev, who was on duty on 21 October 1937 in the office of the secretariat of the chief engineer of Glavaviaprom [Chief Directorate of the Aviation Industry] recalls, four men came in late that evening and told her not to leave and not to answer the telephone. Three went into Tupolev's office, and the fourth remained with her. At about 3 A.M. Tupolev, surrounded by the visitors, entered the reception room.

Passing through the office of the secretariat, silently, as though apologizing, he nodded his head in greeting. As soon as they left Arkhangel'skaya rushed to the telephones, but all had been cut off. Then she ran to the next office and phoned home from there. "Sasha, this is terrible, the ONI have taken him away! Call Kaganovich. My telephones have been cut off."

Mikhail Moiseyevich Kaganovich, brother of the all-powerful L. M. Kaganovich, was chief of the Main Administration. He arrived at the office an hour later, and without taking off his coat collapsed into the sofa. Probably he understood for the first time that he would not find protection from his powerful brother, and in a day or even an hour might follow after Tupolev.

At this point I must go back a few months and tell how I happened to take part in preparation for the flight to America of a third aircraft, the unfortunate DB-A, no. N-209. The history of this long-distance aircraft is closely linked with the N. Ye. Zhukovskiy Military Aviation Academy, where Professor Bolkhovitinov was in charge of the aircraft construction faculty, and Prof. M. Shishmarev, who was my wife's father, worked with him.

The flight specifications of the TB-3 bombers no longer satisfied the Air Force, and they persistently recommended modifications. Busy with creating the future TB-7 (ANT-42, Pe-8), which was to develop a speed at high altitudes twice that of the TB-3, Tupolev refused to make changes. The Air Force leadership began insistently to try to convince Sergo Ordzhonikidze to organize a small design bureau at the factory in Fili, which would be capable of carrying out the modification. In addition, they proposed sending a group of military specialists there. Twenty military engineers, headed by V. Bolkhovitinov, were sent to Fili and began the designing. In order not to interfere with series production, Ordzhonikidze requested that the modification be limited to a small number of changes. The corrugated skin of the TB-3 would be replaced with smooth skin, the landing gears would be made retractable, and the crew's cockpits covered with transparent hoods.

Tests of the modified machine, which was called DB-A (Long-Range Bomber-Academy) began at the end of 1935. Compared with the TB-3, the speed of the DB-A increased by 60 km/h. The ceiling was increased slightly and the cockpit accommodations were greatly improved. When it got to the point of long-range crossings, and a navigator-radio operator was needed, M. Shishmarev recommended me to Viktor Fedorovich, and the latter readily agreed.

On 14 May 1937, pilots G. Baidukov and N. Kastanayev, and I as navigator-radio operator, carried out a record crossing on the DB-A. Having taken five tons of cargo on board, we flew from Moscow to Melitopol and back, some 2,000 kilometers, and set a world speed record of 280 km/h. The results of the tests of the DB-A at the Scientific Research Institute of the Air Force, and this and other long-range flights, convinced S. Levanevskiy that it was possible to fly to the United States in the DB-A, even with a small "commercial load." Calculations showed that the machine could fly approximately 6,500 kilometers, that is, to Fairbanks, Alaska. Attempting to restore his reputation after the unsuccessful flight in the ANT-25, Levanevskiy appealed to Stalin, and the flight was approved.

Of course, the Levanevskiy crew had not had as many flight hours together as had the three previous crews. Those crews of V. Chkalov, M. Gromov, and V. Grizodubova had flown together for a long time and understood each other instinctively, even at a glance. On N-209 the situation was different. The crew finally assembled, consisting of Levanevskiy, Kastanayev, Levchenko, Godovikov, Pobezhimov and myself, was assembled not long before the takeoff. Commander Levanevskiy, navigator Levchenko, and second mechanic Pobezhimov had been in the United States before this, when they were involved in purchasing American airplanes for the Main Administration of the Northern Sea Route, and ferrying them to the Soviet Union. They returned in the spring of 1937.

While they were absent, two of the future crew members carried out the development, refinement, and testing of the N-209 airplane. There was the "shadow commander," as N. Kastanayev's friends called him, and the on-board mechanic N. Godovikov. When the en-route flights were being conducted they brought in a third person, myself, as navigator-radio operator, and when in spring 1937 everyone had finally assembled, six experienced specialists were present, but it was not a cohesive group. One other thing was even more serious: the crew commander had never once taken off with a full load, and the navigator had never accomplished a single super-long-range flight in the cockpit of N-209. The second on-board mechanic, G. Pobezhimov, had also not operated the vehicle, had not engaged in searching out and eliminating defects, and had not studied her thoroughly. Finally, on the eve of the flight, at the request of the NKVD I was removed from the flight, replaced by Galkovskiy, who had not even been in the cockpit before his appointment. There was not enough time to order Galkovskiy another flight suit at the tailor's, and he was given mine.

The date of the N-209 takeoff had been announced, so the airplane had to take off on that precise day. This hastily assembled crew had managed to accomplish only one test flight, and even that was not to the north, but to the Sea of Azov, a total of 2,000 kilometers in clear weather. And it was more than 6,500 kilometers from Moscow to Fairbanks, over Arctic ice, uninhabited tundra, and the forests of Canada.

N-209 took off from Shchelkovskiy Airfield at 1815 hours 12 August 1937—just a month after Gromov had flown successfully over the North Pole in an ANT-25, setting a new world record. The meteorologists informed the commander that the weather en route was poor. According to reports from Papanin, there was strong low pressure activity in the area of the Pole. At 1340 hours N-209 crossed the Pole, and 52 minutes later reported that the outside right engine had failed. This was the last officially registered report from N-209. Later, individual amateur and other radio operators received some fragmentary signals on frequencies close to those at which Galkovskiy was operating. But detailed analysis could not confirm that they belonged to the radio of the N-209.

Apparently the tragedy developed in this way. After the right engine failed, the heavily loaded N-209, even with three working engines, began to lose altitude and gradually sank into the clouds. Ice settled on its skin, especially on the ends of the wing and tail section. The weight of the ice accelerated the decent. Visibility worsened. Judging by the data from the last two radiograms, the average speed of descent can be taken to be approximately two meters per second. Then, at some time around 1420 hours the airplane must have struck the ice, the surface of which in the eastern sector of the Arctic was packed ice. It is highly problematic that an ice field suitable for landing could have been found under such conditions. A landing on the ice-hummocks of a heavy airplane with nacelles of the partially retracted undercarriage protruding out from under the lower wing surface threatened to result in a noseover.

Under such conditions, the cockpit, where the entire crew and all communications and navigation apparatuses were located, must have been crumpled. As a rule, people in such accidents are killed. Specialists agreed rather unanimously that N-209 perished with its entire crew somewhere around 500–600 kilometers beyond the Pole, along the general route of the airplane. Neither the N-209 nor the crew was ever found, even to this day.

I must admit that at first I was very disappointed in having been removed from the flight, especially since no one could explain to me the reason. And now was also the shock from the death of my comrades. . . . You can imagine my state when a month later I was discharged from the ranks of the Army as excess. Where could I find work with such a record? I was without work for several months. I had a feeling that the matter was coming to an end.

I was arrested in April 1938, passed through a "special meeting," and sent to Kuloylag camp in the far north, near Archangel, to cut trees.

IN PRISON | 2

Despite all the draconian measures adopted by the administration to isolate prisoners, in both 1937 and 1938 persistent rumors about the newly organized prison workshops for aviation design bureaus spread along the corridors and among the cells of Moscow's prisons. These rumors slipped by guards and penetrated the innermost recesses of the prison world—through locks, security louvers, and doors. These closed design bureaus (in Russian *sharaga*) worked on high-priority military and industrial problems. The rumors, of course, swirled about and picked up fictitious details, but the essential reality of the prison workshops was never confirmed by those who participated in them. Once an individual was assigned to a *sharaga*, he never returned to his former prison. Hence the whole phenomenon of special closed prisons remained in the shadows.

One winter evening, a vehicle exited the Butyrka Prison gate. This wasn't a paddywagon, but an ordinary van. Inside, three prisoners—myself included—with our belongings sat with bowed heads; we had been brought from the camps by the Main Directorate for Corrective-Labor Camps (GULAG). We had no idea where we were going, or why. After driving all over Moscow, the van pulled up at solid iron gates on

Saltykov Street and gave a signal. A guard wearing the uniform of the People's Commissariat for Internal Affairs (NKVD) came out, spoke with the officer sitting beside the driver, and we were admitted onto the grounds of the TsAGI, the Central Aero-Hydrodynamics Institute, where today there stands a commemorative plaque to A. N. Tupolev.

Passing the traditional monuments to Lenin and Stalin, the vehicle stopped at the door to the Design Department of the Section for Experimental Aircraft Design (KOSOS) at TsAGI. Security guards escorted us into the elevator and took us to an office on the eighth floor. We were treated politely: "Please have a seat, you have come to Central Design Bureau No. 29 (TsKB-29), a special prison of the NKVD. Please read our procedural regulations and sign them." We read: ". . . is forbidden," ". . . is not permitted," ". . . is prohibited," for five or six typewritten pages. Most everything appeared normal, but there were some new items as well. "For consumption of alcoholic beverages [Good Lord, how could one ever get them in prison?] and for any attempt to establish contact with the outside world through free employees, the prisoner will be discharged from his work and sent to hard-labor camp." Second, in the section on punishments, in addition to the standard denial of exercise and store privileges, and solitary confinement, there was an item, ". . . will be deprived of visits." From this it was clear that visits were granted here. We read and signed. We had to sign so many of these agreements to be considered good little boys in those years!

The guard took us to our "places of residence," as he called them. (The strange choice of words did not conceal from us the thought of "going to our cells.") We walked along corridors, along soft, carpeted paths, going right, then left, and finally downstairs—the place was deserted. Finally the "screw" (i.e., the guard; prisoners had several names for "screw") opened a door and politely asked us to enter. We listened carefully, but to our surprise no lock clanged as the door closed behind us.

We were in one of the rooms of the KOSOS. Along the walls were thirty soldiers' cots, covered with flannel blankets. Next to each cot was a nightstand with a pack of Ducat cigarettes placed on it. The window was barred, and there were several chairs. We gathered them together and sat down, only to remain silent for a few minutes. How abruptly our lives had changed—the transformation stunned us. Soon, more mundane impulses overcame us and we felt the urge to smoke. We hand-rolled our cigarettes and discussed in whispers what would happen next. After a brief interlude, the door opened and yet another guard said something like, "Please, supper's ready." From ingrained habit, I untied my food package, got out my mess bowl, and stood by the door. The screw smiled. "That's not necessary. They'll serve you there."

He led us to the dining area. As the door opened, about a hundred men sitting at tables covered with snow-white tablecloths turned their heads at the same time. Someone yelled, someone else ran toward us. There were many well-known, friendly faces, and friendly hands reached out to us. It is difficult to describe this meeting or the emotion it generated in us all. The security detail of about five men asked everyone—politely, but insistently—to calm down and take our seats. The storm gradually subsided, and we could have a look around. At various tables we saw Andrei Nikolayevich Tupolev, Vladimir Mik-haylovich Petlyakov, Vladimir Mikhaylovich Myasishchev, Josef Grigor'yevich Neman, Sergei Pavlovich Korolyov, Aleksandr Ivanovich Putilov, Vladimir Antonovich Chizhevskiy, Aleksei Mikhaylovich Cheremukhin, Dmitriy Sergeyevich Markov, and Nikolai Il'ich Bazenkov—in short, the cream of the Russian aeronautical world. Many friendly eyes looked in our direction, as if to reassure us that everything would be fine now. But I was gripped by fright: seeing all these people meant that it was true, that they, too, had been arrested. What a disaster!

We joined this illustrious group, filling in the empty seats. My mess bowl and spoon now appeared comical and unnecessary. At prison camp you only abandoned them when you were headed for the graveyard. The knives, forks, and plates, which we had grown thoroughly unaccustomed to using, only highlighted the absurdity of my bowl and spoon. A girl in an apron brought meat with macaroni to our table, asking politely, "Would you like tea or cocoa?" (And this was said to me, who only yesterday was addressed as "you piece of crap!")

Most of the men had already finished supper and were leaving their tables when a middle-aged man—it turned out to be the eminent chemist A. S. Faynshteyn, who had been a party member since 1915 and once met with Lenin—remarked with irritation, "The cocoa was cold again. What a disgrace!" The newcomers pinched themselves hard: "Good Lord, is this prison or a dream?"

As the dining room emptied, I glanced around quickly and grabbed several pieces of bread and slipped them into my pocket. From my prison perspective, this represented an important coup. My friends and the guards both probably saw this, but I did not care. One of the principles of camp life was, "If you snag it, it is yours, but if you miss out, you only have yourself to blame and you're a goner. The only road left for you is the one to the morgue."

Following dinner we gathered in the sleeping quarters, which were located in KOSOS's oak-paneled hall. Here old friends engaged in a lively conversation: Aleksandr Vasil'yevich Nadashkevich and Yuriy Vasil'yevich Kalganov, Konstantin Vasil'yevich Rogov and I. M. Kostkin, Georgiy Semenovich Frenkel' and Yuriy Aleksandrovich Krutkov, Vladimir Sergeyevich Denisov and I. N. Kvitko.

But I couldn't believe whom I saw next! Among these "enemies, saboteurs, and spies" was Vasiliy Stepanovich Voytov, a sailor from the days of the October Revolution, a Communist, and, before his recent arrest, director of the huge Aircraft Plant No. 82. I had met him earlier when he served as president of the credentials committee at the Zhukovskiy Air Force Academy at the time I enrolled in 1927. During those formative years, such committees weeded out those deemed unworthy to receive an education. I remembered that fateful interview. Voytov, a stern military man with two diamond-shaped pips on his collar tabs (which correspond to today's lieutenant-general), asked me: "Who's your father, where did he work, and as what?"

"He was a naval officer," I answered.

"Of course, of course, I had the honor to serve with your father on the battleship *Gangut*. I was a seaman, a petty officer second class, while he was the squadron commander."

I understood that, for me, the path to the Air Force Academy was forbidden.[1]

When I encountered Voytov at KOSOS, I couldn't help reminding him. "So tell me, Vasiliy Stepanovich, do you remember the credentials committee you once chaired . . . ?" He avoided this conversation, however, probably because there were several men like him among the prisoners at the *sharaga,* some former members of the Russian Social-Democratic Worker's Party (Bolshevik). These naive people still thought that their arrests were an absurd and tragic mistake. As for us—the engineers, noblemen, merchants, and officers—it was another matter. Perhaps there actually *were* some counterrevolutionists among us!

It took a substantial amount of time for reality to sink in for men such as Voytov. In my own mind, as before, I couldn't find any logic in Voytov's fate: how could it be that someone—not a careerist putting in his time, but a genuine underground Bolshevik—who not so long ago had the right to pass judgment on young men who were striving for knowledge, was now lying in the iron cot next to mine? What were his private thoughts? Later, I asked him several probing questions. This time he didn't walk away from our conversation. "So you remember that interview," I said. "You know, if you'd looked into my application information more carefully, you would have been convinced that my family gave the Russian Navy several worthy sailors."

"So who are they?" he asked.

"Permit me." I began to name them. "My mother's older brother, Vasiliy Fedorovich Shul'ts, commanded our guard ship in Piraeus. On one occasion, the Greek Queen Ol'ga was visited by her brother, a Russian grand prince. After drinking too much alcohol with his breakfast in the wardroom, the prince mistook himself for a naval commander and gave the order to get up steam and weigh anchor. The maritime regulations stated that orders from persons in the reigning family were to be carried out without hesitation. My uncle had to somehow avoid an international incident. In the crowded port, the guard ship could ram and sink any number of foreign vessels.

"My uncle solved the problem by going below decks to the wardroom and committing suicide. Naturally, the fireboxes died out and the ship didn't weigh anchor.

"My mother's second brother, Konstantin Fedorovich Shul'ts, was a flag-officer under the renowned Admiral S. O. Makarov on the battleship *Petropavlovsk*. He perished with the admiral during the Russo-Japanese War when the *Petropavlovsk* ran into a Japanese mine near Port Arthur.

"The third brother, Mikhail Fedorovich Shul'ts, commanded the cruiser *Novik*, the only Russian warship to break out of besieged Port Arthur during that war with Japan and return home. He deceived the Japanese by skirting their home islands on the Pacific Ocean side. The cruiser *Novik* managed to reach the port of Korsakovskiy on Sakhalin Island, but three enemy ships were waiting there. After joining an unequal battle and taking several holes in the hull, my uncle ran the *Novik* aground, removed all guns from it, and set up a potent battery ashore. In this improvised battery he fought until the end of the war. For his exploits he was awarded the St. George Cross. Subsequently, with the rank of vice admiral, he commanded the Siberian Flotilla, and in 1919 was executed in the city of Luga, where he was whiling away his old age.

"My father, too, fulfilled his duty. In World War I he commanded a composite detachment of the Baltic Fleet, led by the cruiser *Ryurik*. He mined all the approaches to the German ports from Memel' to Luebeck, for which he was promoted to the rank of rear admiral and was awarded the St. George Sword.

"So tell me, then, Vasiliy Stepanovich, would you have admitted me into the Academy after learning of all this?"

He answered, "Not at all, dear comrade. In my time we didn't understand what sins for which the Old Bolsheviks were imprisoned. Similarly, you didn't understand how democratic centralism worked. Military regulations forbade us to admit youths from certain suspect classes into Soviet higher educational institutions—and so I didn't admit them. And as for how that influenced the quality of future military specialists, that was none of my business. That's what the higher bosses are for!"

Indoctrination in the *Sharaga*

My life in the prison workshop was not without uncertainties, but I was surrounded by old friends. Andrei Tupolev presided over us as a patriarch. Around beds, chairs, nightstands, and in the passageway he asked the newcomers many questions. We were most cautious and answered his queries in low tones, glancing at the door from time to time to see whether the guards were coming. We learned that the sleeping area was a haven where we could enjoy isolation and autonomy. It was a self-managed territory: the guards were forbidden to enter when the prisoners were present.

Tupolev told us, "We've included you for a long time now on lists of specialists we need for work on Project 103. We have pulled many specialists from the Gulag. They have been looking for you in the camps for a very long time now, from Minsk to Kolyma, and from Dzezgazgana to Noril'sk. Praise God, you were still alive. Others, sadly, were not found alive."

We also asked questions about the *sharaga*. As it turned out, TsKB-29 contained three separate design bureaus: Petlyakov's group, then working on a high-altitude fighter (Project 100); Myasishchev's bureau, assigned to work on a long-range, high-altitude bomber (Project 102); and Tupolev's bureau, at work on the development of a dive bomber (Project 103). In addition, a fourth design bureau—one associated with Dmitriy Lyudvigovich Tomashevich—had just been organized and would soon be assigned to a new fighter design (Project 110). At the helm of this unusual enterprise was NKVD Col. G. Ya. Kutepov, a former electrician and fitter. The first prisoner-based aircraft design bureau, TsKB-29 of the Combined State Political Directorate, had been formed in the early 1930s. Members of the so-called Industrial Party of Designers and Engineers constituted this first prison workshop. They included Nikolai Nikolayevich Polikarpov, D. P. Grigorovich, B. N. Tarasevich, Aleksandr Vasil'yevich Nadashkevich, I. M. Kostkin, V. L. Kerber-Korvin, V. S. Denisov, N. G. Mikhel'son, Ye. I. Mayorov, and others. They designed several aircraft, including the I-5 fighter. As a joke, someone inscribed the letters "VT," meaning "internal prison" (*vnutrennyaya tyur'ma*), on the star insignia on the tail of the experimental model. As a consequence, Kutepov, the future chief of our *sharaga,* joined the "work" with the imprisoned designers.

Kutepov oversaw three deputy "directors" of design bureaus. Balashov "directed" Tupolev, Ustinov "directed" Myasishchev, and Yamalutdinov "directed" Petlyakov. Ten NKVD officers served as junior chiefs to round out the administrative structure of our *sharaga*.

We learned these things that first day at the *sharaga*. The prisoners started to disperse to their sleeping areas at about midnight (*about,* since prisoners weren't supposed to have watches). The lights were turned off, which also was pleasant, since the ceiling lights burned all night in cells, barracks, and in packed Stolypin railroad cars. For the first time in three years, we were going to bed in normal human beds, with sheets, pillows, and blankets. The sleeping quarters grew quiet except for the occasional scraping of streetcars which could be heard outside the window as they turned into Volochayevskaya Street and the park. They were hurrying to spend the night in Apakov Park, formerly Bukharin Park.

That night only the newcomers stayed awake: it was too great a shock after the prisons, the camps, and the transit points to lie in a clean bed, anticipating a return to your old work, the access to foreign technical journals, the use of a logarithmic slide rule, the availability of finely pointed pencils and white Whatman paper stretched on the drawing boards. I had arrived here from the filth of the labor camps where prisoners possessed no rights, where life was punctuated with the shouts and curses of the guards, the gnats and cold in the *taiga,*[2] the heat and tarantulas in the desert, the petty disputes of prisoners over a serving of vile gruel or over worn-out shoes. What a delight to sit down at Culmann drafting instruments, to draw a centerline, and to think about aircraft design!

Just one week before, in my former camp, in the predawn darkness when we marched off to work, a guard with the vacant face of a degenerate shouted, "Sit down! Hands behind your head! I'm warning you: Take a step to the left or a step to the right and I'll treat it as an attempt to escape. I'll open fire without warning." But now it was, "Please step into the dining room." What a contrast in environments. Only those who experienced it could understand the metamorphosis we underwent.

For three days newcomers did not have to work (we were quarantined), Instead, we read, ate well, slept, and strolled in the "ape house," the iron cage built on the roof of KOSOS, where prisoners could take a breather and exercise. This sort of break was a good idea: we needed time to rise from the bestiality of our recent past to the heights of engineering work now before us.

In the evenings, when our friends returned to the sleeping quarters after work, we were deluged with questions. Many of them had come to the *sharaga* directly from prisons and weren't familiar with life in the camps. Most considered their stay at the *sharaga* to be temporary, believing that when we completed the aircraft, we would all be sent to the camps, or worse. As S. P. Korolyov, the future leader of the Soviet space program, used to say, laughing, "No one's immune to Themis.[3] Her eyes are blindfolded: She might take you mistakenly; today you're solving differential equations, but tomorrow you could be in the grim Kolyma camps!" That's just how it was. We would notice that someone had disappeared, completely unexpectedly, usually per some standard procedure. A screw would approach someone at his workstation and ask him to come to the office. The guards had already collected his belongings in the sleeping quarters. Then the curtain would fall. Where, why, what for—they never answered these questions. It was possible that the authorities told Tupolev and Petlyakov something, but no one told us. Such arbitrary acts punctuated our lives in the prison workshop.

Our comrades gave us no peace until we shared our experiences. We felt compelled to tell them about the reality of the camps: we lived with criminals, experienced guard dogs biting those who lagged behind on forced marches, the utter arbitrariness of the camp administration and the guards, the horrible food, the heavy work norms, and the lack of mail. All this made such a strong impression on our comrades that some concluded that it would be impossible to exist there, and that the only way out would be to die. Objectively, it was impossible not to agree with this. However, we believed truth would ultimately triumph, it would come to light. Despite all the privations of camp life, I came across only a few suicides, in fact no more than two or three times in all my wanderings in the Gulag.

We newcomers, in turn, were interested in the history of the *sharaga*. Tupolev stopped by to see me one evening as I sat in an empty room, working on some technical problem. He told me how the prison workshop had taken shape: "During our embryonic phase, they took us to Bolshevo. Just about everybody was there: shipbuilders, tank specialists, artillerymen, chemists. . . . A few days after we arrived in Bolshevo, they called me in to see the prison authorities, and I received my first assignment. I was ordered to compile a list of imprisoned aircraft specialists whom I knew. To tell the truth, I was extremely perplexed. I knew all those who were arrested before me, but what about those after me? Perhaps they would use my list to throw many more people into prison? After thinking things over, I decided to list everyone I knew. After all, they really couldn't have imprisoned the entire aircraft industry, could they? This seemed to be a rational approach, and I wrote out a list of about 200 men. It turned out that, with rare exceptions, all of them were already behind bars. You know, they did things on a grand scale! These lists of the imprisoned aviation specialists expanded constantly. Whenever some newcomer arrived, he was told, 'Sit down and write down whom you saw from the aircraft world.'

"In the end, the police authorities identified about 200 aircraft specialists for work in the *sharaga,* but they did not know where to put them. At the time, there were no calculators, and blueprints were duplicated by hand. Each engineer required as many as ten technicians, draftsmen, copiers, and so on. It turned out that our prison workshop needed space for about 800 to 1,000 personnel."

The only facility on that scale in existence in Moscow at the time was what had been Tupolev's design bureau. Without its leaders, however, it was barely functioning. In order to create the illusion of experimental aircraft construction, then–People's Commissar M. M. Kaganovich[4] transferred a group of second-rate aircraft designers—V. N. Belyayev, Shevchenko, Gudkov, N. P. Gorbunov—to the former Tupolev design bureau. It is possible that they actually were capable engineers; unfortunately, they didn't create anything sensible. This was to be expected, for under those conditions, in addition to engineering ability, one had to possess a strong will and discipline to make it to the top.

The state system required order and productivity. To keep the designers from fighting among themselves, as they had in the past, the authorities appointed Lenkin as director over them—a sort of obedient junior officer. Lenkin stifled the squabbling, but they still didn't produce any good aircraft. This situation began to cast its shadow on Stalin's head of police, Beria. As Stalin's "best friend," Beria assumed the role of guiding light for all science and technology.

After several meetings between the NKVD and the People's Commissariat for the Aviation Industry, there was an edict with Stalin's blessing to subordinate the Soviet Union's aeronautical establishment to police control. Within days, bars were installed on the windows at Moscow-area aircraft plants, and the entire eight-story KOSOS building became a prison.

Tupolev revealed that some of the second-rate chief designers were driven out, others curtailed. As many as 150 "enemies of the people" were transferred from Bolshevo to Radio Street, where several hundred free workers were put under them. The newly created TsKB-29, now under control of the NKVD, commenced its creative work.

After our three days of acclimation, we newcomers were summoned, one by one, to meet with Kutepov, who was situated in Tupolev's former office. For the first time since our arrival, we descended the main staircase from the sixth floor, where our sleeping areas were, to the third floor, which held the offices of the "directors." Oh, how short-sighted you were, citizen Kutepov, reading only the NKVD's notices and thinking that they were actually the Book of Genesis! If you had read something like *Parkinson's Law* (even earlier there existed works that were no less insightful before this *Law* appeared), it would have become clear to you that they would toss you out of someone else's building, and that you would finish your path in life as a mere rank-and-file economic planner or manager.

The "boss" spoke with us, informing us in no uncertain terms of the abyss that separated us. I was ordered to "change into anything more appropriate." (He even winced at the specialists' outer appearance.) "Get a haircut, get a shave, you're assigned to such-and-such a design bureau, you'll be issued your personal identification stamps, or facsimile." I was told, "You'll get detailed instructions from Balashov, the chief of the design bureau."

We didn't know yet that there were no last names in such prisons; they were replaced by the notorious "facsimile"—a rubber stamp with three digits which you affixed to blueprints and calculations instead of a signature. The sum of the digits indicated for whom you worked. For example, Tupolev was 011, his deputy N. I. Bazenkov was 065, and the chiefs of his designer brigades were 056, 074, 092, and so on.

Next morning, in the office of Major Balashov, chief of Design Bureau No. 103 (KB-103),[5] I was issued a set of drawing instruments, a logarithmic slide rule, a chit for the technical library, and my stamp.

Balashov then escorted me to my workstation and introduced me to the free workers as a leading designer and manager. And so, only four days after any screw could treat me as he saw fit in prison, I became the manager of several engineers, technicians, and designers. From that point on, I could order them to work late into the evening, grant or refuse them leave, and decide whether they were worthy of any awards. At first this was a serious concern for me: Won't we get inflated opinions of ourselves? Most important, how would our free subordinates act toward us?

Everything turned out to be much less difficult than I had imagined. We were received not as enemies of the people but as people who had been hurt by life. In the morning we would find signs of touching concern from the free workers in our desk drawers: a flower, piece of candy, pack of cigarettes, or even a newspaper. After getting used to us, they would tell us candidly, "Be careful around so-and-so, he's a stool-pigeon." Lack of rights, rudeness, and denunciations coexisted with tenderness, all-forgiving love, and a willingness to make sacrifices. Others spread talk of our "machinations": how Tupolev had sold the blueprints for an aircraft to Messerschmitt, how Korolyov had planned to flee abroad, and how Bartini—Mussolini's personal agent—had infiltrated the chief designers. Our more generous coworkers were finally threatened that if their liberal attitude toward us continued or—God forbid—if they were caught passing some notes to the free world outside, they would be quickly converted from free workers into prisoners. How could we not remember our "free" comrades with a few good words, how could we not be grateful for their touching signs of concern, and, finally, how could we not be proud of our fellow countrymen who, despite the corrupting propaganda of that era, remained human beings?

Let us return to our story. The "director" of KB-103 was Balashov, a major in State Security. His assistant was Kryuchkov, also a major. Both were rather dull-witted people. What did such "directors" from the security organs understand of technology? Two imprisoned engineers, Otten and Naumov, went to Ustinov, who "directed" the Myasishchev Design Bureau, with a proposal to create a small two-stroke emergency gasoline engine for supplying an airplane's electrical system in case the generators failed.

"So tell me, what kinds are used currently?" asked Ustinov.

"Four-stroke engines," they answered.

"It's risky to switch to two-stroke engines right away," noted Ustinov. "Wouldn't it be better for you to work with three-stroke engines at first?" Thereafter, we referred to Ustinov as "Three-stroke." The reader will smirk, "This is just a funny story!" For us, it was more—this anecdote merely mirrored our grim reality. We were controlled by untrained, often ignorant people.

The equipment brigade included two prisoners: engineer I. S. Svet and me. Both of us were interested not only in producing blueprints for Project 103, but also in everything that increased the airplane's combat effectiveness.

The military navigators told us confidentially, "You know, we bomb mainly at night. During the day, the German fighters shoot down our Il-4s easily, since they are only lightly defended with three machine guns. But at night, when we can't see our targets properly, our accuracy suffers." We tried to solve this problem. For nighttime forced landings, our Air Force was using PAR-2 parachute-retarded illumination flares. Once ejected, the flares descended slowly, illuminating the ground and enabling the pilot to choose a good location. We thought it would be a good idea to fill the hatch of the lead airplane in the regiment with large numbers of PARs instead of bombs. After reaching the target zone, this airplane, which we nicknamed "the Torchbearer," could use them to determine target location precisely and vector its wingmen there via radio. As a result, bombing effectiveness would be increased markedly. Bombs would no longer fall on residential areas, destroying homes and killing civilians, but on defense plants, weapons-storage dumps, and airfields.

After preparing our calculations and blueprints, we reported to Major Balashov, the "director" of the experimental design bureau (OKB). After familiarizing himself with our material, but without evaluating it, he said, "Good, I'll report it to Kutepov and General Danilov [the chief of the NKVD department]. They'll call you."

Kutepov called us in and asked angrily, "Who ordered you to do this?" We didn't return to the "Torchbearer." All our work probably went into the file with the standard marking, "Protect forever."

Did they really think that everything would remain the same for all time?

After the war, sadly, we learned from foreign aviation magazines that English engineers had been concerned with the very same problems and had arrived at a similar solution. They immediately tested it and put it into series production. They named their "torchbearer" aircraft "Pathfinder." The meticulous Germans performed calculations, while the English confirmed the results: the accuracy of night bombing against illuminated targets had increased severalfold. This method kept hundreds of bombers in formation that would have been shot down during daylight raids. And so, here you can see a second benefit—not an anecdotal one, but a real-life one—from the NKVD "management" of the work of specialists and designers.

Prisoners at Work

Andrei Tupolev led as chief designer for Project 103 (KB-103); Nikolai Il'ich Bazenkov was his deputy. The chiefs of the design brigades for the various areas were as follows: A. M. Cheremukhin, strength; A. Ye. Sterlin, aerodynamics; N. A. Sokolov, aeronautical engineer; academician A. I. Nekrasov, theoretical computations (previously, this had involved all the employees at the Central Aero-Hydrodynamics Institute); I. G. Neman (the former chief designer of Kharkov Aviation Institute [KhAI] aircraft), fuselage; V. A. Chizhevskiy (the former chief designer of the Bureau of Special Designs aircraft and gondolas for stratosphere balloons), wing center section; D. S. Markov (former chief designer of the OSOAVIAKhIM plant), tail surfaces and controls; S. P. Korolyov (future head of the Soviet space program), who was soon replaced by Boris A. Saukke (the former chief of the design bureau for construction of the *Maksim Gorkiy* aircraft), wing design; M. N. Petrov, pressurization and air conditioning; Aleksandr Romanovich Bonin, hydraulic equipment; G. S. Frenkel' (the former primary navigator for the Scientific Research Institute of the Air Force, NII VVS), instrumentation; myself, L. L. Kerber, electrical and radio equipment; A. V. Nadashkevich, armament; T. P. Saprykin, landing gear; S. M. Yeger, overall design; and S. A. Vigdorchik, production engineering. The brigades for the powerplant and its equipment were headed by free-worker engineers A. P. Baluyev and B. S. Ivanov. From the old workers at the Central Institute there was also G. A. Ozerov. At Tupolev's request, Ozerov assumed all administrative and financial duties. Obviously, the management staff for our Project 103 was highly qualified. The *sharaga*'s other department managers were of about the same caliber.

An incomplete list of the most famous prisoner-specialists of the NKVD's *sharaga* would begin with academicians and corresponding members of the Academy (former and future), some 21 people, a group that included Robert Bartini, Vladimir Myasishchev, and Sergei Korolyov.[6] The remainder of my list, compiled from memory, includes fifteen professors and doctors of science, fourteen directors, chief engineers, and chief production engineers of aircraft plants, and five chiefs of series-production design bureaus. All in all, there were more than 150 of us.

It must be kept in mind that there were two other prison design bureaus for the aviation industry: one for engines and one for missiles. In all likelihood, we won't be far off if we estimate the total number of specialists from our ministry who were pulled out of camps by the triumvirate of secret police chiefs Yagoda, Yezhov, and Beria at 280 to 300 men of the very highest qualification (the other two prison design bureaus were on a somewhat smaller scale). One has to admire those who, even after such blood-letting, were nonetheless able to deliver thousands upon thousands of aircraft to our heroic army during World War II. Most likely, no other nation could tolerate something similar. The stories of several prisoners are compelling. In fact, all of them are noteworthy. But no memoir can touch upon every memorable person or episode.

For instance, I remember Bartini, seated and lost in thought, with his Roman patrician's head bowed. True, he was not dressed in a toga, but in a prisoner's robe. The son of well-to-do parents, he was attracted to Marxism, joined the Italian Communist Party and, when Mussolini came to power, emigrated to our country. Later, they "unmasked" him, and "for passing state secrets to Italian intelligence" they gave him ten years in prison. A talented design engineer, he designed several types of aircraft. Without a doubt, he would have created even more if he had not had a penchant for very original designs. Our bureaucrats viewed many of his creative ideas as fantastic. Speaking with an accent (and it must be mentioned that in our country, as in pre-Petrine times, foreigners were regarded with great suspicion), he overcame all the hurdles of the bureaucratic system. Eventually, he was appointed chief designer of the Gol'tsman Plant. Gol'tsman, an Old Bolshevik and a member of the Aeroflot board, died in an aircraft disaster while flying to Crimea with P. I. Baranov. He was buried next to the Kremlin wall with honors. During one of the purge trials, someone dragged the deceased Gol'tsman into the "organization." With purely Oriental refinement, Stalin ordered his remains exhumed from the Kremlin wall, burned, and his ashes scattered to the winds. The plant's name was changed to the Vodop'yanov Plant in honor of the renowned Soviet aviator. Like Gol'tsman, Bartini fell out of favor and was thrown in prison. The bomber he designed—in fact an excellent design—was renamed the Yer-2, after the surname of his deputy engineer, Yermolayev. Whenever Bartini was accused of selling something to Mussolini, he would become excited and speak in rapid Italian, constantly repeating the word *inconceptible* (meaning "incomprehensible").

They said that Karl Szilard was a distant relative of Leo Szilard, who had fled from Hungary to the United States. Later, Leo Szilard and Albert Einstein signed the famous letter to President Franklin Roosevelt in which they expressed their fears that Hitler, using the work of Fermi and Meitner, would force Heisenberg, Hahn, Weizacker, and Bethe to create atomic weapons ahead of the United States.

Neither Szilard desired to live under tyranny, and Karl emigrated to the USSR. Both found freedom, but of different types: Leo found Western-style freedom, and Karl Eastern-style. Usually modest and shy, Karl would assiduously pore over aerodynamic computations. When anyone came up to his small table, he would invariably stand up and answer questions with a smile. Sometimes he would succumb to fits of Magyar madness. In a black Russian shirt, pants of the same color, and huge prisoner's boots, gripping his face in his hands, he would walk up and down the corridor. When the bell rang, he would run into the sleeping quarters and throw himself down on his bed. The next morning, all this would be gone and, with a guilty smile, he would plunge into his logarithms. We later learned the reason for Karl's flare-ups: His wife and two young children remained free. As a Hungarian who knew no Russian and was ill-equipped for our reality, Karl was almost certain she would perish. Karl was unable to understand what was happening in our country, and was convinced that Stalin had reached an agreement with Hitler, that fascism was afoot in our country, and that we would all be shot. In spring 1944 he, Krutkov, and Rumer disappeared. Many years later we found out that they had gone to work on the atomic bomb. After the bomb was detonated, they were freed. And so the two Szilards, traveling rather convoluted paths, arrived at one and the same problem in two rather different countries. Verily, "Your ways are inscrutable, O Lord!" We have only to add that Karl's family was supported by the prisoners' relatives: A. R. Bonin's wife took in young Misha, while his little sister was taken in by another prisoner's family, and our families helped his wife. I really don't know which made us happier (perhaps this will sound cynical): the fact that his relatives were adopted or the fact that these families were not intimidated by the police. According to rumors, when he returned to Hungary, first President Mátyás Rákosi, and then his successor, János Kadar, placed Karl on a pedestal, and made him an academician. Have you found peace of mind, our dear Karl?

Yuriy Borisovich Rumer was a mathematician, physicist, and intellect with broad interests, one of the potential Russian Oppenheimers. They brought him from the Mariinsk camps: when he arrived it was May, and warm. He entered our prison workshop in worn-out shoes made from felt boots, and draped in a mattress cover. Tall, with bluish-black hair and broken glasses on a large nose, he looked like a Jewish prophet. He used to work in the Abakan valley in Eastern Siberia, "not, however, to create anything worthy of remembering, as had the great exile, but to prove to the local inhabitants the worthiness of the system of punishments founded by Messrs. Krylenko and Vyshinskiy, who met and were personally acquainted with Lenin and his wife. I don't think that Vladimir Il'ich ever shared their interpretation of the presumption of innocence," Rumer would say with a grin. He worked in the computation brigade, and won over our librarian, Fat'ma Rasstanayeva, by reading all the technical literature in English, French, and German in one month. "He never once used any dictionary!" she said admiringly. An interesting incident subsequently befell Rumer. He was arrested for the second time, and following the usual scenario, he was tried through an OSO, a special military tribunal. They gave him ten years. While he was bouncing along in a cattle car, where the "criminals would win everything I had on me at cards, one thing after another, up to and including my eyeglass frames, and would threaten me that I would to show up at my destination in the suit I was born in, a misunderstanding came to light." It turned out that he had been confused with someone else. "I rode back in a passenger car on a fast train—but still with an escort," said Rumer. "Apparently they were afraid that the respectable free civilian workers would win everything of mine." In Moscow they apologized for excessive force used against him and put him back at his work at an atomic prison design bureau, but as a free worker.

Then there was Aleksandr Ivanovich Nekrasov, the author of an important work on theoretical mechanics. While he was on a business trip to the United States, he barely survived a car accident. He returned to his homeland an invalid, where he learned that he was an FBI agent and was given a standard ten years. As a result of the shock, he preserved only two areas of his memory: he did impeccable work in theoretical mechanics and recalled the distant past—Easter matins, girls from high school, Filippov's pies, Russian poetry, the magazine *Capital and Country Estate*—but mainly he grieved over the kitten that remained at home when they took him to the Lubyanka. Tupolev saw to it that he was protected (as much as was possible under our conditions) from the prison administration and did not get involved in any hands-on work. Nekrasov would sit in a tiny, separate room at a huge, antiquated desk and follow his own course. When Nekrasov was fired because he was an old man poorly adjusted to life, Tupolev took him home. They finally gave him an apartment on the Gor'kiy Embankment, delivered antique furniture from the NKVD's warehouses, and gave him a kitten and a housekeeper, both of whose loyalty was beyond question.

Sergei Pavlovich Korolyov, the future creator of space missiles, came to us from Kolyma, where he had been digging gold with a pick. He was short and bulky, with his head attached at an angle and distinctive hazel eyes. Always a skeptic, Korolyov viewed the future with absolute gloom. "They'll shoot you without an obituary" was his favorite phrase. He labored with Tsander on rockets and was convicted because, as his investigator said, "Your pyrotechnics and fireworks are not only not necessary for our country, but also dangerous. You should do real work and build airplanes. But the missiles are probably for an attempt on our leader's life, aren't they?" It took Wernher von Braun, Peenemünde, and the V-2 for them to remember him. Subsequently, after being honored with the highest adulation, orders, and ranks, he kept in touch with his old friends. In later years, at his dacha near Ostankino, over a glass of cognac he would look around and then whisper to us, "Well, boys, do you remember 'Three-stroke,' Grishka Kutepov, the monkey cage, our rendezvous, and so forth?" Because he was an academician, the police maintained a constant guard around him, even then. The security at his dacha, which was provided by the same types of screws as at the *sharaga,* lent this a special irony. "You know, the most tragic thing now is the fact that there's still so much in common between this situation today and the situation then. Sometimes you wake up at night, you lie there, and think, 'Maybe there's somebody, and he's already given the order, and these very same guards will come in here rudely, and bark, 'Well, scum, collect your belongings!'" His first marriage was to K. M. Vintsentini, a surgeon at the Botkin hospital (named after a famous Russian physician in the nineteenth century). It was she who amputated the leg of our comrade N. A. Sokolov when they found a sarcoma.

And there was Yuriy Vasil'yevich Kalganov, son of a peasant from Orlov, who made his way to the top with difficulty. When the October Revolution broke out, this young man became commissar of a Red Army division. People who suffered the same fate included the prisoners A. Ye. Sterlin, K. Ye. Polishchuk, A. Yu. Rogov, and V. S. Voytov—all commissars of divisions, corps, or armies during the Civil War after the October Revolution. The war ended, and when the party called, they joined the "golden ones," the elite who studied at military academies; afterwards, they were sent into industry. Soon thereafter Kalganov became director of a huge plant in Irkutsk. Someone "accused him of being a subversive, being part of a spy organization." The zealous investigator kept Kalganov standing at attention for days. When they brought him to his cell, his legs had swollen so much that his boots had to be cut off. Kalganov said, "You understand, I told him to go to hell right away. The accusation against me was a slander or a provocation— well, anything you want, only not true." They tormented him for a long time. When they saw that they had come up against an iron will, they decided to turn him over to the special committee. "You know, it was only here in the *sharaga,* after seeing all of you, including such directors and old Communists as Leshchenko, Abramov, Polishchuk, Voytov, and Chizhevskiy, that I came to my senses and understood this play-actor Stalin and his band of henchmen. Their hands are stained with the blood of hundreds of Communists. While I was in the camps, I still had doubts."

After coming to his senses, Kalganov didn't become a counterrevolutionary, however; he became an extreme cynic. With his sharp analytic mind, he threw himself into computational work with delight, went into raptures over the elegance of mathematical solutions, and, rubbing his hands together, would repeat, "Here everything is done honestly, without demagoguery!" He was never separated from his logarithmic slide rule. Whenever he read the newspaper, you could hear his running commentary on the information contained therein: "Aha! Coal production in Hungary has reached unheard-of heights! Let's take a look. [Figuring on his slide rule] Yes-s-s-s, an average of 1.5 kilograms per capita, monthly. That's not a lot." After working in the Scientific Research Institute of Aviation Technology for a short time after he was freed, he took to his bed with dropsy of the legs—a consequence of his Irkutsk interrogations. When he died his wife, Benita Anatol'yevna, carried out his request and buried him at the Vagan'kov cemetery. ("I got used to it. I used to go there to check whether Ekaterina Furtseva[7] was lying in her claim that rising living standards had brought longevity. I'd take a walk and figure. She was lying, like all of them, by about 15 to 18 percent.")

Vladimir Sergeyevich Denisov was in prison for the third time. In 1924 he had been arrested, sent to the distant Solovki camp as part of the first wave of arrests of the intelligentsia. By 1932 Denisov worked in Central Design Bureau No. 39 with Polikarpov and Grigorovich. He was a member of the intelligentsia in the Chekhov mold and a political recidivist—was that really possible? Tall and bald, with a squeaky voice, he would lecture newcomers with mock gloominess: "Yes, gentlemen, from the very founding of our state decent people have always sat in prison, interrupting this condition, which is the natural one in the Soviet state, by brief visits from their families." And when asked if they hadn't conducted investigations humanely in the past, he would reply, "Of course, in 1924 they used to put people into a sauna, into a room with a temperature of 36°C, without water, until the NEPmen confessed where the gold was hidden. But we won't talk about where and how they were injured. In 1932 they would put people on an ordinary pail in a draft. 'Sit there until your ears stick into your sciatic nerve, or until you tell us who recruited you and where, as a spy on behalf of some intelligence service (the investigator would politely permit the sitter to determine which intelligence service).' Which one is more humane, gentlemen, I ask you to determine for yourselves." Denisov was one of the greatest Russian production engineers in wooden aircraft construction, and as early as World War I he was helping to set up production of Farmans at the Duks plant in Moscow. After the revolution he organized production of a huge lot of R-1 (i.e., de Havilland DH-9A) planes with Liberty engines.

Lev Termen was the inventor of the Termenvox and other electric musical innovations. The irrepressible news media trumpeted him to the entire world. He and his instrument were "demonstrated" to Stalin and treated fondly, although the very same Stalin banned *Katerina Izmaylova* and Shostakovich's *Eighth Symphony*. Termen was sent to tour Europe, and then refused to return home. He went to the United States and married a beautiful Creole woman. Rachmaninoff and Stokowski didn't think much of his inventions, and the matter went downhill. Having "seen the light," Termen applied to come home. They gave him permission. He arrived and brought the government all his instruments as a gift. They didn't accept the gift and sent the inventor to Kolyma. By agreement with the administrators, the Creole woman went there as well; she then disappeared. Later they brought Termen to the *sharaga,* where he worked under Myasishchev on instrumentation for Project 102.

Aleksandr Vasil'yevich Nadashkevich was an epicurean and devotee of the "weaker sex": even in prison he took care of his nails and beard "à la Henry IV." He was among the greatest specialists on aircraft armament. This was his second time in prison. "You know, with every time this becomes more and more plebeian. At TsKB-39 I had a separate office with a telephone, and Paufler, our chief—please note, he was no match for our Grishka [Nadashkevich always called Kutepov 'Grishka Otrep'yev']⁸—left it with me on condition that I promise not to call home. I also had a pass, and could walk around the facility grounds without these idiotic babysitters." He would tell an interesting story about demonstrating for Stalin the I-5 fighter which had been built by designers later identified as "saboteurs." "This took place at Khodynka, next to the two hangars in which we both lived and worked. Nikolai Nikolayevich Polikarpov gave the briefings on the aircraft, while I briefed on armament. When we had finished—and we were standing off by ourselves then—Stalin asked, 'They don't make it oppressive for you here, do they?' For ten years this phrase hasn't given me any peace. Was he being a hypocrite or an actor?"

When instructing us in the mysteries of existence, Nadashkevich addressed the prisoners as "sirs." "You must note, sirs, that in the Soviet Union [Nadashkevich was of Polish extraction] it is impossible for the cause to advance without tyranny. Remember Ivan the Terrible, Peter the Great, or Nicholas I, the *oprichnina,* or the Okhrana, the tsarist secret police. They were fanatically cruel to the Decembrists, the revolutionaries, and more recently, the Old Bolsheviks. They beat them with cudgels, rods, and rubber truncheons. Just wait, one of those who comes later will leave the Molotovs, Kaganoviches, and Malenkovs to rot in prison. The arbitrariness will be even more frivolous, and the demagoguery even more trenchant." Unlike Szilard and Korolyov, who acquiesced to a sad outcome, Nadashkevich was an optimist, and was deeply convinced that our situation was not a threat. "They'll be executing the apostates, otherwise the emperor will continue to have no clothes. As for us, since we know how to make good airplanes, which the country can't do without, they won't touch us. Moreover, sirs, remember that they'll cover you with medals. But if necessary, one fine day they'll remove the medals and send you back to the Lubyanka."

There was also Georgiy Semenovich Frenkel', a professor at the Moscow Aviation Institute, a true intellectual who valued classic and modern poetry and the arts. He was a subtle diplomat. Tupolev's Talleyrand (without selling anyone out, of course), he was referred to among the prisoners as "the educated Jew with the Governor General" (*eminence gris*). He derived his definition of what was taking place from confusing sources of Jewish mysticism and Russian black magic. His mix of the apocalypse, Christ, and Stalin proved attractive to no one. Since he was physically frail, and lacked the persistence and insolence necessary to survive in the camps, he buried himself deep in his soul. He subsequently became a hypochondriac and spent all his free time sleeping. Jokingly, the prisoners called a unit of sleep equal to 24 hours a "frenk." "The prisoner sleeps, and so his time passes," he would joke in reply, but even when he made jokes, his eyes remained those of a doomed man.

Yuriy Aleksandrovich Krutkov was a corresponding member of the Academy of Sciences, who looked like Voltaire as portrayed in the sculpture by Houdon. An erudite man and an encyclopedist, he charmed everyone with his keen mind. He was brought to us from the Kansk camps, where he worked as a janitor in the criminals' barracks. "Not bad work, you know," he told us. The subtlety of his remark was striking: "Sometimes they beat you, sometimes they let you have a smoke. I must note that the students at my university were less demanding and never once beat me during a lecture; moreover, they'd even give me a cigarette without complaining, and it wasn't just a butt." He used to tell how, during one fierce winter, he and the janitor from the barracks next door were ordered to cut some firewood. The two middle-aged men, wrapped in rags, filthy, and with several days' growth of gray stubble, slowly pulled the saw back and forth as they conversed:

"Where are you from?"

"Leningrad. How about you?"

"I'm from there, too."

"Where did you work?"

"At the Academy of Sciences. How about you?"

"The same place."

"I knew everybody there. What is your last name?"

"Krutkov."

"Yuriy Aleksandrovich? Good Lord, don't get mad at me, I didn't recognize you! I'm Rumer."

"That's enough of 'who knows whom' here, Yuriy Borisovich. Don't get mad at me, either: I'm going now, to stoke up the barracks, or else—you know yourself—they'll beat me up."

I didn't know Krutkov before his arrival at our prison design bureau. But we all know how people from St. Petersburg become friends readily because of their love for their city. The two of us would often spend our evenings recalling our memories of that city: Vasil'yevskiy Island, the Peter and Paul Fortress, the monuments and the majestic boulevards.

Once I mentioned the name of the shipbuilder A. N. Krylov during a conversation. Krutkov was unusually interested. "How did you know Aleksei Nikolayevich?" "Well, perhaps you can't call it exactly knowing him. After the war in 1904–5, Russia set out to rebuild its fleet. As part of his official duties, my father, Lev Fedorovich Krutkov, the former chief of staff to Admiral N. O. Essen, often had to meet with A. N. Krylov, who at that time headed the model basin. They would discuss the combat qualities of the battleships then under construction. The Krylovs lived nearby, and so, to avoid wasting time, Krylov would stop by our house. He and my father would go off by themselves to my father's study. Both of them were highly temperamental, and soon, from behind the tightly closed doors, we would hear them begin to argue, their voices growing louder and louder. My frightened mother, Olga Fedorovna, would get my brother and me and take us into the kitchen, and, covering her ears in horror, would whisper, 'Good Lord, they'll start fighting any minute now.' They would get together on Fridays, which prompted her to refer to these gatherings as 'black Fridays.'"

Krutkov interrupted me. "Pardon me, you'll understand right away why I'm so excited. Since the day of my arrest, I haven't known a thing about my wife's fate. She used to work in one of the scientific research institutes of the Academy of Sciences and knew Krylov's daughter, Anna Alekseyevna, whose married name is Kapitsa. You can help me learn what happened to my wife, since the Germans have already blockaded

Left: L. L. Kerber, 1903–93.
Above: Andrei N. Tupolev
(1888–1972) as a young man.

Andrei N. Tupolev (standing next to cockpit) with his first airplane, the ANT-1, a
single-engine monoplane powered by an Anzani 35-hp engine, ca. 1923.

The ANT-2 under construction. The Soviet Union's first all-metal aircraft, the ANT-2 was powered by a Bristol-Lucifer 100-hp engine.

The ANT-3 (R-3) was an all-metal, two-place reconnaissance aircraft with a range of 750 kilometers. In 1926, the ANT-3 *Proletariy* made a 6550-kilometer flight from Moscow to European capitals including Berlin, Paris, Rome, Prague, and Warsaw. The following year another ANT-3 completed a round-trip flight from Moscow to Tokyo, covering 22,000 kilometers.

The ANT-4 (TB-1) was a twin-engine bomber Tupolev designed in 1928. Pictured here is an ANT-4 equipped with skis for cold-weather flying, an area pioneered by the Soviets. Powered by two M-17b (500–680-hp) engines, the ANT-4 represented the Soviet Union's first in a series of bombers built in the 1930s. The bomber had a crew of six with an effective range of 1,600 kilometers.

The first Tupolev-designed trimotor aircraft, the ANT-9, ca. 1930.

Andrei Tupolev's five-engine ANT-14 reflected the growing Soviet interest in giant aircraft during the 1930s. The *Pravda,* shown here in 1931, easily surpassed contemporary Western airliners with its capacity for 36 passengers.

The ANT-14 *Pravda* flew to many European cities to showcase Soviet strides in commercial aviation. The *Pravda* is shown here in Bucharest, Rumania, in October 1935.

The ANT-20 *Maksim Gorkiy* became the Soviet Union's most ambitious effort to build giant aircraft. With its red wings and eight M-34RN engines the behemoth aircraft first flew in 1934. The *Maksim Gorkiy* weighed 42,000 kilograms and could carry 85 passengers. This ill-fated aircraft crashed after a midair collision in 1935.

The ANT-40 (SB) went into series production between 1936 and 1939, being deployed in the Spanish Civil War and World War II. Fast and maneuverable, the twin-engine SB-2 bomber became one of the most effective Soviet warplanes, although by the start of World War II the aircraft had become obsolete.

The ANT-42 (TB-7 or Pe-8) was the Soviet Union's first genuine heavy bomber, although it would be manufactured in limited numbers (fewer than 80 entered service). The Pe-8 variant was equipped with a sequence of engines, the last being the 1540–1700-hp ASh-82, ca. 1938. The Pe-8 operated with a crew of 11 and with a range of 4,500 kilometers. This Tupolev design saw only limited use during World War II.

One of Tupolev's seaplanes was the ANT-22 (MK-1), which entered service in 1934. This aircraft was unusual with its six M-34R engines. The ANT-22 served as a bomber with the Soviet Navy.

Andrei N. Tupolev with wife Yulenka Nikolayevna and daughter Yuliya in the 1930s.

Tupolev (extreme left) with the crew of the ANT-25; (left to right) Alexander Belyakov, Valery Chkalov, and Grigoriy Baidukov. Pilot Chkalov is shown with Belyakov and Baidukov after their historic flight from Moscow to the Far East in 1936. This same crew made the first transpolar flight from Moscow to Vancouver, Washington, in June 1937.

Valery Chkalov's ANT-25 at Pearson Army Air Corps Field, Vancouver, Washington, after his historic transpolar flight, June 1937. (Courtesy of the National Air and Space Museum.)

Mikhail M. Gromov (left) with Andrei B. Yumashev and Sergei A. Danilin flew Tupolev's ANT-25 on a transpolar flight from Moscow to San Jacinto, California, in July 1937, breaking a long-distance record. The Gromov flight represented the apogee of world-record breaking flights by the Soviet Union in the 1930s.

L. L. Kerber (right) with transpolar flier Georgiy Baidukov (left) at the time of the historic flights of the ANT-25, ca. 1936–37.

Vladimir M. Petlyakov

Boris A. Saukke

Vladimir M. Myasishchev

Aleksandr A. Arkhangel'skiy

G. A. Ozerov

Tu-2 over Moscow. Tupolev worked on the development of the Tu-2 during the time of his imprisonment.

The Tu-4 was the Soviet copy of the American B-29 Superfortress. At the close of World War II the Soviet Union interned three B-29s, which served as models for the Tu-4, a Soviet copy of the American bomber. Tupolev's design bureau produced the Tu-4 in the course of one year, providing the Soviet Union with a strategic bomber.

The Tu-95 Bear, shown in flight in 1956, became the most famous Soviet bomber during the Cold War.

The Tu-70 became the first important variant of the Tu-4 bomber. Pictured here is a 1946 commercial version which carried 48 passengers.

Tupolev's Tu-114, ca. 1957, one of the Aeroflot's most popular airliners. The Tu-114 had a range of 10,500 kilometers and could carry 176 passengers.

The Tu-124 became one of the Soviet Union's most successful jet airliners in the 1960s.

The Tu-154 entered service in 1968. This three-engine jet could carry 152–58 passengers, becoming a mainstay of Aeroflot service in the 1970s.

Tu-105, Prototype Tu-22 Blinder.

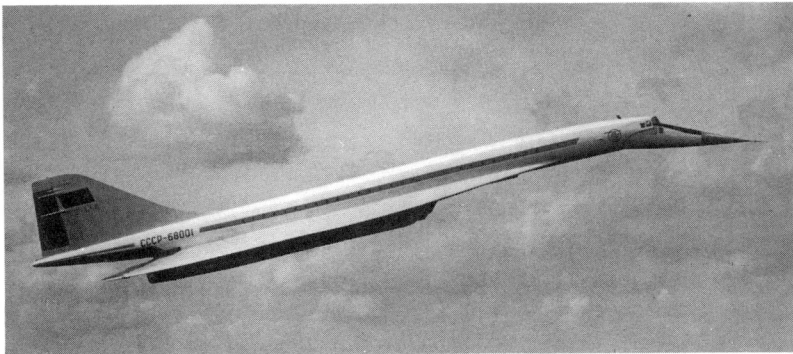

Tu-144 in flight. The Tu-144 was the Soviet rival to the Concorde in the race to build a supersonic commercial airliner.

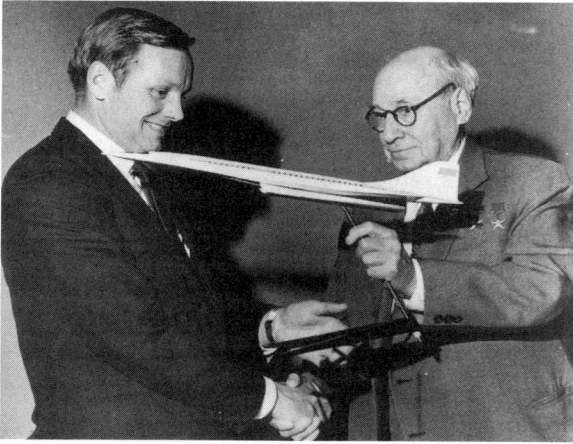

Andrei Tupolev presents model of the Tu–144 to American astronaut Neil Armstrong.

Andrei Tupolev at the Paris Air Show in the late 1960s.

Andrei Tupolev with family *(left to right):* M. Tupoleva, daughter-in-law; A. A. Tupolev, son; Yulenka, wife; Andrei Tupolev; Yuliya Andreyevna, daughter, with her husband, V. M. Vyl' and granddaughter.

Andrei N. Tupolev in military uniform, taken in 1945, at the time of the Tu-4 project.

L. L. Kerber in postwar years.

St. Petersburg. I must know if she's alive. You're already a freed serf, while I'm still a slave. It's easier for you to write a letter to Krylov."

Of course, I agreed. The reply arrived surprising quickly. "I hasten to inform you that I asked my daughter to conduct inquiries. Please set Krutkov at ease: His wife has been evacuated to Novosibirsk, where she is at work and in good health. Please give my deepest regards with greatest respect to Lev Fedorovich and dear Ol'ga Fedorovna, whose life we so unceremoniously spoiled on Fridays with our verbal battles." Krutkov's spirits lifted, and I replied to Krylov that both my parents had already left our transitory world. My father, a vice-admiral, died during a kidney operation in 1921, and my mother perished in besieged St. Petersburg [Leningrad] in 1942.

I called on the Krutkovs in Petersburg after the war, and his wife told me that they had reestablished contact with Anna Alekseyevna. A fine plot for a contemporary O. Henry, if one appeared, wouldn't you say? Krutkov worked in the computation section of the Central Design Bureau and was a consultant and an arbitrator in all types of complex technical arguments. He was a splendid storyteller, and we heard amazing stories from the lives of his colleagues, academicians S. F. Ol'denburg, A. P. Karpinskiy, A. F. Ioffe, and A. N. Krylov, whom he knew well. They freed him after the atomic prison design bureau, and he returned to his beloved St. Petersburg and the university.

Josef Grigor'yevich Neman was an aircraft designer of the Kharkov Aviation Institute. He pioneered work in the Soviet Union on retractable landing gear. Working in this sphere of technology was risky, for if the landing gear didn't lower, you could be sure of being accused of sabotage. Neman, by nature, was an effusive person who continually sought new ways to do things. Always good-natured, forgiving, and responsive, he was idolized by many young workers. He possessed enormous credibility and respect. He was perhaps the only person who could say, as he evaluated a design approved by Tupolev, "You know, Andrei Nikolayevich, to tell the truth, this one's unfortunate; let me think it over, and I'll try to propose something more elegant," and have Tupolev take it calmly. Tupolev, Myasishchev, and Petlyakov valued Neman highly, and considered him a rising star. Before his arrest, Neman lived in Kharkov, where he worked as a chief designer and lecturer at the

Kharkov Aviation Institute. Being separated from his family for several years, he had many female admirers who were attracted to his youthful vitality and good looks. Of course, under the *sharaga*'s conditions, these feelings were purely platonic. However, one of these admirers evidently attracted him seriously, and in Omsk, after he had been freed, he married her. When the Germans were nearing Kharkov, his wife and children miraculously escaped from the city, and reached Omsk after a month-long ordeal. When she learned of her husband's fate, she poisoned herself. Similarly, the lives of many families were ruined in the aftermath of arbitrary arrest. After returning to Kharkov following the war, Neman contracted leukemia. And so perished a talented designer who, in the opinion of many, belonged in the pantheon of great Russian aircraft designers, to join the company of Igor Sikorsky, D. Grigorovich, and Andrei Tupolev.

Vladimir Antonovich Chizhevskiy was an intellectual who saw some secret, higher dynamic at play. "The human mind cannot comprehend everything; just wait, several years will pass and the causes unseen by us will be revealed!" Time went by, we lived to see Stalin's death, and everything came to light. The emperor had no clothes, we discovered, and everyone saw how vile and loathsome were his reasons for everything—what had been supposedly inaccessible to the human mind. Chizhevskiy had been the chief designer of the Bureau for Test Components (BOK) in Smolensk. He had created several interesting BOK aircraft and designed gondolas for the stratospheric balloons *OSOAVIAKhIM-1* and *USSR-1;* like the majority of people, he turned out to be short-sighted.

Aleksei Mikhaylovich Cheremukhin was a man with the most tender of souls, but with cunning. He was also our consummate specialist on aircraft structures. He was an expert on the history of aviation, a military pilot during World War I, and a one-time associate of N. Ye. Zhukovskiy. Moreover, he was a talented artist, and he maintained clandestinely an illustrated chronicle of TsKB-29. All this was lost during the evacuation to Omsk. Like Yu. A. Krutkov, he was a talented storyteller, who regaled us with stories about his student days at the Moscow Higher Technical School. After the war, when he died behind the wheel of his car near

Palanga, we lost a good friend, who occupied a leading position in his field. I remember him vividly as a great chronicler of Russian aviation.

As we see, the conglomeration of prisoners at the *sharaga* was rather diverse and interesting.

The actual prison, in which our life outside work took place, occupied the top three floors of the Design Department of the Section for Experimental Aircraft Construction (KOSOS), and its windows faced the inner courtyard. The sleeping areas were here—three large ones and one small one, always poorly lit—as were the dining area, kitchen, and monkey cage for walks on the roof. The many rooms of the administration and security had windows facing the street. These three floors were connected to the other floors on which we worked by a single inner stairway. We didn't have our own individual cells.

Our schedule was tightly scripted and controlled. We were awakened at 0700, and by 0800 we and our sleeping areas were in order. Breakfast was from 0800 to 0900, after which we worked until 1300, when we broke for lunch. We worked again from 1400 to 1900, then rested until 2000, and had supper and free time until 2300. Then it was lights out. They conducted a head count at night, while we slept. As the war drew closer, they extended the workday to ten hours; in the spring of 1941 they extended it to twelve hours. They fed us well: in the morning we had kefir [fermented milk], tea, butter, and porridge; lunch consisted of two dishes and dessert compote; supper was a hot dish, kefir, butter, and tea. They would bring yogurt and bread into the dining area for those who worked after supper.

This diet and the general lack of physical exercise soon allowed us to gain weight, to take on human form after being reduced in the camps to mere skeletons. There was a shop attached to the prison where, once a week, we could use the money sent to us by relatives via the Butyrka to buy toilet soap, cologne, candy, cigarettes, and even razor blades. The cologne explained the meaning of the item in our rules of conduct on punishment for consumption of any alcohol. Three of our young people—V. Uspenskiy, I. Babin, and L. D'yakonov—for drinking their extra cologne were beaten unconscious and sent into solitary confinement.

The prisoners' insulation from the outside world was superbly thought out. We were under a watchful eye both day and night. Two sets of security guards oversaw our prison workshop. Within the *sharaga* there were the professional prison guards from Butyrka; outside was plant security. Screws were always on duty at the entrance to the sleeping areas. They weren't so much guarding us as making sure that some free worker didn't accidentally wander into the sleeping areas. A second strong screening force consisted of three armed screws who stood at the only door that connected the *sharaga*'s spaces with other areas of the large KOSOS building. In addition, screws dressed in plainclothes walked the corridors of the *sharaga* all day long, glancing into the work areas from time to time. From 2300 to 0800, they left one screw stationed on each floor, but set up posts at each sleeping area. A third line of guards secured all exits and entrances to the grounds of TsAGI, and patrolled within the courtyard and along the fences. After getting used to this, we noticed yet a fourth line, of "gentlemen in plainclothes," who hung around in Radio Street and along the banks of the Yauza River.

Of course, a force one-tenth this size would have been sufficient to guard us. I must confess, we didn't think about escaping, and if someone had thought of escaping, where would he go? Whose help and whose shelter could he seek out? What would happen to his family, and to all his friends and relatives for such a bold act? No, this elaborate display of security was not intended for us, but for those who were free. It really was necessary, so the people would come to believe that we were, in fact, "enemies of the people."

At first the prisoners were distributed among the sleeping quarters according to when they arrived. Then the bosses permitted each sleeping area to contain workers from a single design bureau. Middle-aged men—those inclined toward solitude and quiet—were in the fourth sleeping area. Although they took us for a shower once a week, toward morning the air in the sleeping rooms nevertheless reminded you of something between an army barracks and a passenger rail car with unreserved seating. After stopping by the sleeping quarters for Project 100 and seeing the dozing Petlyakov in tattered socks, one newcomer was shaken. The man who created the Russian Flying Fortress, the ANT-42 (or Pe-8), two years before the Americans created their Boeing

B-17 Flying Fortress—and here he was, a prisoner in torn socks. . . .
In 1936, four years before the war, the Pe-8 flew at an altitude of 10
kilometers, cruised at a speed of 420 km/h, and delivered three tons of
bombs to a target at a range of 2,500 kilometers. This bomber entered
into production four times, and four times it was taken out of
production.

It occurred to me that if Molotov had said to German Foreign Minister Ribbentrop during the latter's visit to Moscow that a thousand
similar bombers were at our airfields, ready for action, would the
Germans have launched Operation Barbarossa in 1941? As to whether
our aviation industry could have produced them in four years, the major
plant directors sitting with us—Leshchenko, Abramov, Voytov,
Kalganov, and Usachev—answered clearly in the affirmative. And, as
a matter of fact, it wasn't even necessary to answer. The answer lay in
those tens of thousands of combat aircraft delivered to the Soviet Air
Force during all the war years.

Tupolev's Design Team

Trains carrying specialists began to arrive at Bolshevo in late 1938 and
early 1939: shipbuilders, artillerymen, tank specialists, rocket experts,
and communications specialists. Tupolev was a center of attention for
the aviation cadres. Soon the nucleus of the future experimental design
bureau coalesced around him: S. M. Yeger, G. S. Frenkel', A. V.
Nadashkevich, and two young men who, although they could draw,
clearly ended up in this group due to bureaucratic confusion in the
Gulag: soundman V. P. Sakharov, and graduate of the Machine-Tool
Manufacturing Institute I. B. Babin. This core group was augmented
by engine specialist A. P. Alimov, a capable technician and master of all
trades. In particular, it was he who built the wooden aircraft models at
Bolshevo.

In the corner of the oak-paneled hall in the sleeping quarters stood
Tupolev's bed; his neighbors were Yeger and Frenkel'. Almost every
evening, this corner turned into a "technical council" where the creation
of aircraft 103 [Tu-2] was discussed. Tupolev usually sat on his cot with

his legs tucked under him, Turkish-style, in his favorite Tolstoi shirt and warm socks. He was surrounded by the meeting's participants. This sight probably looked a lot like the Moscow Art Theater's staging of Maksim Gorkiy's *The Lower Depths*.

A sheet of plywood would be dragged out from under the bed (to avoid information leaks, paper was forbidden in the sleeping areas) and some sort of component sketched on it. With a soft pencil (Tupolev hated hard pencils; if he happened to pick one up, he would stare at it intently through his thick glasses, spit, and toss it over his shoulder), Tupolev would correct the drawing while explaining his thinking, from time to time resorting to strong Russian words in order to help his audience understand. "No, you still haven't found a solution here; you can't glue the engine nacelle to the wing with dung. He [pointing to the designer] thinks that this shaft here will take the load. That's silly: This isn't a shaft, it's a nozzle, and it will twist without loads, all by itself." Biting his hangnails (the sharp pocketknife that had always been in his pocket had been confiscated: we weren't allowed to have anything sharp in prison, although we could buy Gillette razors in our shop) and laughing infectiously from time to time (his large stomach shaking), the chief would instruct his flock.

Sometimes Myasishchev or Petlyakov would drop by our sessions. These two were almost polar opposites. They called Petlyakov "the great one who has taken a vow of silence," because at any meeting he preferred to sit silently, with only his inquisitive, wise eyes giving any clue to his thoughts. Listening attentively to the group, he would analyze the comments, discarding all verbal chaff (and at meetings, particularly those at ministries, this could be about 80% of what was said). He would then precisely and succinctly communicate his opinion. Most people did not understand this process, and considered Petlyakov a simple executor of Tupolev's decisions, but this wasn't so. Petlyakov was the chief of the first brigade at KOSOS, which designed the TB-1 and TB-3, the ANT-14, and the ANT-42. Of course, the overall design of the airplanes, and decisions such as the multispar wing and duralumin corrugated skin, were Tupolev's. Petlyakov, however, carried out the detailed development. Moreover, he produced the blueprints for the

Pe-2 without Tupolev's participation. In short, he was a completely independent and talented chief designer.

Petlyakov brought to mind a figure from the American Wild West. He was short, solidly built, with energetic facial features, a strong chin, and cold, gray eyes. On the inside, however, he was a gentle, shy man who couldn't shout at his subordinates, much less curse them, just as he couldn't complain about them to his bosses. The group knew his traits, respected him, and worked amicably and—most important— with great enthusiasm.

After he converted his high-altitude No. 100 fighter into the Pe-2 dive bomber, as Stalin had ordered, Petlyakov was freed from custody and transferred with the rest of his group to supervise the series production of the aircraft. In fall 1941 they were evacuated to Kazan', where Petlyakov became the plant's chief designer. His life ended tragically in early 1942 when he was summoned to Moscow. He set out in one of the Pe-2s being ferried from the plant to the front lines, but en route, near Arzamas, the airplane caught fire and crashed. Of course, sabotage was suspected, an inquiry ordered, and several persons were subsequently arrested. Petlyakov was buried in Kazan'. About fifteen years later, fearing demonstrations and the arrival of hundreds of Georgians, they quietly lowered the coffin of Vaso (Stalin's drunkard son) into a grave next to Petlyakov's.

Vladimir Mikhaylovich Myasishchev was the complete opposite of Petlyakov. Even in a crowd, his appearance was striking. Handsome, with head set at a proud angle, and always dressed in refined fashion, he looked like an actor. At TsKB-29 Myasishchev designed a long-range high-altitude bomber, aircraft No. 102. After Petlyakov's death, Myasishchev was appointed chief designer of the Kazan' plant, where he managed production of the dive-bomber for the remainder of the war. This took him away from finishing No. 102, and in the end, due to the lack of the necessary engines, interest in it disappeared. This was unfortunate, for Myasishchev had incorporated many interesting innovations in aircraft No. 102, including an undercarriage with a steerable nose wheel, box-tanks in the wing, weapons operated by remote control from the pressure cabin, among other innovative features.

Myasishchev's credo let him down on two occasions. He used to say, "I will attempt any mission and will fulfill it, if our industry gives me the necessary components, that is, engines, instrumentation, and raw materials." However, time after time our industry failed to deliver these on time, and Myasishchev's experimental design bureau was eliminated.

Although Pavel Osipovich Sukhoi was never arrested, it is interesting to compare him to Myasishchev, since they held a common philosophy. Like Myasishchev, Sukhoi was unquestionably a talented designer. At KOSOS he managed the construction of the long-range, record-breaking ANT-25 aircraft, which made the famous transpolar flights by Valery Chkalov and Mikhail M. Gromov from Moscow to America. He also designed the *Rodina* aircraft, a twin-engined experimental design, in which the female crew of V. S. Grizodubova, P. D. Osipenko, and M. M. Raskova made their record flight.[9]

Sukhoi's other credo was as follows: "When I make an airplane, I satisfy all the customer's requirements. After that, it is none of my business; if such an airplane is necessary, let the ministry and the plants organize its production. I'm a designer—not a production controller, not an organizer, and not a fixer." The result was always the same: good aircraft were built in single copies, and were not put into series production. This continued until a capable go-getter and forceful organizer—Y. A. Ivanov—appeared at Sukhoi's side. He took on the difficult burden of state testing for the splendid Su-7 airplane, which he adapted successfully for series production. From then on the Army received several models of excellent airplanes with the "Su" designation.

In comparing Sukhoi and Myasishchev's work styles, we must note that Sukhoi was more of a realist, while Myasishchev could be distracted from practical designs to pursue highly experimental concepts. Myasishchev's instincts led the aeronautical industry to make less than rational decisions on occasion, and Myasishchev and his associates suffered as a result. Both made important contributions to Soviet aviation in the post-1945 period.

Aleksandr Aleksandrovich Arkhangel'skiy avoided arrest, but it would be inappropriate to overlook Tupolev's loyal friend and comrade-in-arms. During all those years of the purge he lived in constant fear, waiting each night for the knock at the door. His wife, Natal'ya Dmitriyevna, daughter of Prof. D. N. Ushakov, the famous linguist, would recall how he would upset her by not eating and not sleeping, and how he became quite unstable mentally. Arkhangel'skiy played a special, irreplaceable role in Tupolev's design bureau. He is well known for his leadership in the design work for the ANT-9 and the SB bomber. No less important was his advocacy of Tupolev's design team before various groups such as state planners for the aviation industry sector, the Main Directorate for Northern Shipping Routes. His persistent and gentle manner served Tupolev well. I think the authorities decided to let Arkhangel'skiy remain free precisely because someone had to monitor the operation of various Tupolev-designed aircraft.

Tupolev's approach to aircraft design remained concise and practical: "Our country needs airplanes the same way it needs black bread. One could propose delicacies such as pralines, cakes, and pies, but to no avail, for we lack the ingredients to make them. Accordingly, we must fashion a doctrine that directs our energies toward the possible. We need aircraft designs suitable for large-scale series production, to be based on our actual engineering and production capabilities. If, as a consequence, some of our designs turn out to be somewhat behind Western designs, to hell with them: we'll go with quantity. No unjustified rift between quantity and quality, however, should develop. To prevent such a rift we must encourage sophisticated engineering capabilities to allow for the development of experimental aircraft, even as we stress series production of proven designs. Our experimental design bureaus should pursue two missions: new models ready for series production; and more experimental types that anticipate the future, to allow us to be at the cutting edge."

Tupolev implemented this approach throughout his career. The R-3, TB-1, TB-3, and ANT-5 (I-4) aircraft reflected this shrewd approach. The ANT-25, SB bomber, and the ANT-42 (TB-7) reflected his timely focus on experimental types. In addition, he believed that "given the conditions in the USSR, pygmy design bureaus, even if they're headed by talented designers, can't achieve much. They don't have what it takes to get through the bureaucratic obstacles. We need powerful organizations like KOSOS, and only two of them—or a maximum of three— would be necessary." During those years he believed that the solid design bureaus created around Polikarpov and Il'yushin could be this type of design bureau. "During the ten years from 1927 to 1937," he argued, "our design bureau created ten large-scale production aircraft, and they met the requirements of the Air Force and Civil Air Fleet. Meanwhile, Polikarpov's design bureau created five such aircraft and Il'yushin's another two aircraft. The other "pygmy design bureaus"—and there were several—didn't create a single combat aircraft for our Air Force." Clearly, in Tupolev's mind, many of their talented designers could have been more productive if placed in the context of a larger design bureau. Tupolev's approach made sense in light of the economic conditions in the Soviet Union. Our country was short on funds. Just think how much it cost to maintain all these design bureaus!

It would be naive to portray Tupolev as a real Tolstoian and nonresister. Regardless of his sphere of activity, a major talent always intimidates those who are less gifted. This phenomenon of talent generally manifests itself as early as childhood, certainly by the time of young manhood. However, it is to Tupolev's credit that many "pygmy design bureaus" existed even when he held the twin posts of director of KOSOS and chief engineer of the Main Directorate for the Aviation Industry of the People's Commissariat for Heavy Industry, that is, when he could have closed down any micro-design bureau with a single stroke of the pen. He did not do this. However, Tupolev's views were well known; he did not conceal them. A critical attitude toward the activities of others is everyone's inalienable right. Most aviation specialists recognized this right; however, some of them—in particular A. S. Yakovlev—started rumors that Tupolev couldn't tolerate the existence of aviation design bureaus other than his own.

When it came to technology, Tupolev was open-minded, listened to others' opinions, and would change his mind for a superior idea. He didn't tolerate hare-brained schemes, however, and would dismiss those who proposed such schemes very quickly. He never punished his associates if they defended passionately ideas that were erroneous or foolhardy; instead, he took the time to change their minds. It's also of interest that in his personal life he adopted a lifestyle that was above reproach, bordering on the puritanical; at no time did he depart from these virtues, although there were many temptations associated with a person of his stature.

It wasn't easy to make one's way into Tupolev's inner circle since his standards were so high. One had to love one's work above anything, almost ascetically forgetting all else, have initiative and persistence, be just, not worship authorities and bosses, seek new paths boldly and, after finding them, not yield in the face of opportunist ideas. A person who met such standards would gradually win Tupolev's trust. His experimental design bureau could be characterized as an association of capable engineers, each granted autonomy to solve his own problems in his own specific area. As an orchestra, each associate played his part in the score precisely and in tune, and in the key chosen by the director. This type of organization and these principles contributed to the growth of creative abilities. It is not surprising that the family tree of Soviet aircraft design sprang from the design department of TsAGI.

To return to the *sharaga:* M. A. Nyukhtikov was designated test pilot for the prototype of our Project 103 aircraft. At this time, Petlyakov's Pe-2 had been flown and was undergoing tests. After beginning test flights for the Pe-2, P. M. Stefanovskiy, the test pilot, complained that if he reduced the throttle as the airplane made its descent, it fell toward the ground rapidly, "like a stone." Meanwhile, our design work on Project 103 (Tu-2) had just begun. The search—tormenting at times— was under way to find the optimum configuration for the Tu-2, to assure that it would not display any undesirable flying characteristics.

One night Tupolev sat in Yeger's team with one leg tucked under him (his favorite position). On the blackboard was a wing profile—exceptionally well designed, in our opinion—similar to an American prototype. With a soft pencil, Tupolev made corrections to the tail section of the wing profile. Our aerodynamic engineer, A. E. Sterlin, sitting next to him, clearly did not approve of these changes. But Tupolev said, "There, we'll squeeze it in a little here." Turning to Sterlin, he added, "And don't you start making farting noises: the airplane will be stable near the ground. We will land smoothly nice as you please, like this . . ." (here he gestured as if pulling the controls toward himself, then smoothly let them return).

Nyukhtikov, who happened to be present during this scene, told us later, "At the time I thought to myself, 'Okay, let the Old Man chatter away, it'll make an impression on us, but I know a thing or two, and you can't fool me with such primitive nonsense.' During the first flight of the Tu-2, however, everything happened just as Tupolev had said it would, and I vowed to apologize to him." And in fact, when we gathered in the hangar of the Scientific Research Institute of the Air Force in late fall 1940 and—with bated breath—heard the evaluation of our work, Nyukhtikov repeated word for word what Tupolev had said a year before at the blackboard.

Apropos of this, at the very same time pilot F. F. Opadchiy detected a certain directional instability in the Pe-2, particularly noticeable at slow speeds. The bosses became angry. They took Petlyakov to the Lubyanka headquarters of the NKVD, to see General Kravchenko (who at the time oversaw all special prisons), and then to Beria himself. Ugly and threatening conversations followed, with the word "sabotage," if not uttered, certainly in the air. Petlyakov's design team began to suffer depression and anxiety, arguing with one another. The conflict was clearly coming to a head.

Interestingly, one trait common to almost all prisoners in the *sharaga* was a tendency toward depression. It only took some insignificant incident, such as a rumor or a few ill-chosen remarks from the "management," and deep anxiety would spread among the prisoners with lightning speed. At times you could walk into the sleeping quarters and find dozens of people lying on their beds with their faces buried in their pillows. They would say, "They won't let us go . . . Everybody's going to the prison camps . . . I heard that they've already kicked out some men," and so on. It was striking to see the speed with which our society of a hundred people could lapse into complete despair and pessimism.

Petlyakov made futile attempts to find a simple solution to the Pe-2's chronic instability. There was talk of a major redesign of the tail surfaces, but that would have been perceived as a big scandal, with new deadlines, and the honor of the "managers" would now be suspect. Kutepov summoned Tupolev, who walked around the tail surfaces for a long time, muttered, wheezed, chewed on his nails, and then said, "Volodya, extend the elevator surface beyond the two outside rudders;

In Prison | **185**

otherwise you will have negative interference. The additions will stabilize the air flow, the efficiency of the controls at low speeds will increase, and, I think, you will manage without major reworking. Do it like this in series production . . .'' (he continued in highly technical terms). That night they made extenders at the plant, the next day they installed them, and on the following day Opadchiy reported that the instability had disappeared.

Another example of Tupolev's quick solutions to problems occurred in 1947. We were working on the Tu-14, when the Rolls-Royce Nene engine, which had a large diameter and a long nozzle, did not fit properly with the streamlined wing. Tupolev was called in. After staring at it for a while, he began to increase engine nacelle diameter ahead of the wing, while simultaneously reducing it under the wing. He could be heard mumbling, "Here's where the flow is compressed, here's where it straightens out, and so we'll avoid interference between the wing and the engine nacelle." It was as if he saw the airstream flowing over the airplane, compressing around the engine nacelle, and then, after straightening out, flowing over the wing. Clearly, none of those present understood his reasoning; moreover, everyone agreed that he had made the engine nacelle worse. After consulting among themselves, our overall design experts decided to invite in our leading aerodynamics expert, academician S. A. Khristianovich. He took a look, shrugged his shoulders, and stated skeptically, "Well, something's not right. However, it might be better if . . ." And he went on in very general terms. Giving in to the authority's skepticism, Yeger, Sakharov, and Babin decided to correct Tupolev's design somewhat. The lightning struck that evening. Tupolev became extremely angry. "You fools, you snivelers, you bums"—he made abundant use of such phrases—"you're driving me crazy with the ideas of Khristianovich: he's a theoretician, and pretty far removed from practical measurements of designing. Moreover, he's working on different tasks, such as looking for general principles. What did he do, tell you something specific or suggest a design? Hell, no! So why did you . . ." and so on, in the same vein. But the Tu-14 flew just fine, and its rather hideous-looking (at least in our opinion) engine nacelles turned out to be essential for optimal performance.

These three examples—on the Tu-2, the Pe-2, and the Tu-14—demonstrated Tupolev's problem-solving capability, based on personal intuition. It is worth mentioning that at the time intuition was regarded as something like a witch in old Salem, and was treated as non-Marxist, a manifestation of thinking. In the precise sciences, students were routinely taught that everything had to be expressed in formulas, which allowed little room for Tupolev's style of aircraft design.

Incidentally, the Tu-14 was the first time that I encountered a conflict with the generally accepted formula, "If it is beautiful, then it makes good sense." It is doubtful that the Concorde itself would have pleased Louis Bleriot, Farman, Duperdussin, or Voisin—they probably would have spit in disgust.

The Tu-2 and the PB-4

About the time that Tupolev's circle was forming in the *sharaga,* or possibly even earlier, in the Butyrka prison (another city prison), as he would later say, Tupolev's idea for an "aggressor" aircraft—a dive bomber capable of carrying the heaviest bombs and flying faster than the fighters of that time—was taking final form. Yeger, Sakharov, and Babin, working from morning to night on three drafting boards placed on rickety tables, endeavored to make a general scheme. Despite the tragedy of interrogations and torture Tupolev had just undergone, and despite the surroundings—which brought to mind more than anything else a train station during a natural population migration—his thought processes operated clearly, and the Tu-2 gradually took shape. According to Tupolev's concept, the airplane was to have a crew of three. There was one pilot; for firing he had a forward battery of four ShKAS machine guns in the nose of the airplane and two ShVAK cannons in the root of the wings. Immediately beneath the cockpit there began long bomb-bay doors, behind which up to three tons of bombs, including one 1,000-kilogram bomb, could be suspended. The forward area of the doors was canted for bomb exit during nearly vertical dives. Behind the bomb bay sat the navigator and the gunner, who had two machine guns for protecting the rear hemispheres. The airplane was compressed as much as possible: according to rough calculations carried out by

Tupolev himself, with twin 1,400-horsepower Mikulin engines the air-plane could reach a speed of 600 to 630 km/h, or more than fighters at that time.

Tupolev used to say during the war years, "First and foremost is speed, speed, and more speed. The Messerschmitt Bf-109 fighter can now reach speeds of 500 km/h. And we know that its designer isn't sitting idly by, but is attempting to increase its speed. We must achieve 600. There's no such thing as miracles: you can't get speed out of thin air. A fighter has one engine, one man, one cannon, and two machine guns. It is true that we have two engines, but the crew, bombs, cannon, and more complex equipment create a greater load. So where will we gain our increase in speed? From aerodynamics! Take a look at the rough lines of the Bf-109. For our Tu-2, by contrast, we have to fashion a more compact, streamlined airplane, and reduce its weight drastically. Only then will we beat the German designers.

"Speed, however, is not a panacea for all ills. Remember how one designer decided that speed was the single goal, and created a twin-engine bomb-carrier without any defensive armament. What came of this design? The military scrapped it. We won't take this route, and have outfitted our Tu-2 with potent guns. Of course, this means additional weight, but without armament, the airplane is highly vulnerable. And now on to bombing accuracy: no sort of hocus-pocus will help you here. Let's take a look at a gun: it is always aimed at its target. But what about us? We're flying forward horizontally, but drop our bombs vertically. Accordingly, we have to turn our Tu-2 into something like a rifle, but with an extremely long barrel aimed only at the target. After all, when we make it dive at this target, isn't it like an extremely long rifle?"

Dive bombing requires a special sight, but it had not been created yet. Tupolev was sure that if he waited for it from outside, he'd never get it. Although it consisted of precision optico-mechanical instruments whose design and production lay completely outside the area of expertise of his design bureau, Tupolev decided to create the bombsight himself. He called in Georgiy Semenovich Frenkel' to perform this essential task.

"Zhorzh, our aircraft is doomed without a bombsight. Can you design one and produce it under our conditions?" Frenkel' agreed. An automatic diving pilot was required so that after the pilot had picked up the target in his sights, he could maintain the aircraft on target automatically. A. A. Yengivaryan and I. M. Sklyanskiy completed the design of this autopilot on schedule. Highly complex technical problems associated with the Tu-2 were often solved this way, by using absolutely extraordinary measures. The Tu-2 would be built in a huge production run; accordingly, it had to be simple to produce by workers with average skills. To achieve this, Tupolev divided the airplane into a huge number of smaller components that could be assembled individually.

From time to time, high-ranking NKVD personnel would appear, walk around and silently look over the drawings, then move away with serious looks on their faces. These visits prompted great anxiety because our uninvited guests gave no hint as to the purpose of their visit or their conclusions. At one point, Tupolev disappeared. One night the guards carted him off to Moscow. He returned a day later, stern and angry, and informed us that we must prepare a report on the Tu-2 aircraft for the Lubyanka in three days. On that occasion they took three of us, with blueprints, to the Lubyanka: Tupolev, Yeger, and Frenkel'. First the group was received by General Davydov, the chief of all the prison design bureaus (Davydov would be thrown in prison in 1939, and replaced with Kravchenko, who was himself thrown in prison not much later, in either 1941 or 1942). The general approved the concept and announced that Tupolev would be taken to report to Beria on the following day. In the meantime, so as not to "trouble" them with another ride, they were put up in solitary cells in the inner prison of the Lubyanka.

The meeting took place in Beria's huge office, with windows overlooking the historic Lubyanka square. The gathering involved great pomp. The blueprints had been spread out on a table in advance. At the end of the table, nearest Beria, then known as the "closest assistant and best friend" of the great leader, sat Tupolev; next to him was an officer, and across from him was Davydov. Farther off, next to the wall, between two officers, stood Yeger and Frenkel'. After hearing out Tupolev, Beria said, "I've told Comrade Stalin of your proposals. He agreed with my opinion that instead of an airplane like this we need a high-altitude, long-range, four-engine dive bomber. We will call it the PB-4. We don't intend to inflict pinpricks [here he pointed disapprovingly at the blueprints for the Tu-2]. No! We will smash the beast in his lair!" Turning to Davydov, he added abruptly, "Undertake measures [he nodded toward the prisoners] to see that they prepare proposals for the PB-4. That is all!"

Tupolev returned as mad as a thousand devils: Beria's plan was clearly untenable. A high-altitude airplane required a pressurized cockpit and a restricted field of view. Creating such a multiengined aircraft with sufficient range would required a large, ungainly design—and an excellent target for the enemies' air defense batteries. In addition, Beria's foolhardy proposal would bring in its wake enormous problems for series production. Pressurized cockpits made it impossible to use reliable defensive armament. At that time the USSR still had no remotely controlled armament that would permit firing from the pressurized cockpit or the aiming of cannon outside the cockpit. In short, there were numerous "cons" and not a single "pro," except for a primitive idea: Since the Germans and Americans had single-engine dive bombers, we had to outdo them and create not a Tsar Bell but a Tsar Dive Bomber!

That evening Tupolev gathered his group. "This is very crucial. It is possible that this dilettante has already convinced Stalin, and it will be difficult for them to give up the PB-4. I know Stalin a little: he doesn't like to change his decisions. We have to choose the airplane's overall appearance in good conscience and calculate the approximate weight; it is too bad that we don't have Petlyakov here, since he knows all the data on the ANT-42 better than I do, and the PB-4 will have to be about the same size. Take the ANT-42 as the baseline, pressurize the cockpit, come up with a bomb exit from its bomb bay during dive bombing, and calculate the weight beyond that of the ANT-42. The nominal overload for the ANT-42 will have to be increased for the dive bomber. Frenkel' and I will write the explanatory note."

Tupolev's well-reasoned explanatory note laid out four basic points:

1. A high-altitude, long-range, four-engine heavy bomber had already been created: the ANT-42. To "smash the enemy in his lair," it was only necessary to organize its series production.

2. Since the percentage losses due to air defense batteries would always be high, a dive bomber had to be a relatively small aircraft, yet large enough to perform its task.
3. In order to aim during its dive, the bomber had to be highly maneuverable, which is difficult to achieve with a heavy four-engine airplane.
4. Tupolev could guarantee the projected performance data stated for the Tu-2, while he could not for those required for the PB-4.

Approximately one month later, the NKVD ordered Tupolev to the Lubyanka alone. This time he disappeared for three days, and his colleagues became extremely worried about his fate. Finally, he rejoined us and told this story: "My report made Beria openly angry. When I finished my presentation, he looked at me spitefully. He had evidently told Stalin a lot about the PB-4, and, perhaps, had convinced him. This amazed me, since from past experience I had the impression that Stalin—even if he didn't understand the subtleties of aircraft design—nevertheless had a good grasp of reality. Beria said that they would look into matters. For days I sat in a solitary cell, worried about my fate. Finally, they summoned me again. Beria announced, 'Comrade Stalin and I went over the materials once again and made a decision: make the airplane with twin engines, and quickly. Once you've finished, then begin work on the PB-4 because we will need it in the future.' Then the following dialogue took place between us:

> *Beria:* 'What speed does your 103 [Tu-2] aircraft have?'
> *Tupolev:* '600.'
> *Beria:* 'That's low, it has to be 700! What range?'
> *Tupolev:* '2,000 kilometers.'
> *Beria:* 'That's not good enough, it has to be 3,000! What kind of load?'
> *Tupolev:* 'Three tons.'
> *Beria:* 'Too low—it has to be 4. That is all!'

"Turning to Davydov, Beria added, 'Order the military to quickly prepare the technical requirements for the twin-engine dive bomber. Correct the parameters stated by Comrade Tupolev to match my instructions.'

"At this point the audience ended, and we walked out into the secretary's area. Davydov nodded to Kutepov and Balashov. On tiptoe, they subserviently disappeared behind the sacred doors and soon reappeared, now in the form of messengers, loaded down with blueprints and calculations."

Many years later, Tupolev shared the rest of the story. "I had only a few fateful conversations in the government, a small number of encounters in which our fate was determined. This was one of them. It was utter madness to design the PB-4. The military would have scrapped such an aircraft, and they would have been right, for it was inconceivable to dive at point targets in a four-engine bomber. Beria would have characterized the military's negative attitude, of course, as sabotage, since he would have had to justify himself before Stalin, too. Recalling his malicious look, I'm inclined to believe that he would have sacrificed us without a moment's hesitation, and what would our fate have been?"

When Tupolev returned and described the events during those three days, everyone sighed with relief. A terrible storm had passed. We now looked forward to genuine work. Wasting our talents and energy on the PB-4 would have been like the line in the revolutionary song, "We dug our own grave, the deep hole was ready." A weight indeed had been lifted and Tupolev's design group could once again work and create.

Soon thereafter Tupolev's design team was transferred from Bolshevo to the KOSOS building in Moscow, where the group for Project 103 was being formed, and our work became hectic. However, before things became busy, Tupolev had the following conversation with General Davydov: "In order to put all my energy into working on the airplane, I have to be certain that my wife and children are alive and well. I can't do it without that. Let them bring me a note from my wife, Yulenka Nikolayevna."

The bosses viewed this request with great alarm. After all, Yulenka Nikolayevna was also in prison, and the children were being looked after by their grandmother, Yenafa Dmitriyevna. They were huddled together in a single room left to them in the apartment where the family had lived in Kalyaevskaya Street, without anything to live on. At first the bosses ignored Tupolev's demand; however, after encountering his iron will, they brought the note. It is hardly necessary to mention that the note was written in the office of the investigator who was handling Yulenka Nikolayevna's case, as he dictated it, and "in the interest of the possible release of your husband."

A Journey Outside the *Sharaga*

Our country's information service was always shaky, particularly in the defense industry, where everything necessary and unnecessary was classified. After being separated from life for two years' worth of camps and prisons already, we were completely unaware of what our industry had produced recently. The three who were responsible for outfitting the Tu-2 aircraft with the newest and best armament—Nadashkevich, Frenkel', and I—went to Tupolev and said that we needed to visit numerous factories. "Hmmm," said Tupolev. "Prisoners visiting secret experimental design bureaus—that is grounds for war. I'll try to have a talk with Kutepov."

Although they had some difficulty with the idea, our "managers" eventually understood and one fine day, escorted by Major Kryuchkov and two guards, we set out for the plant visit. The "circus" began at our first stop, where the guard at the outer gate demanded our passes. Kryuchkov showed him the certificate that theoretically opened all doors and informed him that the others were "specialists under him." "Nonsense! Specialists, indeed! Maybe even spies!" Things got livelier, people began to gather around, and we soon found ourselves in a confrontation.

Kryuchkov disappeared, but soon reappeared with the chief of security, and they let us in. When the equipment developers entered the reception area, Act 2 began. First came the long silence, then the detailed questions: "Where did you go, where are you working, and on what?" Kryuchkov's uniform and the guards with their pistols helped fit the pieces together. The "free ones" took great pains to explain what it was that we needed. They bade us farewell with warmth. Our first contact with the free world turned out to be our last. You have to suppose that it made a painful impression, if not on Kryuchkov, then at least on the screws.

After we returned to the *sharaga,* we told our team about our adventure in detail. A common thread was the point that at the plant "they understand everything." Henceforth, NKVD officials, at our request, began to obtain the information we needed, but they often mixed things up, brought back the wrong blueprints, and returned with items we didn't need. In short, this became a game of "Whisper Down the Lane," and it became very difficult to design the best aircraft in the world. Nonetheless, we gradually began to acquire materials and start work on the aircraft.

Work continued on Project 103. For the mockup of the Tu-2 we made countless modifications of the crew compartments. Our work moved forward slowly, largely because we were hampered by imprecise data. Finally, Tupolev told Kutepov that it was time to request an Air Force commission to evaluate our mockup. The prisoners expressed concerned about the outcome of this meeting. Most of us had known these "bosses" for many years. They were politically trained people, and devoid of doubts. We expected that they would have a very biased, guarded attitude toward the aircraft designers/prisoners. As luck would have it, the appointed chairman of the commission was not just an officer but an engineer, Gen. P. A. Losyukov, a wise and perceptive man.

The commission met in Kutepov's office. (Only a short time before, the very same officers had come to the very same office for similar meetings, but it had been Tupolev who had received them.) When everyone was present, the prisoners were brought in. General Losyukov immediately set the mood: he stood up, shook hands with Tupolev, and they exchanged bows. After Yeger gave a substantive report, everyone moved to the model shop on the sixth floor, where a full-scale model was surrounded by people in medium-blue Air Force uniforms. For two days, the prisoners answered the officers' questions, trying to show that the planned aircraft was worthy of defending our socialist state. Finally, everything had been examined, climbed over, felt, measured, understood, and evaluated. At a plenary session, the military—as usual—expressed their maximum requirements, while the prisoners responded with more realistic real-life ones. Emotions sparked conflict, but gradually everyone yielded to reason, and a compromise resulted. The advent of the positive evaluation of the aircraft became the occasion, per tradition, for a modest dinner with wine. Even here the spirit of compromise ruled: prisoners could not give toasts or clink their glasses. After the meal we were led away, to allow the military and the NKVD to sit down for their own celebration.

Many years later, Losyukov admitted to Tupolev how difficult this whole episode had been for him. Several commission members had attempted to create a conflict between the representatives of technology and politics. There was no escape from that entrenched principle of socialist bureaucracy—"just to be on the safe side." The commission could not make an unambiguous evaluation of the aircraft, for the obvious reason that the saboteurs had built the mockup. It was customary to write a flowery conclusion, and then let the generals from the People's Commissariat for Defense and the generals from Internal Affairs figure things out. Not so in this case.

We found the process extremely difficult, too. We expected that resolving even minor issues would bring the response: "Aha! Since you don't want to satisfy our lawful requirements . . ." Such a threat could entail far-reaching consequences for us such as an isolation cell, being sent to a prison camp, time added to your prison term, or even worse. Everything worked out, however, in this situation for us.

In recalling the work of the commission, I must mention Tupolev's attitude toward certain military specialists who later claimed that they had "improved" our airplane. He was open to recommendations that actually improved the design, and usually accepted them. However, when anyone made clearly self-serving proposals that did not improve the airplane as much as permit their originator to assert what an uncompromising champion of progress he was, Tupolev usually subjected him to moral debunking through ridicule.

Later, we were at the commission meeting for the Tu-14, which was the first Soviet aircraft outfitted with ejection seats. A military doctor at the meeting demanded that the aircraft be fitted with a latrine in the cockpit, to allow the crew to remain seated. Tupolev almost fell off his chair from laughter. "I can't stand it, he's killing me!" he cried in his high-pitched voice. "Yeger, give him a toilet instead of a catapult.

He'll pull on the chain, and out he'll fly!'' The poor doctor didn't know what to do. Naturally, Tupolev's ridicule ended the conversation.

A second incident comes to mind, again to illustrate Tupolev's wit. When we met to evaluate the Tu-16, one of the commission's officers, a Lieutenant Colonel T., demanded that the top gunner's field of vision include both the upper and lower hemispheres. In vain, Nadashkevich tried to explain that this was impossible, but also—and more important—unnecessary, since the two tail gunners had an excellent view of the lower hemisphere. But the lieutenant-colonel persisted. Then Tupolev said to him, with a smile, "When an eye grows on your ass, then I'll give you your field of vision." A huge outburst of laughter erupted. The officer could not finish his presentation, and the issue was dismissed. This sort of humor mixed with disdain was quite effective, and people gradually stopped making ridiculous demands: no one wanted to be ridiculed by Tupolev.

Ebb and Flow of *Sharaga* Life

Petlyakov's Project 100 was the first to see light. Once they finished assembling the Pe-2 at the plant, flight tests began under P. M. Stefanovskiy's supervision. The order had been given to show the Pe-2 over Red Square on May Day. We were overjoyed because we would see an actual flight by an aircraft from our monkey cage. Then the prison administration decided—"just in case"—to keep us under lock and key during the celebrations. In the face of pressure from his colleagues, Petlyakov went to Kutepov to complain. Finally, they reached a compromise. They would open the monkey cage during the air show, but two screws would be on duty.

On the day of the flight, so many people crowded into the monkey cage early in the morning that the unfortunate screws were pressed against the bars. They begged us to make room: "We're suffocating!" The day was clear, and the silhouettes of the Kremlin towers could be seen clearly on the horizon. Airplanes first came into view in the distance from the Belorussian Station side; they approached Red Square flying in formation. We could see the silvery Pe-2 tearing along beneath them, passing them. But the airplane's silhouette was unusual—some type of dark objects stuck out on its underside. Soon the airplane passed the formation and then shot skyward like a Roman candle. For three days, the prisoners fretted about what it might have been on the airplane, while the screws on duty said nothing. On the morning of 3 May, the free workers informed us that the pilot had forgotten to raise the landing gear. We had been lucky, of course: Stefanovskiy had been flying at high speed. If the undercarriage doors had disintegrated and fallen—God forbid!—on the viewing platform, what would have happened to Petlyakov and his group then?

The Politburo, the central leadership of the party, liked the airplane, but Stalin proposed to "revise its intended use somewhat." This meant that series production would begin on a dive bomber, not a fighter. Petlyakov now faced the challenge of redesigning his high-altitude fighter into a dive bomber. Considering the success of the German Junkers Ju-87 at the outset of World War II, this was probably wise, but things did not get easier for the prisoners. Petlyakov's team was hoping to be released after the demonstration flight, but the call for further work postponed their freedom for at least a year. We were also concerned by the lack of a solid, well-thought-out doctrine from the military leadership. Our work had been punctuated by arbitrary moves, such as making a dive bomber out of a fighter, which was no easy task and one where many of the original qualities of the aircraft would be lost. At the same time they had forced Tupolev to design the huge high-altitude PB-4 instead of the lower-altitude dive bomber he had proposed. Moreover, these abrupt shifts took place on the very eve of World War II. Even we civilians understood how these policies compromised our military preparedness.

The change in the Pe-2's design affected the Tupolev team. From a purely selfish perspective, we feared a second dive bomber might mean cancellation of the Tu-2, and subsequently then we would lose the chance for freedom which hovered in the future.

In our prison workshop we worked day and night, and as the Russian proverb says, you cannot harness a timid doe with a workhorse. Everyone displayed a certain degree of nervousness, snapping at each other.

Our anxiety prompted us to curse Petlyakov for his short-sightedness, only eight days after the air show when they had silently tossed him aloft in the sleeping quarters to honor him joyfully. Now they cursed the entire hierarchy from Kutepov to Beria, to themselves or under their breath. After about two weeks it became clear that the dive-bomber adaptation of Petlyakov's Pe-2 could not compete with the Tu-2: His airplane could not carry large bombs, it had a shorter range, its defensive armament was insufficient, and its top speed was 100 km/h less. Perhaps this wasn't good for Petlyakov's team, our partners in misfortune, but we cheered up and the output of blueprints for our airplane returned to the required level.

Ye. P. Shekunov, the chief production engineer of the *sharaga,* introduced something new on the blueprints for the Tu-2: the "assembly number." Only he issued these numbers. On one occasion, when he was busy with something urgent, he jokingly remarked to one of the engineers who was pestering him about the assignment of an assembly number, "Write down 'Gordian.'"[10] And so the term "Gordian subassembly" was used until it reached a vigilant plant production engineer, who went to the bosses, suspecting sabotage. No matter how hard Shekunov tried to explain to Balashov, chief of Project 100, that this was only a joke, and that there also existed such phrases as "Augean stables," the "Procrustean bed," and the "sword of Damocles," Balashov gave Shekunov the order: "Use no foreign terminology!"

This incident spurred us to find out the cultural level of the free workers. At the time, everyone was interested in A. N. Tolstoi's *Peter the First.* It was well known that our leader liked this novel, and gossips were claiming that Tolstoi had added "Loyal, without flattery" to his family coat of arms. The novel's action unfolded in the area in which the design bureau's building stood. To the left of our building, the Kukuy stream flowed into the Yauza River; opposite it, beyond the river, at the ponds of today's park of the Ministry of Higher Education, Peter used to take Anna Mons for a drive. Not far away, the roof of the Lefortovo palace could be seen. During the KOSOS building's construction, they found the grave of Bryus, a grandee and inventor of the perpetual calendar, in the wall of a demolished Lutheran church. It soon became clear that these circumstances held no associations for the majority of the free workers; moreover, they did not connect names such as Lefort *rayon,* the German market, the Semenov Church, Soldiers' Street, Hospital, and others with the history behind them. Such was the historical erudition of the Soviet intelligentsia, which was packed with dates of party congresses instead of "that pandemonium," as one of the free workers put it, related to the history of our great nation. It was as if we had no history.

We prisoners felt sorry for many of these free workers. Perhaps they weren't to blame after all, since the ubiquitous political propaganda shaped them day and night. All past achievements of world civilization became subordinated to the official ideology. Everything was viewed from the present. There was little place for Aristotle, Seneca, Erasmus, Confucius, et al. Today the issue of the life or death of the first socialist state and, more important, the fate of this state would be decided by airplanes, tanks, guns, tractors, locomotives, rails, metal, coal, and oil. . . . If production increased by a certain percentage, then we have a right to shout "hurrah!" Let's assume that, to achieve all this, it is necessary to resort to sacrifices, including using gold from our cultural patrimony to buy lathes, presses, and rail-rolling stock. Let's do that, too. After all, there's lots of gold around—in the Hermitage's museum storerooms, in museums, at the Tret'yakov Gallery, in the church patriarchs' robes, in the monasteries, and in the churches. There is a lot of it collected there!

As a consequence, Russia lost much of its heritage and cultural treasures. We offered the Hammers, Morgans, and Rothschilds their choice of canvases by Rembrandt, Rubens, Leonardo da Vinci, Van Dyck; icons painted by A. Rublev and F. Grek; many priceless items of gold, platinum, and silver with diamonds and pearls confiscated from our churches; and, finally, the family jewels of the Romanovs, Morozovs, Ryabushinskiys, Mantashevs, and Putilovs.

After all, the proletariat has "nothing to lose but its chains," and you can't help much with them! What are the groans of a few little members of the intelligentsia, professors, scholars, and even academicians? Let them moan: eventually they will calm down. They will never stand against the justice of our bright, progressive, one true doctrine, not by one iota.

It seemed to those of us imprisoned that all these things were indeed tragic. Moreover, the need for an inner, spiritual life was proved to be ineradicable: Neither the camps nor the slave labor in the TsKB destroyed it. Naturally, we understood the importance of building the next blast furnace or mine and increasing the daily output of cars or tractors, but we thought that this should be compatible with music, poetry, and, ultimately, moral philosophy.

In the evening, when we became "free," groups of people joined by some common interest would gather in the corners of the sleeping areas. A music lovers' club arose. With the approval of the "management," they made a violin, viola, and cello with their own hands, using bakelite-finished plywood, which was used extensively in aircraft construction. They didn't sound like Stradivariuses, but the trio of Bazenkov, Bocharov, and Borovskiy performing Offenbach or Strauss on Sundays invariably attracted enthusiastic listeners.

There was also a club for poetry lovers. Our library received additional materials from the holdings of the Butyrka prison, including the poets Fet, Pleshcheyev, Tyutchev, Nadson, Blok, and (naturally) Pushkin and Lermontov. They didn't keep any Aleksei K. Tolstoi there, of course, because he was an insufferable fault-finder. We also ended up with books from confiscated private collections, sometimes with inscriptions that had slipped through out of police carelessness: "Ex Libris Bukharin," "A. I. Rykov," and so on (they weren't inspecting bookplates at that time) provided reminders to us of victims of the purge era. One of our bibliophiles hid two such books, but lost them during the evacuation of 1941. Many of us amused ourselves by making various things such as cigarette holders, tobacco pipes, or cigarette cases. We sawed brooches, monograms, or ladies' belt buckles out of plexiglas, or we used it to make dolls or toys for our children.

We had artists, too: A. M. Cheremukhin drew typical scenes from our life in the prison design bureau, while T. P. Saprykin drew the rooks on Savrasov Street. Several men wrote poetry, and one eccentric even wrote a novel on the construction of an aircraft plant in Siberia, in the style—particularly favored at that time—of books such as *Steel and Slag, Hydroelectric Dam,* and *Far from Moscow.*

The issue of relations with women was a special one. Our headquarters doctor, a doctor's assistant from the Butyrka, let on in strictest secrecy that bromine or something else was put into our food with the intention of reducing sexual ardor. It is possible that this was true, but the issue isn't just physical. Men need a woman's caress, her sympathy, her support, and, finally, friendly contact, which we value so little until we have been incarcerated. After remaining within the confines of the *sharaga* for weeks, months, or years, when prisoners encountered friendly concern, a spoken caress, or, at times, deep understanding, it gave rise to serious feelings, albeit platonic. At times these encounters fell on fertile soil. It was a well-known fact that some wives, particularly party members, had repudiated their arrested husbands. Unfortunately, some even did this publicly, at meetings. It was difficult to explain their reasons: perhaps fear, perhaps there was a biological basis, that is, the urge to protect one's children and kin. Nevertheless, it made sense to be proper. It was impossible to act otherwise with prisoners, who had developed their own code of honor and decency in relations with representatives of the prison administration, guards, and investigation. Their souls were too easily wounded. Naturally, a woman who had publicly disavowed her husband could not count on his returning to her. After learning of such betrayal, meeting another woman who unselfishly displayed tenderness and attention, who had no right to receive anything in return, who was constantly risking being thrown out of the TsKB or even Moscow, or worse—being arrested, could make one ask, "Who would cast the first stone?"

Building the Tu-2

Work on the Tu-2 progressed. Blueprints gave rise to parts in the shops, these merged into assemblies, and then completed components of the airplane appeared. Gradually, not yet born, the airplane was spread here and there on storage shelves. Assembly jigs for Tupolev's pride and joy were being prepared for final assembly in the huge assembly hall.

Taking shape among the girders of the assembly jig were the stream-lined nose of the Tu-2's forward cockpit, the graceful tail surfaces, the mighty caisson wing center section with its long bomb-bay doors, the detachable portions of the wing, the engine nacelles, and the well-designed landing gear. It can be said without exaggeration that the external lines of the test model of the Tu-2, in an engineer's opinion, were elegant. This result was not accidental: it was a consequence of careful streamlining, to squeeze the maximum speed out of the air-plane. Tupolev became vexed by any component of the aircraft that protruded from the smooth, sleek fuselage into the airstream. The radiocompass loop antenna—a flattened ring 250 millimeters in diam-eter and 30 millimeters thick—was one such thorn in his side.

He asked me, "Kerber, why can't you put it inside the fuselage?"

"That's metal, and it screens the antenna. The reception range for the airfield and homing radios would be reduced."

"Well, what if the fuselage were wooden?"

"Then you could."

"Sveshnikov and Vigdorchik [designers] have argued for such a mon-ster, a fuselage about three spaces wide, and wooden."

"Andrei Nikolayevich, that means new materials, new production engineering, glue, and assembly jigs."

Tupolev didn't spare us, but he didn't spare himself, either. All Tu-2s produced had a nose portion of the fuselage with up to four frames' worth made of wood.

There was not a single extra inch in the crew compartments; despite this, Tupolev demanded a well-designed interior space: "What a screwed-up mess! Everybody stuck in his own two cents' worth! This isn't an interior, it's crap! A man has to work here, and rest sometime, too, and instead of making it comfortable, they made it God only knows what!" At these words, the next panel or set of controls bade farewell to the cockpit, screeching mournfully as its nails were ripped out, then traced an arc through the air before landing on the floor.

There was consternation, of course, with any call for the redesign of any part of the airplane. During one such call for modification, the lion's share of the reworking of blueprints fell on the head of our armament specialist, A. V. Nadashkevich. His tasks were indeed numerous and burdensome. He finally complained to the boss. Never prone to depression, our youths put together a song on the spot, in which Tupolev, fighting for improvements, sang:

> And so for this we need
> Only a single change.
> But it's a tiny, nonsensical thing,
> Only just one set of rivets.
> You have to replace them,
> Thicken the skin next to them,
> And lengthen the longerons.
> And shift the engines just a little.

The recalculations ended thus:

> But your blueprint
> Is only on the drafting board, after all,
> So, Comrade Nadashkevich,
> Everything's fine, everything's fine!

When he heard the song, sung using motifs from a French song popular in those years, "Everything's Fine, Beautiful Marquess," Tupolev burst out laughing, and in the enthusiasm of the moment promised, "To hell with you! I won't make any more changes." But afterward, everything followed the usual trend. During these incidents, if anyone said, "Andrei Nikolayevich, remember there's a plan, deadlines, and blueprints . . . ," Tupolev would interrupt brusquely, "Does the plan really state that we have to make the thing hideous?"

On another occasion we were modifying the cockpit interior and needed the blueprints from production to complete our work. When Balashov learned about it, he called us in: "What sort of new innovations are these? No reworking! And what kind of 'interior' is this? You're making up some sort of words! What's the blueprint number for this 'interior'? Bring it to me!" Tupolev, however, did not waver: he demanded frequent refinements. Worrying ourselves sick, working into the evening and part of the night, we came up with new designs to respond to Tupolev's requirements. One might say that prison wasn't the best place for seeking out original solutions. Yet, even in this environment, Tupolev was not ready to deviate from his principles.

The year 1940 was memorable. Time was an hourglass, a tiny, relentless stream of sand that counted off the months and years. The Wehrmacht's columns had already marched across Poland. Ribbentrop was in Moscow. The dive bombers were sowing death and destruction, military bands were thundering out their melodies, Himmler was building Auschwitz and Treblinka, and on the borders of Flanders, Alsace, and in the Ardennes, German generals Manstein, Guderian, and Kleist ordered their tank corps forward as wedges. We, alone—outcasts, prisoners, scum—understood nothing, for we had neither newspapers nor radios, but we felt that the world was headed for the abyss. A ten-hour workday was introduced. Many of us were happy, for it meant less time for onerous thoughts. The Tu-2 came off the assembly jigs, they mated the components, and they began to fit it out. Day and night, dozens of prison designers, each with his own "guide" [guard], swarmed over the airplane so thickly that you could not get close to it. A shortage of "guides" developed while they were working out systems on Project 100, fabricating parts for Project 102, and mounting engines and equipment on our Project 103.

When we returned from the shops, we would tell our police overseers that the workers were complaining. "You bring us in in the morning, but you don't arrive until evening." The three chief designers—Petlyakov, Myasishchev, and Tupolev—made a proposal: one guide for every two prisoners. This wasn't accepted. A guide took a prisoner on receipt in a prison, and he could not look after two of them. Then our bosses requested another group of guides. It turned out that this was impossible, too. All the qualified ones were taken, and training—like that for working dogs—lasted six months. "Management" met us halfway with its own plan: compress the time the prisoners spent in the shops. This decision brought inevitable confusion and unpredictable consequences. On the Tu-2 they mixed up the hydraulic lines for lowering and raising the landing gear: one strut would be lowered while the other would be raised. It looked like the airplane was running like a man. Normally, you would laugh and reconnect them, but here that wasn't the case. A guide summoned Balashov and Kryuchkov. The "managers" traced the diagram with their fingers . . . tried to figure it out . . . looked askance at Aleksandr Romanovich Bonin as he began to get anxious . . . time passed . . . the workers became angry. When he could not take it any longer, one of them said, "Comrade Chief, I've been working on things like this for about 10 years now, and I've gotten pretty good at it. Please go have a smoke, and we'll fix things in an instant! Otherwise it's a disgrace to keep Bonin away from important work." Once that crisis passed, mysterious things began to happen with the flaps. For no good reason they would arbitrarily shift to their fully extended positions. Once again the "managers" laid out the diagrams and, wrinkling their foreheads, tried to figure them out. This time the matter turned out to be more complex, and they kept me in the prison chief's office "just in case." I became a hostage. Things began to look bad. The workers in the shop saved me. They noticed that several idle guides, standing on the scaffolding, had leaned their elbows on the skin of the cockpit, bending the metal and shorting out the contacts. On the lunch break, when the prisoners were taken to the design bureau and no guards were around, the workers pressed on the skin and duplicated the failure. They informed the shop chief, who informed Kutepov, and they let me go.

At this point, it is appropriate to relate the working class's attitude toward the so-called saboteurs. The mechanism for summoning an imprisoned designer to the shops was as follows. After noting the designer's personal number on a blueprint, the shop master went through the plant dispatcher and the TsKB to summon the designer to the shop. The prisoner was supposed to be addressed as "Citizen Designer," but workers didn't address us by anything other than our first names and patronymics. In addition, the workers were supposed to write reports to the prison administration concerning all changes we made to the blueprints, and all errors they noted. It was bad enough that they didn't do this, but they also said, in friendly fashion, "Petr Petrovich [or Ivan Ivanovich], on the basis of my experience, I'd recommend this or that here. It will be simpler, more reliable, and cheaper." I must add that if a mistake was made by an ordinary designer and not a prisoner, the free worker would file an efficiency-improvement suggestion out of which he would get a small award. When the free workers did not submit such suggestions about our designs, it meant that they were refusing such awards in the name of friendly sympathy. They would greet us by shaking hands and treat us to cigarettes, whispering in our ears, "Let's go into our shop, we've set aside 100 grams, drink it—you'll feel better off in your soul." Finally, by offering to carry some news of us home, the advance guard of the proletariat showed that they understood the "class enemies" splendidly.

It made us wonder: Who were our "managers," the Kutepovs, Balashovs, and Ustinovs? Actors or complete fools? Every time that an incident like those described above occurred (according to our free colleagues), the "management" would hold meetings and discuss whether it was an error or malicious intent. The specter of "sabotage" haunted us.

It was memorable how the ill-fated flaps atoned for their guilt and made both the prisoners and the prison officials laugh heartily. While typing the description of the aircraft, a young woman typist had, in every instance, typed "gonorrhea" (*tripper*) instead of "flap" (*trimmer*). When she was called in to "management," she informed them una-bashedly that a flap was "terra incognita" for her, whereas she was at least acquainted with gonorrhea, if only theoretically.

One evening they summoned Konstantin Vasil'yevich Rogov to the shop. Getting into the car, he hit his head on something, and blood began to gush out of a wound on his forehead. His terrified guide began to yell, and the workers took Rogov to the first-aid station. When the prison duty officer telephoned the Butyrka, the order came back to deliver Rogov to their hospital immediately. Rogov told us later how they interrogated him about whether he had tried to commit suicide.

The funniest thing in this incident was the guide's behavior. After finding Rogov and assuring himself that no one was listening, the guide asked him, "Citizen Designer, please don't write me up, I'd get in big trouble with the higher-ups, and they'd take away my award! My wife gave me orders: 'Ask them not to complain.'" "They" didn't complain.

Departures and Greetings

Finally, we reached the point where everything on the Tu-2 had been checked out, fitted, and tested. The only thing that remained was testing the engines. The Tu-2's wings were detached and it was moved out of the assembly hall into the yard. Gas and oil were added, and tanks with compressed air were brought for starting the engines. The free-worker flight engineer, M. F. Zhilin, ascended the ladder into the cockpit, and the command "Clear the prop" rang out. The air hissed. First one and then the other propeller reluctantly began to move. The propeller slipstream tore off the yellow fall leaves, which flew around the yard in a cloud between the shop and the KOSOS building. From open windows, hundreds of people watched the first test of their off-spring. The Tu-2 shivered like a thoroughbred race horse before a start. A smiling Tupolev, in a warm overcoat, shook our hands.

The next morning the Tu-2 was covered with a tarp and tied down. Several trucks loaded with parts and airfield gear lined up; then the towing vehicle pulled up. While we were sleeping peacefully in our barred sleeping areas that night, they transported the airplane to the airfield.

This was a huge landmark for our work, and the "managers" re-minded us that they had both a stick and a carrot. Some of the prisoners were granted visits. We were ordered to make ourselves presentable; then we were taken to the Butyrka. They made sure we were aware that in the prison we weren't creators of war machines, but prisoners. "Hands behind your back, face the wall, no talking!" We had become unaccustomed to such sweet, familiar forms of address at the *sharaga*, but the prospect of a visit kept us quiet. Accompanied by the rapping of his keys against his belt buckle or the stairway railing (to say, "At-tention! I'm bringing a prisoner!"), the screw led me down long corri-dors to a tiny, windowless room that held a table, three chairs, and an hourglass. After several minutes of torturous expectation, the door opened, and a second screw brought in my wife and child. This was my son, whom I had never seen before. We greeted each other, I kissed the boy, and he looked at me as if I was a stranger. The hourglass was turned over, and time began to run out. We faced each other across the table while a guard sat the end of the table. My wife told me about her life. The screw would sometimes interrupt: "You can't talk about that." I said almost nothing about myself; the reality was under lock and key, and who needed something fabricated? Time ran out; my wife said,

"Sonny, say goodbye to Papa." He stretched out his tiny arms toward the screw. This was bitter, but understandable; the guard had uniform collar tabs and shiny buttons. When the screw corrected the child, he extended his arm toward me with the very same indifference. The second screw led them away. We expected so much from such visits, and they yielded so little. Under a watchful eye, we sat as if bound, and the 10 minutes out of the 1,019,800 spent in isolation flew by as if they had never occurred. We returned to the *sharaga* in silence, lost in ourselves. Evenings after those visits were always difficult; everyone disappeared into his corner. Each of us wanted to be alone, to relive and re-experience this gift of fate. For several days afterward we weren't ourselves.

Another shock came when we heard they were freeing a group from the Petlyakov team! Petlyakov himself was first; they drove him home right after his progress report on the testing of the Pe-2 dive bomber. There had been rumors about his pending release for a long time, and when he did not return, those rumors increased. In all the sleeping areas, people discussed who and when until well past midnight. They calculated their odds, built hypotheses, and expressed their assumptions. This excited not only the Petlyakov group, but also everyone else, since there had as yet been no precedent.

An ordinary morning dawned. After breakfast we dispersed to our workstations. At about 10 o'clock it spread like lightning around the *sharaga* that Petlyakov—now a free man—had arrived and gone into Kutepov's office. About 11, when they began to call those who would be released one by one into the office, the excitement peaked. Those who had been summoned were not returning. Under various pretexts the prisoners went down to the third floor and walked around near the doors of the "management," hoping to learn something, but it was in vain.

During the lunch break, as we sat in the dining hall, those who had been released were taken into the prison offices, and our contact with them came to an end. We hadn't congratulated them or bade them farewell. While this was cruel, it was even more cruel that the freed men didn't say a word to their former colleagues in Petlyakov's team. It was terrible to see those who remained in prison, for they were completely dazed. Freedom, a chimera hovering invisibly nearby, had evaporated. What would happen to us now? Where would we work, or would we work at all; would we eventually be freed, or would we ever see freedom?

Every group of people develops defensive reflexes to adapt to these arbitrary conditions. The same was true of us. Although no one had given any promises to release us after the Tu-2's flight tests, everyone considered this a foregone conclusion. And if this was so, then we had to work and live, live and work. And, indeed, we lived, created, argued, cursed, read, built, complained, and laughed. At times it was gallows humor and, at times, genuine laughter. After all, it is really impossible to "stand before one's fatherland in mute reproach" forever.

On the day those men were released, the well-ordered life in the *sharaga* fell apart. Most of the prisoners were Muscovites, and somewhere nearby our families lived hard lives, without the earnings of the main breadwinners. If the family was not half-starving, then they denied themselves almost everything, living an austere life because of the arrest of a loved one. For us, release was not so much a moral category. Instead it meant in practical terms the right to work and provide education for our wives and children; they would no longer be called son, mother, or wife of an "enemy of the people." Finally, they would be able to sleep peacefully, without jumping up at night at the sound of a knock at the door. We yearned for a normal life.

The day after the releases proved difficult, too. Dead silence filled all the rooms, as in a family that has lost one of its loved ones. The tragedy of Petlyakov's colleagues who remained in prison became clear, not only to those of us who were working on other projects, but also to the free workers, and, I think, even to that portion of the guards who most resembled human beings. The remainder of the Petlyakov group was showered with attention because everyone wanted to ease their load at least a bit.

Two or three days later, at precisely 9 A.M., the freed Putilov, Izakson, Minkner, Petrov, Yengivaryan, Rogov, Kachkachyan, Leshchenko, Stoman, Shekunov, Abramov, Shatalov, Nevdachin, and with them, our N. I. Bazenkov, shining, happy, and looking younger, appeared at their workplaces. Bazenkov ended up among them because the "management" was convinced that, until they freed Tupolev, Project 103 needed at least one capable person for trips outside. Without such a person things would have more been difficult for the prison management, since they had a poor understanding of the mysteries of technology. Release did not affect the freed men. They continued to behave as they had previously toward those who had remained in captivity, that is, friendly and, I would say, brotherly. However, this "thaw" didn't last for long. Even on the second day we felt a sense of estrangement. They clearly avoided lengthy tête-à-tête conversations, they appeared downcast and restricted their movements. What had happened? We learned later that the plan called for them to be transferred immediately to the series-production plant. As usual, something went wrong, and the move was postponed for several days. The prison administration became alarmed. The proximity of free men to prisoners was always the Achilles' heel of the isolation system. Now it turned out that Achilles had another heel: the freed men. The administration gathered the freed men together and instilled in them the idea that freed prisoners were not as much vile criminals as those who remained in captivity; accordingly, extraneous contact with them was unnecessary, and not in a free person's interest! Since the administrators themselves understood that their words were utter nonsense, they turned them into their own jargon: "Hold conversations only on official topics. Excessive contact is not recommended. Be vigilant!" That's how it was.

This difficult and, for many, tragic episode passed and soon was forgotten. After a week the rhythm of activity in the *sharaga* returned to normal. The abandoned beds were removed from the sleeping areas and a few tables were removed from the dining hall. Calm prevailed once again. Nothing gave a hint that a tempest had passed over our "quiet backwater." It was disturbing when the freed prisoners would slip every so often and say, "Remember how well we used to be fed? You don't eat like that on the outside."

The newspapers that friends sometimes brought to us hinted at all kinds of riddles. The outside world had changed without explanation. Molotov seated next to Hitler in the chancellory, Ribbentrop and Stalin in the Kremlin. How could we make sense of this altered reality? Your head would open from thinking about it, but to no avail. There was now an alliance between Germany and the USSR, and after years of rivalry we learned of our mutual interests. Yet we were being rushed to complete our project because the country needed dive bombers. A bomb shelter, in fact, was being built in the basement, and tearful young women would come to work saying that their boyfriends had been called up into the army. In time, they brought gas masks into the storage area at the prison design bureau. Those who lived at dachas in Byelorussia and Riga regions complained about the roads. It was impossible to sleep at night because columns with tanks and guns were on the move! Rumors, rumors, rumors, while the government remained stubbornly silent!

Our pilots ferried several aircraft from Germany to the Scientific Research Institute of the Air Force: the Junkers Ju-87 and Ju-88; the Messerschmitt Bf-109 and Bf-110; the Dornier Do-217; the Heinkel He-111 bomber; the Henschel ground-attack aircraft; the Fieseler Storch courier airplane; and the Focke-Wulf reconnaissance airplane. It is possible that these were given to us with some ulterior motive, as if to say, "Take a look at what we intend to use to smash you." We were taken to the Scientific Research Institute to examine these German aircraft with swastikas emblazoned on their tails. This was the combat equipment that had routed Poland, Denmark, Norway, Holland, Belgium, and France, but was only breaking its teeth against England. There was much that was interesting about these aircraft, and we could emulate certain kinds of advanced technology without remorse. We examined them in great detail, flew them, and talked with personnel who were familiar with them. German aircraft designs revealed numerous clever details.

While inspecting the Ju-88 I had an unexpected encounter. After climbing onto the wing via a stepladder, I was examining a power-supply unit located in the fairing of the engine nacelle. Suddenly, the ladder shifted: obviously, someone was climbing it. I figured it was my guide, and continued to study the unit. When the person appeared over the wing, I turned pale: here was my brother, B. L. Kerber! He was working in Yakovlev's experimental design bureau, and they, too, had been brought in to examine the German equipment. When I recovered from the shock of our chance meeting, I peppered him with questions. He began to tell me feverishly about his wife, children, and the older generation. Everyone was alive and healthy, they were waiting, and hoping, he reassured me. . . . The guide called up to me, "Citizen designer, the group is getting ready to go to lunch!" There *is* such a thing as good luck! That evening in the sleeping quarters everyone envied me.

We were taken to the institute's mess hall for lunch. In our country, dining halls—including military ones—are divided into endless categories. The Scientific Research Institute had five, for sergeants to generals. Whether it was because the highest-category mess hall was isolated, or because they ordered our security to treat us as equivalent to generals, that is where they took us. When we entered, three generals were quietly talking at one of the tables: P. A. Losyukov, S. A. Danilin, and N. P. Shelimov. Upon seeing Tupolev, the generals rose, exchanged bows with him, and began to do their best to take care of us. They seated us, then expressed interest in our impressions of the German equipment. Their actions, however, revealed an awkwardness due to the situation, an obvious restraint due to Kryuchkov's presence (the screws had been left outside the door), and overall embarrassment at this sudden encounter. Again there was a conflict between reality and the rumors with which they made fools of people, like the fairy tale about Tupolev selling the plans for our Tu-2 aircraft to the Germans. We were having a general, rather relaxed conversation when Tupolev, turning to Losyukov, said, "Well, Prokhor Alekseyevich, I was honored: I examined the Messerschmitt Bf-110, and saw 'my' airplane." Everyone fell silent. His meaning was clear to everyone. Kryuchkov clearly became upset. An oppressive silence fell over the small room. Lunch came to an end, the generals stood up and Danilin and Losyukov went up to Tupolev and shook his hand warmly.

This incident was a signpost of the times. They saw, they were horrified, but they remained silent. In the dramatic setting of an unexpected meeting, however, sincerity had the upper hand; people were unable to conceal their feelings and showed them without thinking about it.

The story of the plans supposedly sold by Tupolev to the Germans was a falsehood, of course, but the story in all its fantastic detail did explain Tupolev's arrest to the people. Military aircraft with twin tail construction, like the Bf-110 and Tu-2, were similar in silhouette and on occasion during the war our antiaircraft batteries became confused and downed Tu-2s.

Success and Disappointment

Two weeks after examining the German aircraft we again traveled to the airfield. Our Tu-2 had been reassembled, and it was time to prepare it for its first flight. Tupolev created an operational group for flight testing, which included the chiefs of the design teams and their leader, A. M. Cheremukhin.

Our trips outside followed a fixed scenario. A small bus was brought into the KOSOS courtyard. The screws sat in the last row of seats. Then, counting heads, they let us in. A second group of screws sat in the front seats, and Kryuchkov sat next to the driver. Tupolev had a permanent seat in the right front corner. Between him and the door, on a hinged seat, sat yet another guard. During the entire trip we eagerly scanned the city and countryside. In the bus there was a sense of freedom, for we stared out windows without bars, as if we were making a passage into the life of free people. But there were always reminders of our situation. During one of our trips in the spring, when streams ran through gullies, patches of sunlight played on the bus ceiling, sparrows chirped furiously, and the invigorating spring air poured in through the open windows, we had a flat tire at the Preobrazhenskiy Outpost Square. A complicated situation arose since, after all, they could not let us out onto the street! After consulting, the screws got out and surrounded the vehicle. Although they were all in civilian clothes, their backs, with pistols protruding, did not escape the attention of the young boys. A herd of children, at first timidly, then more and more brazenly, began to circle nearby. Finally, one of them ran up to the door and shouted impudently, "We know who you are!" Sitting near the door, Tupolev showed some interest. "Well, who are we?" Unperturbed by his patriarchal appearance, the boy tossed back, "Crooks!" After this, Tupolev would frequently say to us, with a sad smile, "Well, crooks, let's go!" or "Let's think this over, crooks!" This mischievous child had evidently gotten to him. Many years later, we were riding in his limousine. It was a March day, and the sparrows in the trees along the Yauza River were chirping, flying around, and fighting. "Tell me, do you remember how they called us 'crooks'? They noticed real well, didn't they?"

We were testing the Tu-2 in the fall of 1940. At the airfield, the airplane was brought out of the hangar; the wheels left clear tracks in the soft newly fallen snow. M. A. Nyukhtikov was in the cockpit. When he tried the engines, snow flew from the prop stream. Once everything had been checked out, Tupolev yelled, "Good luck!" Nyukhtikov gave it some throttle, the airplane shuddered and disappeared in clouds of snow. After several minutes, the Tu-2 appeared at the start of the takeoff runway. It began its takeoff run, braked, and again began its run. After repeating this three times, Nyukhtikov taxied to the hardstand. Our group was arguing about whether or not it actually took off. I had seen dozens of aircraft perform their first taxiing, and such arguments arose every time. Nyukhtikov taxied up and killed the engines; the exhausts crackled noisily as they cooled in the freezing weather. He descended the ladder. "Everything's in order, it is ready for its first flight."

A first test flight requires a mountain of paperwork; the Tu-2, created by "saboteurs," understandably required a mountain of paperwork. For several days Tupolev tore down mountain after mountain, calling in one of us, then another, pulling out paper after paper to put the proper "facsimile" ID number on each page. He became more and more ill-tempered and finally threw the stamp down on the table. "Put them wherever you want, even on the rear ends of the 'managers,'" he cried in a falsetto. Just as children playing "pretend" believe that, if God is willing, even a stick can shoot, so the grownups from the NKVD, when they suspect something and then, suddenly, it really does shoot, brazenly insure themselves by requiring the stamp "011" on all papers [i.e., Tupolev had to sign for each stage and assume all responsibility].

Finally, all formalities were completed, the weather forecast was checked, and the day for the flight was set. Extremely worried, we rode to the Scientific Research Institute. One after another, the luxurious cars of the "management" overtook our small, unhurried bus. On this occasion, all decorum was cast to the wind and, like fat flies, both designers and all higher-ups "involved" in creating the Tu-2 gathered around the sweet pie.

Nyukhtikov, the pilot, and Akopyan, the navigator, donned their parachutes; silently they took their places in the aircraft. Although they later denied this, they were no doubt extremely nervous about a test flight for the Tu-2—an aircraft designed in a prison workshop. Then again, perhaps they never really believed the fairy tales told by Yezhov, Yagoda, and Beria? The soul of another man is always murky and shrouded in darkness . . .

The engines were started. Soon the Tu-2 strained to move forward into the air. Nyukhtikov raised his hand, the engine mechanics pulled the chocks away from the wheels, and the airplane slowly taxied for takeoff. Behind the Tu-2 walked Tupolev, slowly, unhurriedly. Such was his inner strength that no one made any effort to stop him and no one went to escort him. He cut across the field, and we knew for certain that somewhere near the spot on which he stood the airplane would leave the ground. That's how it was, for every such flight. What was going through the mind of this lone middle-aged man at this juncture? His wife was in prison, his children were probably suffering, and his creation, upon which so much depended, would now take flight.

The engines' roar increased. Nyukhtikov released the brakes, the Tu-2 gathered speed, left the ground, and disappeared in the fall haze. At this point, the tension that had kept us wound up all week was released, and everything became amazingly ordinary. So what? Another airplane had taken off . . .

Twenty minutes later the airplane landed smoothly and rolled up to the assembled group of NKVD and Air Force chiefs. We were off to the side: they asked us not to approach them. After listening to the crew, the chiefs set off for Moscow in their limousines, hurrying to report to even higher-up chiefs how they had handled the important mission entrusted to them and had created a combat aircraft. Then Nyukhtikov walked over to us, smiling, and told us in friendly fashion that he was satisfied with the Tu-2. He reported that the aircraft was simple to control, it was stable in flight, and there was no tendency to yaw or stall. He shook Tupolev's hand cordially, and Tupolev embraced him.

The stage emptied, they loaded us in the bus, and we set out for the prison. "No ovations, please!" as Ostap Bender used to say. We returned and found that the prisoners had been waiting for us. Our report caused much joy. Late that evening, when emotions died down, all those who had worked on the Tu-2 gathered in the oak-paneled sleeping quarters. There were no speeches, wine, or toasts. We all wished to congratulate Tupolev. Although there were bars on the huge window, and a screw silently walking the corridor, I cannot recall a freer, more cordial, warmer gathering. How could it happen that the Tupolev group celebrated the birth of its fifty-eighth airplane under such unusual conditions? Who "helped" in this?

Not all official accounts of Soviet aviation were accurate. Let us turn to the well-known books by A. S. Yakovlev, *Aim of a Lifetime* and *Soviet Aircraft,* which made the charge that Tupolev:

> did not prepare a replacement for the outdated, slow-moving TB-3 bombers in timely fashion;

> did not attempt, in timely fashion, to increase the speed of the SB bombers, which began to suffer noticeable losses from German fighters during the war in Spain; and

> did not create a high-speed bomber needed by the Soviet Army at the outset of World War II.

Is it possible that Yakovlev, who was an aircraft designer and—as of January 1940—a Deputy People's Commissar for Experimental Aircraft Construction, could be so ignorant? Is it possible that he didn't know that:

> In 1936, five years before the war, KOSOS had prepared a replacement for the TB-3, and had created the TB-7 bomber, which had twice the speed of the TB-3?

> In 1937, Tupolev was fired from his position, subjected to punitive measures, and thus was unable to modernize the SB during the war in Spain?

As early as 1940, Tupolev developed the best assault bomber in our war with Nazi Germany—aircraft 103, that is, the Tu-2?

Yakovlev was more than just an author of aviation books. He was also the "leader's" [Stalin's] consultant on aviation issues and naturally shared his views with him. Given Stalin's well-known suspicion and his mistrust of the old intelligentsia, this would have been fertile soil for the question, "Yes, but is Tupolev completely loyal?" (This question will be addressed later.)

The testing of our airplane was going well. During the fourth or fifth flight they endeavored to reach maximum speed. Inside the hangar a small room had been set aside for our work. Here people came in for meetings and to analyze the ongoing test flights. The room was always crowded, in part because each of us had his "very own" guide. While we were measuring maximum speed, the "management" ordered the guides to withdraw, since the maximum speed tests could not be entrusted to such guard dogs. Colonel Miruts, the lead engineer for the military, was excited and, as usual under such conditions, reported with a slight stutter, "643 kilometers per hour." This was a fantastic number: the LaGG, Hurricane, and Bf-109 fighters were all slower than the twin-engined Tu-2. You should have seen Tupolev's face—boundless joy, pride, and personal satisfaction—as a mischievous kid, in the work of his team, "That's our guys!" This was his triumph, the success of a man who knew how to master Beria, and how to work on a genuine, necessary military design rather than the imaginary PB-4. Of course, there was skepticism about the accuracy of the measurement; could the pilot have been descending? The runs were repeated over the measured course several times, and the result was soon confirmed.

The prison design bureau's bureaucracy went into action. After meeting with Balashov, Kutepov went to see Kravchenko, who then reported to Beria. And he, according to the rumors, immediately went to the big "boss," Stalin himself. Trembling, we waited for the next development, but the "management" did not say a word. Several days later, production engineers from the aviation plants appeared at the *sharaga*. Only from them did we learn that Stalin had decided to put the Tu-2 into large-scale series production. And that he had imposed a year's deadline to get the Tu-2 into production.

We walked around like birthday boys. Clearly, in a day or so the curtain would rise for us, too. After all, they released Petlyakov's group immediately after the decision to put his Pe-2 aircraft into series production! Freedom soon would be a reality, we thought. Depression and despondency were replaced by happiness and optimism. Some of us washed and ironed our clothing, those from other cities got train schedules from the free workers. Joy overcame our free colleagues, too. They congratulated us, invited us to their homes, and told us what wine and food their wives would prepare for us.

A day went by, then another, then a week, but everything remained the same; in addition, rumors persisted that the military wanted certain modifications on the designs of the Tu-2. After examining the German aircraft, they insisted that the crew be consolidated in a single compartment. Such a change of doctrine was common in that context. Only recently, at the Tu-2 mockup, the point had been made that if the crew was dispersed, the aircraft possessed a greater chance of survival in combat. Now, "to ensure survivability," the Air Force insisted that the crew be consolidated.

Meetings were held, somewhere someone was deciding something, while we waited. Finally, a sort of Solomon's verdict was reached. So as not to offend Petlyakov's group, so the wits said, we too (just as they did) had to create a second airplane, the Tu-2U, as a standard for series production. In the Tu-2U the navigator was shifted forward into the pilot's cockpit, new AM-37F engines were to be used, one more defensive position with a machine gun was added, fuel-tank capacity was increased, and external pylons for two one-ton bombs were planned. "But everything else, fair Marquess, is just fine, just fine . . ."

The most poignant aspect of all this was that the deadlines for submitting blueprints to the plants had not changed. "We could not postpone the production of series aircraft, we could not 'disappoint Josef Vissarionovich [Stalin].'" Simply put: They were afraid to report to him since, after all, heads could roll, and it wouldn't be ours, which were cheap, but their own, and "that's always unpleasant," as they sang in another song that was popular at that time.

Understanding what a blow this was to the prisoners, the "management" floated a rumor that we would be released as soon as the Tu-2U team produced its flight data. Despite these promises, the euphoria that had prevailed during recent days was replaced by an equally acute depression. It was as if we were different people. Trifling issues that had been solved on the fly the day before became cases of squaring the circle, where we were asked to design an entirely new airplane. The cockpit wasn't fitting together, the contours were changed every hour, the engine wouldn't fit properly, the center of gravity had shifted, a new landing gear was necessary. . . . The changes multiplied and snowballed. Tupolev sat in Yeger's team until late at night, technical conflicts grew into personal ones, a friendly society of capable people turned into a collection of neurasthenics. Everything was going to hell in a handbasket, and the threat of a complete breakdown was staring us in the face.

Realizing the tragedy of the situation, Tupolev resorted to an unprecedented measure: one evening he called a meeting of all the prisoners participating in the reworking of the airplane. Of course, this was done illegally. Sentries were posted. Tupolev described in detail all the meetings to which he had been invited, explained the reasons for the changes, and ended thus: "They don't inform us, they order us; however, only a fool could avoid seeing that we're moving toward war. It's equally clear that no one except us can design the bomber the country needs. I will probably be correct if I say that we love our motherland no less than others and, in all probability, more than those who have brought us together here. The conditions are difficult and, if we can renounce personal bitterness and see beyond it, even tragic. And so, understanding all this, I'm giving you a mission that no one but you can fulfill. As for you, I know you will fulfill it, because you are who you are. We must put all our ability and knowledge into the Tu-2U, and all our talent, too. Let's grit our teeth for the last time and complete this mission. We have barely enough time, but we have to succeed. Therein lies our guarantee of liberation; we can't remain prisoners during a war, for we can't fight in chains."

We departed in silence. The responsibility given to us was a heavy burden. Two days later Kutepov asked Tupolev with a smile, "What sort of meeting did you hold? Were you electing people to a trade union?" Someone had already managed to sell us out.

From that point on, we worked until late at night. The "management" didn't protest; in addition, around 10 P.M. they brought yogurt, tea, bread, and butter to the mess hall. The free workers were put on a mandatory ten-hour workday, and they worked most Sundays, too. They could not say anything in front of their bosses, but they would complain to us: "It's getting harder and harder to live; food is disappearing, you have to stand in lines, and there is no time." The people were growing certain that war was inevitable. They understood this intuitively, and waited for some persuasive words from the party or government, but there were none.

On one of these evenings, when the work had been going unusually well, the door creaked open, but, instead of a screw, Tupolev walked into the large, empty room. He approached the lighted table, pulled up a chair, and, tucking his leg under him, sat down.

"Working?" he asked.

"Yes, I'm finishing up the power-unit diagram."

"You're good at everything, only your damned tobacco stinks."

I should mention that Tupolev could not stand tobacco smoke, since he had lung problems his entire life. To spare his health we never smoked in his presence, and I put out my cigarette. Shifting on the chair, he made himself more comfortable and fell silent. This occasion allowed me to pursue a more private conversation with Tupolev. For a long time all of us had wanted to learn a bit more about his arrest, how his investigation had gone, whether they had tortured him, and what they had demanded of him—or had they simply ordered him, as they had most of us, to think up something appropriate. For this type of conversation it was necessary to cross a certain line or, if you will, a certain closeness or intimacy had to be present. In the hubbub of our daily life, such moments never occurred. But now it was as if a quiet angel had appeared, and I risked asking about these things. At first he began to speak as if he were unwilling and then, with increasing trust, "What cell were you in?"

"No. 57."

"How about that? I was in 58. A fateful number for me: I was accused under Article 58, cell No. 58, and for my 58th airplane." Tupolev went on to explain the circumstances of his arrest. "They kept me at the Lubyanka for a long time, in solitary. Finally, they transferred me to the Butyrka. It got easier and harder; all the same, there were people around you. No, they did not beat me. They just kept me standing for a long time, but that was hard, since I'm heavy-set. You're standing there, and the investigator keeps hammering away at you, 'Write, you bastard! Who did you sell the blueprints to?! How much did they pay you? Write, don't be bashful! Your buddies Arkhangel'skiy, Sukhoi, Petlyakov, and Myasishchev broke down a long time ago. They turned you in. You're the only one being stubborn. Give up, it'll be easier for you.'"

Tupolev sighed. "You know, that stupid maniac kept repeating this over and over, and I was standing there, and my legs were hurting, and my eyes were starting to close since I wanted to sleep, and so I'm standing there, thinking, 'It seems all I've done my whole life is build airplanes for them—no, not for them, for my country.' Of course, there were some mistakes, and not everything worked out, but that's the way life is. You know, I really like to whittle sticks. You whittle, and whittle, and sometimes you whittle a real monstrosity, so that it scares you, and you spit in disgust and throw it away. That's how it is with whittling. But an aircraft design is more complicated. Also, they give you an assignment, but then aviation leaders say, 'Let's refine it.' Baranov says one thing, Rukhimovich says another, Alksnis, the air commander, gets his say, Voroshilov has his, Ordzhonikidze says something yet different, and finally they report to 'him [Stalin]' [Tupolev raised his eyes and pointed toward the ceiling], and then you get something else completely unexpected from 'him.' And then, after all this, when you're looking at the airplane when it's out in the field, as you see that they've whittled it down to the last straw, then there's just one thing left: create a new airplane.

"Of course, there were miscalculations." He suddenly grew animated. "Do you think that Mitchell, Focke-Wulf, and Messerschmitt didn't make them?" At this point, his look became subdued. "So, okay, you get up and you console yourself with, 'Forgive them, for they know not what they do.' No, you can't forgive them for this. You can't!" he remarked with conviction. "I believe that all this will be out in the open, and even within my lifetime. And you—do you believe this?"

"I wish it were so, Andrei Nikolayevich, but it's not happening."

"You have to believe, you can't live without that. Otherwise you can't stand it, and you commit suicide." He finished, and stood up, smiled sadly, patted me on the head, and walked off with his tiny steps, with his right shoulder jutting forward.

It was hard, to the point of tears, to watch the designer's disappearing figure. For many years his aircraft had made up the main design of our Air Force—a man who had been slandered and declared a criminal. Now he worked in a prison workshop.

We were all tormented by the question of who had helped in Tupolev's arrest. Even now this troubles many aviation specialists. Without a doubt the arrest could not have occurred without Stalin's approval, and to make the arrest the police would have had to compile materials. Individually they wouldn't convince anyone of Tupolev's evil intentions. When compiled, however, these materials were viewed differently, as part of a large plot.

One of the first records to discredit Tupolev appeared in the NKVD's file after a Kremlin session that resulted from Sigismund Levanevskiy's ill-fated polar flight in 1937.

In 1976, when he was already seriously ill, S. V. Il'yushin persistently sought a meeting with the Minister of the Aviation Industry. Since he was extremely busy with urgent matters, P. V. Dement'yev asked his deputy, A. A. Belyanskiy, to meet Il'yushin, who was gravely ill. With difficulty, but articulately, he said, with his voice breaking at times,

As I leave this life, I want to confide in you. We were having difficulty finishing off the DB-3. We could not get the oil cooling on the engine to perform correctly, and the oil was overheating. Overall, we could not have managed without the assistance of TsAGI and the test data they provided for our project. No less important had been the studies at the Central Scientific Research Institute for Aircraft Engine Construction [TsIAM]. The deadlines were getting closer and closer, and the authorities kept telling us, "Faster! Faster!!" Although the Chief Directorate of the Aviation Industry [GUAP] and Tupolev, its chief engineer, were helping, it seemed to me that it was too little and reluctant. That's when I wrote to the proper authorities, wrote bad

things, wrong ones, and evil things, mentioning Tupolev. Now it's tormenting me. I want Dement'yev to know what happened."

Belyanskiy retold the conversation in detail to the Minister, and shared
it in confidence with N. I. Bazenkov, with whom he was on friendly
terms. Bazenkov, naturally, didn't keep anything back and informed
me. I wrote down his words then and there. Thus, at some point during
those years, the doubts of another great man regarding Tupolev's con-
scientiousness appeared in the files of the KGB.

The most active informer on the "doubtful" aspects of Tupolev's
activity was A. S. Yakovlev. He had his own, original method. His
denunciations and slanders were generously recounted on the pages of
many books. Subsequent "facts" were borrowed from them. Before
this, however, let us trace this author's path from obscure student to
Deputy People's Commissar for Aviation Industry. After the university,
he was sent to Factory No. 39. This young, capable man soon became
bored with the routine work in the factory administrative offices. He
dreamed of more creative work, and in the evening would build light
airplanes with a group of friends. One light airplane, the AIR-7, had
an accident. It was a miracle that the pilot, Yu. Piontkovskiy, avoided
disaster. Convinced that it was impossible to avoid such incidents under
those conditions, GUAP gave the order to cease amateur work at the
defense plant. Yakovlev's group was transferred to a separate area of the
plant. It was this incident that was the grounds for that author's slan-
derous libels against Tupolev.

In the fall of 1933, the indignant Yakovlev wrote a letter to Politburo
member Ya. E. Rudzutak, complaining that GUAP was preventing him
from working on the future of Soviet aviation. There weren't very many
Politburo members, and the origin of the AIR model was known to
Rudzutak. So he summoned Yakovlev, who unintentionally mentioned
the newly created "air taxi," with a cabin for four persons. Intrigued,
Rudzutak invited the designer to fly to his dacha at Nikolina Gora for
a visit. At the roar of the engine, Voroshilov and Mikoyan, who had
been strolling nearby among the hills, approached. The name of the
young designer became known to four members of the Politburo at
once. This was an important benchmark for Yakovlev.

At the Tushino Air Show in 1935, Yakovlev met Stalin. The leader greeted him favorably, as was recorded in a photograph that Yakovlev placed in his book. Yakovlev told Stalin of his flying activities. Soon thereafter, Yakovlev organized a design bureau which was attached to a plant on the Leningrad Highway. Here he was appointed chief designer. Feeling that he was now on solid ground, Yakovlev hastened to promote himself at the expense of others: "The heavy Tupolev bombers on which the pilot A. B. Yumashev set aerial records," he stated, "are obsolete and slow-moving; they fly low and have insufficient range."

Yakovlev's cynicism was apparent. Such a statement was false, as judged by the fact that, as early as 1937, the new Tupolev TB-7 (the ANT-42) demonstrated a speed of more than 400 km/h at an altitude of 10,000 meters; such speeds were unattainable by any fighter at that time. Moreover, it wasn't Tupolev's fault that there was only one regiment of TB-7s available at the start of World War II. Someone interfered with their construction. Who, one might ask?

Having postured in this way, Stalin summoned Yakovlev. The young designer had opened Stalin's eyes to the real "state of affairs." In a confidential conversation, Stalin requested, "Say what you think, and don't be embarrassed. Although you're young, we trust you. You're an expert in your field, you're not linked to the mistakes of the past, and so you can be more objective than the old specialists, whom we trusted greatly, but who led aviation and us into our current morass."

It was 1937. In Spain, the Messerschmitt Bf-109E fighters—with their more powerful engines that Germany had improved under operational conditions—began aggressively to down Tupolev SB medium bombers. Stalin viewed these losses with anger. Once again, it appeared, the airplanes of this "old specialist" proved ineffectual. Why didn't Tupolev improve his SB design in time? After all, Yakovlev had proved it was possible. The twin-engine Yak-4, which Yakovlev had created with unusual rapidity, and which used the very same engines as the SB, flew 100 km/h faster!

"Put the very similar Yak-4 or the BB-22 [as its designer called it] into series production immediately!" The military didn't need the Yak-4 because it was defenseless and therefore insisted, "Arm it!" However, when everything that made it a bomber was installed, that is, two machine guns, a radio, instruments, sights, bomb racks, and so on, the gain in speed had disappeared! Our pilots had a sober assessment of the Yak-4 in the war with the Finns: the airplane had stiff controls, the cockpit was cramped, instrumentation was insufficient, and it had no defenses. Several hundred Yak-4s were removed from the inventory as unsuitable for the war against the Germans. Everyone awaited with trepidation the fate of its chief designer, a Communist, who had dared to deceive his own General Secretary. But nothing happened.

Tupolev and his colleagues were already in Beria's torture chambers, however.

Final Days at the *Sharaga*

Our work at the *sharaga* continued. We worked as hard on the Tu-2U as we had on the Tu-2. Gradually, everything settled down, blueprint production was on schedule, but the schedule turned red as our planner, K. P. Borovskiy, crossed out line after line on it. Even the very complex issue of the navigator's sight was resolved. In his new location the navigator sat much higher, and the optical sight wasn't long enough. Despite all the efforts of the *sharaga* "management," the People's Commissar responsible for the plant that manufactured the sights categorically refused to lengthen the column. Solomon Il'ich Buyanover, the chief designer at the plant, was summoned to the TsKB to meet with prisoners Tupolev and Nadashkevich, whom he respected greatly. Buyanover became quite upset and declared, "Andrei Nikolayevich, Aleksandr Vasil'yevich! If I'd only known! For your airplane I'll do it in an instant!" And he did.

The formation of the fourth design bureau, that of D. L. Tomashevich, was nearing completion. R. L. Bartini, I. M. Sklyanskiy, and V. I. Siprikov transferred there. They began a design for a fighter, Project 110.

V. M. Myasishchev went far with his project. The project for his long-range bomber, Project 102, was approved, the model commission was held for it, and the construction of the 102 aircraft was in full swing. Everything moved along more or less smoothly. In the assembly hall they rolled out the new nose section for the Tu-2U and the engine nacelles, mated the components, and began to outfit it. Aesthetically, the airplane was less elegant than the Tu-2. The aggressive appearance of the first airplane had yielded to less expressive lines. We even remarked ironically that borrowing the common crew compartment from the Germans—universally recognized aggressors—had made the Tu-2U less aggressive.

One evening we noticed that our ranks had thinned. Krutkov, Rumer, and Szilard had been taken away, to parts unknown. It was rumored that the NKVD had ordered them transferred to prison camps for excessively close contact with our free and female colleagues. The prisoners were indignant: "Good Lord, Krutkov is nearly 70, Rumer is a confirmed misogynist, and Szilard does nothing but talk about his wife and children." Then after a scene involving prisoner I. S. Svet and the chief of our prison, due to some stupid nonsense, "Granddad Svet" was sent to an isolation cell at the Butyrka. The war began several days later and we forgot about Svet. When they moved everyone and everything out of Moscow in October 1941, including the prison, they sent him to the Pechora prison camps, where he subsequently outfitted the railroad to Vorkuta with automatic blocking and signals. Actually, there were probably more significant events, but a prisoner's memory always reacted most painfully to the transport of prisoners.

Finally we completed the Tu-2U. It was spring again, the year was 1941, and the streams were flowing, when she made her first flight. Nyukhtikov and Akopyan praised the airplane, although its speed was 50 km/h less than the prototype Tu-2. Flight tests began. One evening, after a routine test flight, rumors suddenly spread throughout the *sharaga* about a disaster involving the Tu-2U. In the morning we learned that the right engine had caught fire near Noginsk. Nyukhtikov had given the order to bail out, but Akopyan apparently snagged his parachute strap on something and perished. An accident commission began an investigation into the causes of the fire. The illusion of freedom melted away once again. Even the most hardened pessimists, however, did not suspect the blows that would rain down upon the prisoners in the aftermath of this incident.

Petlyakov's Pe-2 dive bomber was taken to the Scientific Research Institute of the Air Force. Pilot Khripkov and navigator Perevalov were assigned as its crew for testing. During one flight a fire broke out in the crew compartment during takeoff. Khripkov made a forced landing just outside the airfield's fence, in a field where, unfortunately, a kindergarten class was taking a stroll. Several children died. Our airplane, however, emerged from the crash with minimal damage. Khripkov and Perevalov, also injured, were arrested and taken to the hospital. An investigation was begun. Petlyakov and his colleagues were subdued. The free workers spoke of malice and sabotage.

Various learned experts were brought in to investigate the crash. They came up with a ridiculous version of events, involving static electrical charges that built up due to supposedly improper complete wiring of the equipment in the airplane. The matter became more complicated, and the true cause was lost in this implausible explanation. Our police overseers—who were always inclined to search out malicious intent and sabotage—for the first time went to the opposite extreme. They computed the potential for static discharges, and slowly but surely drowned in their calculations. Petlyakov's team remained worried. A series of interrogations was launched. There were hints that the electrical charges were not accidental. At the last session devoted to the accident were about 10 NKVD workers, as well as our "overseers," headed by Yamalutdinov, V. M. Petlyakov, Petlyakov's "chief" electrician, K. V. Rogov, and a plain-looking, middle-aged expert from the All-Union Electrical Engineering Institute, a professor with an Armenian surname. Kutepov, the Chekists, and our "overseers" liked the static-electricity version best. They were already imagining subversion. They needed Rogov's confession, but that was a trivial matter. The meeting was drawing to a close when the plain professor, realizing what was happening, declared angrily, "I've given up trying to understand what's going on. It's an absolutely fundamental truth: On an aircraft that has just taken off, there are not—and cannot be—any charges capable of producing a fire. Please record this in the record. My presence here is no longer necessary." He stood up, bowed, and disappeared. When Petlyakov related this to Tupolev, the latter noted, "That's how it happens. Someone will appear to be a sluggard—that's what I thought. But, you know, he'll bring a second case against you, Volodya. And what a brave man he turned out to be."

Soon thereafter the flight mechanic noted a gas leak on the wrecked airplane from the manometer fitting over the switch for the landing gear or flaps. Everything now made sense. The crew was freed, the two fitters who installed the manometer and switch were then arrested, the airplane underwent repairs, and the testimony of learned experts and prisoner-designers was hidden in folders (they might be useful later!). Gradually everything returned to normal, and it appeared that the storm had passed. The suspicions, however, had made their way to the top. Two accidents involving airplanes built at the *sharaga* put the higher managers at the NKVD on alert. Could it be that Kutepov and his assistants overlooked something perhaps?

Events, possibly with serious consequences for us prisoners, were in the air. Everyone was downcast; even Tupolev wasn't himself. Our forebodings were justified. After several days we were dealt a blow, but from an entirely different direction. It turned out that the production of the high-altitude AM-37 engines for the Tu-2 had been halted. The plant's entire effort had been thrown into increased production of the lower-altitude AM-35. The latter were needed for the low-flying Il-2 ground-attack aircraft, whose production was gearing up at several aircraft plants. However, one other engine plant was building the ASh-82 with the same power and altitude rating as the AM-37, so we were ordered to redo our airplane to handle the ASh-82. The problem was that the ASh-82 was air-cooled, which meant its front end was much larger than water-cooled engines. This would cause a further reduction in our airplane's speed. In essence, changing the engine meant a substantial reworking of the airplane: completely new powerplants had to be designed; the water radiators had to be removed from the wing center section; the oil system had to be redesigned; electrical starters had to be worked out for the engines; and room had to be found for larger batteries. This was a great deal of work, and the government decided to set the deadline for reworking the airplane at two months hence. To the question, "And when will they release us?" Tupolev responded, with annoyance, "What do you mean? Don't you see what's going on?"

All of our calculation teams were unbelievably overworked. Unfortunately, their results didn't please anyone, since the aircraft's combat capabilities had been downgraded. In particular, the speed of 643 km/h for the Tu-2 was reduced to 610 for the Tu-2U, and to 560–580 km/h for the Tu-2V, as the modification with air-cooled engines was known. This represented a loss of 80 km/h! Morale among the prisoners working on the Tu-2V was so bad that Kutepov called us together. His none-too-convincing words intimating that we would be freed after we completed the blueprints did not convince anyone. Some sort of *force-majeure* was necessary. On the evening of the following day, Kutepov called into his office the leaders of the teams in Project 103, together with Tupolev, and swore that he had received an order at his People's Commissariat on the day before to assure us that we would be free after the very first flight. This type of "confidential" conversation between the "manager" of the TsKB and his prisoners was unprecedented. Kutepov spoke at length, until Tupolev interrupted him. "Grigoriy Yakovlevich, you don't have to say anything. We understand very well that not everything depends on you. Pass on to your bosses that we'll make the new modification." At this point, he gestured toward all of us, and his voice wavered. "But tell them, too, that you can't deceive people forever. Even prisoners have to believe in something."

He stood up, glanced at Kutepov, and added brusquely, "Let's go, men!" In the doorway he turned back and continued: "We'll build *Vera* [our nickname for the Tu-2V], but after that . . ." Whatever came after that was said in our absence.

Wartime Emergency

Everyone was extremely busy. All the teams were helping the engine specialists. We were not retiring to our sleeping areas earlier than 1 A.M. By correcting all the manuals and canons, Tupolev had production work from the pencil drawings, not from blueprints printed as photostats, but directly from the Whatmans off the drafting tables. No sooner had we found a general design than the production engineers began to prepare the manufacture of parts for an assembly, the loft specialists plotted their outlines on the loft floors, and the fitters began to figure the construction of the assembly jigs. This insane work went on nonstop for two weeks, when TASS's notorious announcement appeared on 14 June.

This announcement did considerable damage. "The announcement dulled the troops' vigilance. . . . Commanders stopped spending the night in their garrisons. Soldiers began to change clothes for the night," writes L. M. Sandalov of the Brest garrison at that time. We at the *sharaga,* however, deep in the rear, sighed with relief. The war wasn't so close after all, and we would succeed in completing the Tu-2V! All of us understood that this wasn't a TASS announcement at all, and that Stalin himself had written it. Faith in Stalin was so strong that, when they arrived at work on the fifteenth, the free workers said, "Thank God! We don't have to stock up on food; now we can put everything into the Tu-2V."

While writing about all these events, I almost omitted what was most important for me, personally. On 5 May they called me into the prison office, with all my belongings, as if to be transported somewhere. Worried, I went to Tupolev. He squeezed my shoulder, as if to say, "Don't worry, everything will be okay." I was taken to the Butyrka to the office of the prison chief, where Kutepov was seated. Both men stood up and shook my hand. "You've been freed!"

How did this happen? Literally from the day of my arrest, my wife, Ye. M. Shishmareva, kept trying to obtain a review of my case. Encountering a wall of indifference, she turned to the pilot-heroes of the ANT-25 flights from Moscow over the North Pole to the United States. They knew me very well from the preparation of their ANT-25s for the flights and, naturally, didn't believe in any of my "machinations." G. F. Baidukov and S. A. Danilin, as deputies of the All-Union Central Executive Committee, wrote a complimentary statement to the government about me, with a request to review my case. On 6 May 1941, I became the holder of a document stamped with "Butyrka Prison. 5 May 1941. No. 5," and which ended in the words, "You are ordered to free him from custody in connection with the termination of his case." One week later I became a free worker at TsKB-29. I describe certain details of numerous subsequent events on the basis of my comrades' stories.

A sunny Sunday, 22 June, dawned. Under the influence of the TASS announcement, the majority of the free workers had requested time off and had left the city. The huge halls of the TsKB were nearly empty.

Beyond the windows lay a quiet city on its day off; music rose from the park of the Ministry of Higher Education. There were no radios in prison, and the screws didn't talk about what was happening outside with the prisoners. The prisoners were working quietly.

At noon the music stopped abruptly, and people started running to the loudspeakers on the street and in the park. Movement on the streets froze, and passengers jumped off streetcars. Something impossible had befallen the capital. Grasping the bars, the prisoners strained to hear and understand what was going on. The guards were there, too.

It was war! Nazi Germany had invaded!

It is difficult to convey the state of mind that gripped us: this was the end, everything had collapsed and perished. Distraught prisoners didn't know what to do or what lay ahead of them. They went to bed late, certain that the TsKB would send them to the camps. This wasn't herd-like panic; rather, it was ironclad logic. It was impossible to assume that the latest weapons for fighting fascism would be designed and built by the very same state criminals who had recently "sold blueprints" of these weapons not to just anyone but to the Germans!

The prisoners waited with trepidation: what would the next morning bring? Amazingly, absolutely nothing. Actually, that's not exactly the case. In the morning the screws came in, in uniform, with their weapons and gas masks. Later they distributed gas masks to the prisoners. When the free workers arrived, they didn't know anything new, either. In the newspapers (which the free workers brought in openly from that day onward), aside from Molotov's speech, there was absolutely nothing!

Soon the enemy advanced rapidly on Minsk, Daugavpils, and other areas. It very quickly became clear that nothing had come of "First Marshal" Voroshilov's infamous slogans: "We won't give up an inch of our land to anyone!" And "with little blood and a mighty strike, we'll be on the enemy's territory!" As was typical in such cases, this could be fraught with reprisals against the prisoners.

Rations grew worse. Then they placed paper strips over the window panes. When news of the enemy assaults on Smolensk and Kiev became known, the prisoners were forced to dig trenches in the courtyard and redo the old scrap area into a bomb shelter. For those who understood airplanes and bombs just a little, this drill was clearly senseless. Once the blackouts began, they brought black flannel to the plant, and we fabricated blackout curtains.

Pravda and *Izvestiya* were now posted in the *sharaga* corridor. Even without them it was clear that our army was in retreat. Descriptions of individual heroic exploits by soldiers and officers brought no joy, since the overall picture appeared too gloomy. We understood that they had to print them, but why were we being smashed. Why were we giving up oblast' after oblast'?

Rumors began to fly about plant evacuation. In late July and early August, as the front lines approached the capital, it became obvious. We worked on blueprints during the day. In the evenings we changed into overalls and packed up the *sharaga* property into crates, numbered them, and carried them to the assembly building. There, four older prisoners had to drag heavy boxes while a healthy screw walked behind them. If one of the prisoners faltered, a second prisoner would detach himself from the returning group to help out. Three men could go in one direction, while five went in the other direction—and nothing happened. Only recently this would have meant an escort crisis that interfered with aircraft assembly. The administration's answer would have been, "A guard signs for one prisoner and cannot keep track of two." How could you not be amazed by this charade involving "state criminals" and their guards?

The night of 22 July marked the first raid on Moscow. The prisoners were sleeping peacefully in their bedchambers when the sirens began to wail and the alarmed screws burst into the sleeping areas, yelling, "Alert, get dressed, into the bomb shelters!" This raised a real Tower of Babel: People were rushing about in the sleeping areas and corridors—there was stamping of feet, the cries of guards who had lost their charges, complete darkness, and utter confusion. Somehow we gathered in our shelter-crypt, with its ceiling that a pail of water could go through, not to mention a bomb. Tupolev arrived about 15 minutes later, also half-dressed, but with one bare foot. "I couldn't find one of my slippers. Some scoundrel grabbed it in the panic." He was extremely irritated by the thoughtless cramming of the prisoners into the cellar, and by his humiliating appearance. The screw who had led him in was the very embodiment of fear. His prisoner had clearly given it to him good on the stairs and in the courtyard. I must say, the guards as well as the "management" of the *sharaga* had a healthy fear of Tupolev, probably due to the difference in their intellects. They knew he was an exceptional talent and they were nobodies. His treatment of them also had an influence. Once, after he had climbed all over an airplane in the assembly building and then couldn't find his guide, Tupolev—to the delight of everyone, both prisoners and free workers—would shout, "Hey, whichever of you is mine, get over here real quick, or else I'll be inconvenienced looking for you."

Tupolev disappeared at the end of July. We saw him for the last time after his lunch in Kutepov's office. He did not return by dinnertime, and his bed remained empty that night. Again there were worries and talk. The following morning the "free prisoner" (that's how we had christened N. I. Bazenkov, who had been freed with Petlyakov's bunch) said that Tupolev had been freed, that he was home with his family, and evidently wouldn't be at the plant for a few days. Complex feelings overwhelmed the prisoners. Everyone was glad for Tupolev, but the question remained: What will happen to the others?

Evacuation

The situation in our subunit of the *sharaga* at the time of Tupolev's release could be described as follows. Flight testing was complete for the Tu-2, and while the aircraft was outstanding, there were no engines for it. The Tu-2U was smashed and, again, there were no engines for it. We had completed the blueprints for the Tu-2V, almost all parts were ready, the airplane was being assembled, and would be ready in a month or two. Then it would be flight-tested, and the N series-production plant would build it. Tupolev and Bazenkov would oversee its series production together with the free designers. On these grounds, there was reason to consider the other prisoners to be like the Moor who has finished his business and can then leave. The question was, where can the Moor go? Consequently, it was "finita la commedia, noch einmal," as one of the prisoners used to say. This scenario wasn't too stupid, when you consider that we didn't get any new information. It seemed to us that Tupolev was the only one who could put everything in its proper place. But he wasn't there. These were some of the most difficult days for the prisoners.

At lunch the prisoners were ordered to assemble in their sleeping quarters. People sat on their beds silently, with heads lowered and eyes dulled, passive, ready for anything. In came Balashov, Kryuchkov, and three officers from the NKVD. No one looked up. Balashov informed us, "Citizens, the TsKB will be evacuated tonight. Our KB has been allocated three cattle cars. Divide up into groups of 16 to 18 men, collect your personal belongings, and be ready by 11 P.M. Bedding and meals will be provided. Several free workers from the plant administration and the plant management will be traveling in the adjacent passenger car. You will receive all necessary information from them during the movement."

The timid question, "Where are we going?" remained unanswered. The second question, which was tormenting everyone, was, "What about our families?" The answer came back, "The necessary instructions will be given and they will be given assistance." With dull indifference the prisoners dispersed to pack their belongings. Silent, bent-over figures crammed their goods into their bags.

The screws arrived at 2230. The prisoners threw farewell glances at their sleeping quarters where, all the same, they had spent two years, and with their bags and suitcases set out down the main marble staircase into a new phase in their lives. How would it end?

It was dark, with not a light anywhere. The bus engines were idling at the gate, the guards bustled about, there were escorts and some VIPs. The prisoners were lined up and counted. The guards checked the lists by flashlight, checked the lists again, and divided us among the buses. The soldiers, with rifles, took their places, and the cortege set off. The gates opened, and the building of the Design Department of the Section for Experimental Aircraft Construction dissolved in the gloom. Would we return here? If so, as what—free men or prisoners? Would this be our city, or would occupiers be lording it over us?

Then came the warehouses of the Kazan' Railroad freight station, a train, cattle cars with barred windows, shouting soldiers with rifles, from the NKVD troops. Where were we going? To Magadan, to Kolyma, to the Eastern Siberia camps, or to another set of camps? And how would we go—as prominent specialists and builders of awesome war machines, or as scum?

Searchlight beams swept across the night sky. Our guards were talking in half-whispers, and there was a gloomy solemnity, just like before the liturgy in a gigantic church like St. Isaac's or Christ the Savior. The prisoners stood single-file in front of the rail car doors. New escorts, with dogs, came up. They would guard the prisoners inside the rail cars. With a clang the door slid back, and three screws did a headcount three times over. One thing is for certain: socialism means taking inventory! After the recount, they let the prisoners inside. The door clanged again and the lock bracket banged. Silence. No one said a word.

After shunting here and there, the train departed at about 3 A.M. The wheels clicked at the rail joints; the road to nowhere had begun. As it grew light, the prisoners sat wherever they had found room during the loading. . . . In the half-light of early morning one of the prisoners read a station name through the bars: Kurovskaya.

"So where are we going?" he asked.

"Eastward, gentlemen, eastward," answered Frenkel', our navigator.

At about 10 A.M. the train stopped and the door opened. Outside were some soldiers with their dogs. Two others handed a pail of water for washing into the car, a teapot with tea, and a box containing 18 rations of black bread and 36 pieces of sugar. Two of the "prominent" ones, with an escort of a soldier with a dog, dragged the latrine pail under the cars and poured it out. That was it! The door was closed. No information was exchanged. From a small opening accessible from the upper bunks, a train carrying children could be seen next to us. Their frightened faces stared at the animal-like men behind the bars. It was a train from Murmansk. The city had been bombed, and they took the children directly from the kindergartens to the trains, and then to the east.

It was August, and scorching; there wasn't a cloud in the sky. When the train stopped, the car would heat up. The prisoners would remove clothing and grumble. Kalganov, who had been in the Eastern Siberian camp system, told us by way of consolation that he had traveled to the camp in the same sort of car, but with 48 men, mainly common criminals. They were given a pail of water in the morning, one in the evening, and the temperature was + 30°C. He told us how one man had died, and they left him in the car for two days. "Nowhere to put him," explained the escorts. We fell silent for a while, pondering: Maybe the same thing's in store for us?

The train moved slowly. We crossed the Volga on day 2. The fact that we didn't know where they were taking us and what was happening with our families oppressed us. No one talked. We just lay there silently, with eyes closed. It was easier that way.

Another day went by. At one of the stops, the screws brought us two pails with soup and groats. When Balashov walked up to the car, we peppered him with questions: Where were we going? What was happening with our wives and children? The major's physiognomy was impenetrable and dim-witted, as always, and we got no answers. We set out again, and a conversation began: How did they ever manage to train a soldier to such an extent that no emotion ever showed on his face? We did derive information from the fact that the "management" was here: "Since they're traveling with us, we're not a train of prisoners, but the TsKB train, and that means a lot." On the following day a train of flatcars next to us had guarded cargo covered in tarps. We could make out the lines of the Tu-2V and the Project 102 aircraft under the canvas. This meant that the plant was traveling to the same destination. Our mood slowly began to improve. At night, somewhere near Sverdlovsk [Ekaterinburg], the train with the *sharaga*'s free workers ended up next to us. The free workers were also in cattle cars, but with their families, without bars on the windows or an escort. We talked back and forth. They had been ordered to take warm clothing, but they didn't know where we were going either. We learned about the situation at the front only at stations, from the loudspeakers.

After the Urals came the steppes, and the heat became difficult to bear. The prisoner-aerodynamics specialists used materials at hand— paper, cardboard, and rope—to fabricate air intakes, "scavengers," to direct the air flow into the cars. These helped somewhat, but at the next stop the escorts removed them with their bayonets. As to why they did this, our homespun philosopher answered, "Probably because the scavengers could help us to vaporize out of here!"

Frenkel' became ill from the stuffiness. The prisoners banged on the door and demanded a doctor. At the next stop the door opened wide and our doctor from the Butyrka appeared. Per instructions he couldn't go alone into a car with prisoners, so the sentry put a screw—also one of ours—in with him. The locomotive whistled, the guard closed the door, and both of them remained in our car. Frenkel' was on an upper bunk, and as one of the prisoners helped the doctor up, the prisoner managed to whisper to Frenkel', "Find out where we're going!" The medic listened to Frenkel''s heart and lungs and took some medicine out of his bag, while the frightened screw sat below.

After they had crawled out, Frenkel', looking like Mephistopheles— pale, with a black lock of hair plastered to his forehead, and wrapped in a sheet—worked his hand out, raised it toward the roof, and said, "No ovations, please! We're going to Omsk! And that cretin doesn't know anything more—not a thing, you understand."

Okay, so we were going to Omsk. Then they recalled that there was no aircraft plant in Omsk. "The scoundrel lied to us!"

After ten days the train stopped at the Omsk freight station. Toward evening trucks arrived. Apparently they hadn't managed to evacuate any paddywagons here. Customs were simplified. Prisoners were put on the floors of the trucks, four escorts with rifles stood in the corners, and the cortege set off. This was a provincial city, mainly with single-story houses; the windows were lit up—no blackouts. The road was horrendous and the trucks bounced mercilessly, beating up the rear ends of the "prominent" ones properly. After turning among the streets, the trucks pulled into the courtyard of a two-story building similar to a schoolhouse. We had arrived. Around the yard was an ordinary fence. In the corner was a common five-man toilet; next to the building was a washbasin with ten spigots. There were no bars on the windows. The screws counted out groups of ten and led us to our rooms. Each room had ten military cots almost on top of each other. There was a night-stand in the corner, and a dim bulb hanging from the ceiling. There were doors opening into the corridor that ran the length of the building, and a stairway in the center of the corridor. A door to the outside off the stairway, on the second floor, was nailed shut. The only exit was into the yard, and at the gate there was another guardpost.

So here it was, our new prison design bureau! This wasn't Moscow. . . . It turned out that this building would house all those working on the Tu-2V and those from Petlyakov's group who remained prisoners. The others were taken somewhere else. There was no one whom you could ask anything. Balashov and Kryuchkov weren't there, and the screws were silent. They had been turned into such fools that even when you asked, "What is this, Omsk?" they would answer, "I don't know."

The prisoners were hungry and asked for dinner. The senior screw begged our pardon: no dinner had been prepared. He ordered us to go to bed. There was nothing we could do. Through the window we could hear our neighbors playing Grieg's *Peer Gynt,* and the prison design bureau fell asleep to its strains.

In the morning, the filthy, unshaven "prominent ones" began to make themselves at home. The washbasin and toilet were in the courtyard; the screw on duty sat in a chair at the doors to the building. A second screw patrolled next to the gate to the street. At about 10 A.M. they called us into the dining area. It was extremely cramped, but they nevertheless fed us groats, tea, and gray bread. With nothing to do, the "prominent ones" loitered in the yard. From the house next door the provincial eyes of a Kustodiyev-style prima donna stared at us curiously. Irrepressible children would pull themselves up with their hands, and their faces would appear over the fence, first here, then there. The poor screw didn't know whether or not to drive them away. Apparently the Moscow infatuation with vigilant security for state criminals would have to be simplified. We wondered how the "management" would get out of this situation.

You didn't have to be a Kropotkin, Savinkov, or Camus to leave our "Schusselburg Fortress" in Omsk. Just climb over the fence and you were free. . . . We had to wonder whether the Berias, Merkulovs, Kravchenkos, and all the countless others who had gotten fat on our suffering could continue their foolish games even in the year of the Nazi invasion? Would they really "exterminate sedition" as they had in the past, and would they drag into their "troops" tens upon tens or hundreds of thousands of strong, healthy young men who could have been oh-so-useful at the front?

Life in Omsk

They took us to the sauna. Every room of ten men, with their screw, went to the streetcar stop and, mixing with the crowd of inhabitants who stormed the car, burst into it. The screw caused a stir when he bought ten tickets. "Where are your children?" asked an old woman. But when he again shelled out for ten persons at the sauna cashier's, the curiosity-seekers made his entire face turn red. "Who is it you have with you—is it children, juvenile delinquents, or crazy people from the hospital?" asked the people from Omsk. When we removed our clothes, we all became alike; it wasn't easy to tell who was a criminal, who was a guard, or who was a free person in the steam-filled sauna. Blissful from washing and cleanliness, the prisoners returned to the prison design bureau. In the morning Balashov and Kryuchkov arrived and gathered the "prominent" ones in the dining area. Finally they gave us some information, albeit primitive: "You're in Omsk. Here we will work and build a series of Tu-2V aircraft. You'll learn the details tomorrow; for today, you can rest. They'll feed you, since you're here. Going outside the fence is forbidden."

On day 3 our "free prisoner" N. I. Bazenkov arrived and informed us that there was no aircraft plant in Omsk, but we had been given an incomplete motor-vehicle assembly plant and a plant for motor-vehicle and tractor trailers. We had to use them to organize a new aircraft plant and, as if that wasn't bad enough, we had to begin production of airplanes for the front in December, only five months away. Several minor aircraft plants had been evacuated here from various cities, and we had to create a new workforce from their personnel.

The Experimental Design Bureaus of V. M. Myasishchev and D. L. Tomashevich were housed beyond the Irtysh, in Kulomzin. There, using the aircraft repair shops of the Civil Aviation Fleet, another small plant was being organized for experimental aircraft construction. Tupolev was to arrive in several days. Our work would begin the following day, looking for the crates for the Tu-2 in the chaos of unloaded trains and delivering them to the place set aside for us. Where Tupolev's bureau would be located was still unknown. The free workers were being quartered among the local inhabitants. City organizations were handling this, and for now the free workers were spending nights in the open on their belongings.

"Terribly poignant," as they say in such situations. It was already August. Only five months remained until the start of series production for the airplanes, and there were no walls, no roofs, no electricity, and no water. Nothing! The prisoners became depressed, but Bazenkov was upbeat. "Yesterday I saw the first lathes start up in an empty field. They're bringing criminal prisoners and dispossessed kulaks or peasant farmers from everywhere, and are organizing a huge camp for them next to the site of our future plant. They'll build the buildings around the lathes that are already operating." Say what you will, all this was thought up in style! But we also wanted to know about freezing weather. "They say it's from mid-October on, but we'll cross that bridge when we come to it," Bazenkov answered. Finally, there was the most worrisome question: What about our families? Bazenkov replied that he himself and, on his orders, several of our other colleagues had notified their families. Most of the families were traveling there in one of the trains that followed.

One morning several days later they asked Cheremukhin, Sokolov, Sterlin, Nadashkevich, Saprykin, Frenkel', Yeger, Neman, Chizhevskiy, Vigdorchik, Ozerov, Nekrasov, Saukke, Bonin, and Aleksandrov to remain at the prison design bureau. A bus pulled up, they were put in, and were driven into the center of town. There was a huge square with the theater on it. To the left was a newly constructed large gray building, with NKVD guards at the door waiting for the prisoners. We were taken into a reception area on the first floor, where Tomashevich and Sklyanskiy joined our group. They had been brought from the other side of the Irtysh by steamer; they didn't know why we were there, either. The group was invited into an office. How large and magnificently empty it was! Several minutes later, as we huddled together, the door opposite opened, and out came two unfamiliar NKVD generals, with Tupolev, Bazenkov, Kutepov, and Balashov. Something like the dumb scene in Gogol's *The Inspector General* then took place. The prisoners noticed that Tupolev's eyes were filled with joy. The chief general approached the table, picked up a sheet of elegant vellum paper, assumed the pose of a Roman procurator, and read in a heartfelt, almost Shalyapin-esque voice: "On the basis of a report from the Collegium of the NKVD, the Government of the Soviet Union, taking into account the conscientious work on the Tu-2V bomber aircraft by the specialists named below, hereby decrees that the following are to be released from custody: Cheremukhin, Aleksei Mikhaylovich; Myasishchev, Vladimir Mikhaylovich; Markov, Dmitriy Sergeyevich; Neman, Josef Grigor'yevich; Chizhevskiy, Vladimir Antonovich . . ." He listed eighteen men who, on a hot August day in 1941, became free citizens of their country again.

In spring 1942 they freed the next group of prisoners, and by 1945 the shameful institution had ceased to exist.

Nothing in the difficult past could eradicate our belief in the truth to come, and our faith that this terrible injustice would be uncovered and that people's good names would not disappear from the history of Soviet aviation.

It was only later that the rule of aerodynamics stating that the total area of the cross section of an aircraft must change smoothly over its entire length, was justified theoretically and confirmed experimentally. That is, the fuselage, wing, engine nacelles, and tail surfaces of an aircraft must be narrowed at junction points.

NEW CHALLENGES | 3

Now that all of Tupolev's department managers had been freed from prison, attention turned to increasing the number of front-line bombers needed for the military. Two Tu-2 air regiments were already operational, and a third was preparing to go to the front. The plant production rate was one airplane per day. A plan had been worked out in detail, however, for the plant to produce sixty aircraft per month. This plan would bring the Omsk plant to maximum capacity.

More plant space was needed to increase the production of Tu-2s. At the time, two plants in Siberia were producing obsolete Il-4 bombers, which—due to poor armament—could only operate at night. The slow Il-4s were highly vulnerable to the "Messers" [German Messerschmitt fighters] during the day. We, who were patriots on behalf of our own aircraft, thought that turning an Il-4 plant into a Tu-2 plant made sense; apparently, the People's Commissariat did as well. Before long, representatives of one of the plants came to Omsk to study the design of the Tu-2 and the engineering to produce it. Naturally, we and the plant workers provided all possible assistance. After some calculations we were convinced that within six months one plant would be capable of producing ninety Tu-2s per month. Together with our production, that meant five regiments of excellent bombers. What support for our Air

Force! The plant workers producing the Il-4 endorsed the idea as well. They said that pilots arriving at the plant to pick up Il-4s would complain boldly (you couldn't say this very loudly), "We are the ones who survived, coming to pick up the next batch."

Then a dramatic and arbitrary shift in production priorities altered our tenuous sense of security. One evening L. L. Sokolov, the director, brought Tupolev a coded message from Stalin: "Halt production of the Tu-2 bombers, and set up production of the Yak-9 fighter!" This was incomprehensible! No one in aviation—not the Air Force leadership, not the manufacturers—could understand it. The Battle of Stalingrad was behind us, and Soviet offensives had pushed westward. It would appear that bombers were needed to sustain offensive operations. Even if we admitted that Stalin did not understand all the finer points of wartime planning, he wouldn't make such a decision without conferring with his consultants and advisers, would he? Who were these people who supported such an absurd decision? The plant shops were full of talk about some villainous influence or sabotage.

An order is an order, however, and we—literally with tears in our eyes—watched as the aircraft production facilities that had been created under inhuman conditions, on an empty site, were torn down. At a crucial point of the war, they were closing for two months a plant that was producing the best bombers, instead of one producing obsolete ones!

There was nothing left for us in Omsk, and Tupolev's experimental design bureau was sent back to Moscow. We returned to our home building, which we had bade farewell in 1941 during the evacuation of industrial enterprises, far from certain that we would ever see it again.

After we took up our old places, at times we caught ourselves wondering whether all that had ever really taken place—Omsk, the empty landscape piled high with cargo unloaded from rail cars, the construction of the plant, setting up production, and the triumph of sending the first regiments of Tu-2s to the front. We had participated in the epic evacuation of the aircraft industry.

Tupolev, convinced that production of the Tu-2 would resume, ordered our group to produce two modifications of the Tu-2 design. The first was a photo-reconnaissance version, at the time urgently needed at the front. This modification proved to be a simple task. We only needed to mount several aerial cameras and a fuel tank in the bomb bay in place of bombs.

The second modification became a real challenge, one that anticipated the future. The Il-4 bombers were obsolete in the face of German fighters and air defense. Making uncomplicated changes to the original design of the Tu-2—mounting wings with a larger span and additional fuel tanks, putting a second pilot in the cockpit, and providing the navigator with a more comfortable workstation—would produce a higher quality, long-range bomber. Called the Tu-2D, the bomber had slightly less range than the Il-4. However, it was faster and—more important—it possessed awesome defensive armament: three heavy machine guns instead of the single machine gun on the original Il-4. Unfortunately, due to the interruption in Tu-2 production, this redesigned Il-4 never reached the front in large numbers.

These special tasks did not challenge us unduly. There were many other, more unusual assignments. One that comes to mind occurred after the war, in the summer of 1946. Stalin was vacationing in the Caucasus and, they say, he complained to M. Aminskiy, the People's Commissar for Health, who was vacationing with him: "It's mutton, mutton, and more mutton! When I was exiled at Kureyka, they fed us shaved frozen meat. Now that is food fit for a king!"

Soon thereafter we were ordered to fabricate an insulated chest for carrying frozen salmon to Sochi. The only passenger aircraft at that time was the Li-2, our version of the DC-3, with a range of 1,000 kilometers and a speed of 200 km/h. The official procedure began in the far north: A tamper-proof insulated chest containing the salmon was given to a courier in Arkhangel'sk, who brought it to Moscow on an Li-2. The chest was transferred to another Li-2 and within five hours the special cargo reached Sochi. Without a lot of clever debate, we made two containers. The smaller one, containing the fish packed in ice, was placed into the larger container. Between them were layers of ice. But it was not possible to serve the shaved meat without careful inspection, so a committee of doctors, engineers, and Chekists spent several days feeding on the "flying" salmon. A special route was thus entered into the Aeroflot schedule: Arkhangel'sk–Sochi. The shaved frozen meat was then prepared at Stalin's dacha. Naturally, no one cared how much this cost.

This reminds me of a related incident, from my time in the camps. One night they threw a military quartermaster with a magnificent beard into our cell. The next morning they shaved off his beard. The events that brought him there occurred at the Central Red Army Officer's Club, where a graduation ceremony was being held for students of one of the academies. When Stalin stopped by unexpectedly, there was a flurry of excitement, but the buffet had no Tsinandali or Khvanchkary wine—his favorite drinks. The quartermaster supply officer sent a waiter to the nearest store to buy some wine. When the reception was over, they asked the supply officer where he had gotten the wine. When he replied that the waiter had bought it in a local store, they politely requested that he take a ride with them to the Butyrka prison. The quartermaster was sentenced to ten years for a terrorist attempt to poison the leader.

It seems that Stalin ate only food that had been inspected in a special State Security laboratory. Any food item had a seal on its packaging, and a laboratory record was strictly maintained, showing all sorts of studies of the food item.

Other details of life in those years show the way they protected us against so-called errors and endeavored to maintain moral purity. When Stalin pronounced, "Cadres decide everything!" they created the position of Deputy Director for Cadres at all defense plants. Members of the NKVD occupied these positions. A certain Col. S. P. Pil'shchikov arrived at our plant in Omsk. I can't help providing several examples of how he protected us, not only against spies, but also against deviations from "purity" of belief, and even fought for our honesty.

Four RS82 rockets (Katusha) had been mounted on the Tu-2 to increase its firepower. They had to be fired singly, in pairs, or in a salvo. Since there were no ready-made launchers, we built our own, but they required bronze connectors. When we couldn't find any bronze in Omsk, the military representative warned us that he would stop accepting shipments. Pil'shchikov called me in to threaten me: "You are interfering with the work! I'll start a case against you!" Somewhere they found the bronze, and my "case" did not materialize.

About a month later Pil'shchikov was rummaging through my paperwork and found a note to the effect that I didn't know where my father was. Again there were threats of an investigation, ending with a demand that I become a secret informer. Since this demand interfered with my work (as well as irritated me), I went to Tupolev. It worked. Suddenly, the harassment stopped. Tupolev explained with a grin, "I threatened him, warning that I planned to appeal to Moscow tomorrow, and would say that he was throwing my workers off balance. . . ." Since Pil'shchikov was a coward like the rest of them, he informed his boss that there had been some mixup, and that he would figure it out. When we returned to Moscow, Pil'shchikov came with us.

His influence lingered. During the war, we had been half-starved in Omsk. Ozerov proposed, "Let's add something to what we get ordinarily—say, a ruble apiece—and I'll find a woman in Tsagovskiye Doma; we'll have her buy fresh food at the market and bring it back here." The quality of the meals improved markedly. Pil'shchikov cut out these liberties, however. Some uncleared old lady in the director's dining room? A paradise for information leaks! The "old lady" was more than 70, she was barely literate, and hard of hearing.

The story of Dr. L. Timashuk is instructive (later he unmasked the machinations of the Zionist doctors in the Doctors' plot).[1] He presided over an unheard-of wave of anti-Semitism. Timashuk's campaign against Jews reached us. Several of our major specialists were Jews: Frenkel', Sklyanskiy, and Sterlin, to name a few. Frenkel' had died. Sterlin became bogged down with aerodynamic formulas and remained mostly out of sight. That left only Sklyanskiy, and Pil'shchikov led the attack on this innocent person. But Tupolev found a way to outwit Pil'shchikov. He sent Sklyanskiy to one of the provincial plants and encouraged him to blend in there. Sklyanskiy was a highly qualified electrical specialist. At Kazan', as soon as they started assembling the Tu-16, which was packed with electronics, he went there as the manager. Several months later, when "this idiotic"—in Tupolev's words—"campaign is over, we'll bring him back." And we did, thereby saving Sklyanskiy.

Soon afterward, we had to protect our librarian, Irina Baranova, the niece of Petr Ionovich Baranov. Many years after Baranov's death, it was found that he, too, had been involved in some sort of anti-Soviet activity. Pil'shchikov demanded that she be fired. Tupolev became enraged. "I certainly know better than you how Petr Ionovich worked. If only God were willing that you tried to work as hard! Your quest for his Jewish relatives smacks of Hitlerism. Stop it, or I'll mention having you replaced with someone wiser." This had the desired effect. It wasn't that easy to remove from the group people that Tupolev knew, valued, and trusted.

Pil'shchikov, however, was not yet finished. He latched onto the engineer named Grum-Grzhimaylo. "What strange sort of double last names you have," Pil'shchikov remarked. Grum-Grzhimaylo calmly answered, "And what do you think of Ul'yanov-Lenin?" The police colonel backed off.

We still didn't have a heavy workload in our empty design bureau building. Our experimental plant had begun making fuel tanks for other aircraft. New assignments weren't coming in. While our management was considering our options, Stalin announced a plan to revive production of the Tu-2, but at different plant locations.

Due to such haphazard decisions, the best bombers designs were delayed. For the sake of argument, if we assume that only the Omsk plant had produced them, the Army was short about 800 airplanes. How many soldiers' lives would they have saved?

Tupolev became convinced that four-engine, long-range strategic bombers would be necessary in the future, too, and he began to develop such an airplane, under the code number 64. A. A. Mikulin's aeropropulsion design bureau was already conducting static testing on the new AM-42 engine with special superchargers that made high-altitude flight possible. The potential of such a bomber interested the Air Force, and we were ordered to produce a rough design and build a mockup. Unfortunately, in following the slogan, "Everything for the front, everything for victory!" the short-sighted managers forbade all design bureaus and Scientific Research Institutes to become distracted with developing promising experimental instruments, assemblies, and equipment. All efforts went into supplying the front, that is, into helping the plants that were producing equipment for airplanes in series production.

At the same time, our pilots who had flown the American B-25 Mitchells and A-20 Bostons received under Lend-Lease were convinced that the combat effectiveness of airplanes was incomparably greater with good equipment. How could we carry out high-altitude bombing at night, especially when the targets were cloud-covered? How could we free pilots from difficult physical work of flying a huge airplane? How could we find a point target in fog and clouds and appear at the target? How could we perform a landing under adverse weather conditions after returning dog-tired from a long flight? It was clear that the necessary airplane could not be designed without a new set of instruments to carry out many crew functions by automated equipment and computer. However, these did not yet exist in our country. After we decided what new equipment was required, we turned to our related industries for assistance. They said that it was impossible—for ideological reasons.

Toward the end of the war, rumors from the West reached us concerning a new branch of science that could greatly simplify the complex nature of aircraft design. The creator and head of this new science was the famous American mathematician Norbert Wiener. He called it "cybernetics," from the Greek, *kybernetike:* the art of steering, or controlling. Wiener defined the tasks of cybernetics as follows: the science of general laws on the acquisition, storage, transmission, and processing of information. Wiener elaborated the new cybernetic systems, which he considered to be the formal study of methods of control and communications found in living organisms and machines. His interest included automatic control systems, computers, the human brain, and mathematical programming, among other things.

When those responsible for Soviet official ideology heard about this new area of science, there was panic. What was going on? Marxism-Leninism appeared threatened. This bourgeois cosmopolitan, a class enemy and ideological saboteur, was going after our holiest of holies! If the computer invented by Wiener was capable of replacing the human brain and seemed outside the Marxist-Leninist worldview, then couldn't the computer, controlled by humans, arrive at completely absurd conclusions, up to and including the conclusion that it was possible to live without Marxist philosophy, without dialectical materialism, without the course on the history of the party, and, finally, without the leader's declared wisdom? Total nonsense!

I don't know who the commander-in-chief of ideology was at that time, but this general—clearly a character out of a Saltykov-Shchedrin novel—reported the new heresy to the proper authorities. The order came back from the very top: "Cybernetics is a pseudo-science, and thus it is to be closed and forgotten, now and forever!"

At this point it is appropriate to note Tupolev's attitude toward the party's role in shaping industrial growth. He acknowledged the party's demonstrated skill in organizing the masses, restructuring the people's consciousness and awakening their initiative. The party, in fact, had played a major role in carrying out truly grandiose alterations of our society, the campaign to transform a backward, agrarian country into a leading industrial state. Tupolev resisted the party's interference in science and the solution of complex technical problems. He believed that the party offered little benefit, and only interference.

He loved to repeat, "M. Lomonosov, D. Mendeleyev, A. Popov, I. Pavlov, N. Lobachevskiy, V. Vernadskiy, N. Zhukovskiy, N. Berdyayev, P. Florenskiy, P. Kapitsa, L. Landau, A. Sakharov, and N. Vavilov were all nonparty members, but how high they elevated Russian science! Fortunately, the interference by S. Trapeznikov, P. Pospelov, and I. Mints and their mighty organization is limited to praising T. Lysenko and his agronomic tricks; to rewarding the microbiologist O. Lepeshinskaya, who found spontaneous generation of the simplest single-celled organisms in her apartment bathtub; and the discoveries of Dr. L. Timashuk, who established that all the medical luminaries in our country were ignoramuses and, moreover, criminal plotters." Even in these situations Tupolev's sense of humor never abandoned him. After finishing this long tirade, he would laugh, "You know, it is really amazing. All the leaders of our group are non-Party members, and they haven't disgraced the Motherland!"

When TsKB-29 was in Omsk, Vladimir Myasishchev put the crew into two pressurized cockpits in his DVB-102 long-range, high-altitude bomber. Naturally, the defensive guns remained outside the cockpits. They separated the gunner, with his sight, from the cannon or machine gun: he had to control them remotely. Targeting remote-controlled weapons is complex because the gunner with his sight is separated from the gun or cannon. Aiming properly is a challenge. Engineers S. M. Meyerson, T. M. Bashta, N. S. Naumov, and K. Ye. Polishchuk undertook to solve this problem. At the time they knew nothing of Norbert Wiener and his cybernetics, but they independently reached the same conclusions as their colleague on the other side of the Atlantic. In exactly the same way as he—not suspecting that they were involved in a "pseudo-science"—they were firmly convinced that they were doing something worthwhile in creating improved weapons for their country.

Cybernetics and Automatic Navigation Systems

Several years passed. With the design of large strategic bombers, the number of gun turrets and gunners continued to increase. A gunner needed to be able to fire from any turret, or even from several at once. Both flight engineers and any of the pilots had to be able to control engines with tens of thousands of horsepower. All types of corrections in the control system in the form of automation and computers now appeared in aircraft systems. The traditional autopilot was joined by navigation coordinate calculators, inertial and astronomic correctors, and so on. This equipment was based on an understanding of cybernetics, that is, on the most basic fundamentals of the "pseudo-science."

Cybernetics, however, remained unacceptable to the state authorities. A certain metaphysical Gordian knot had been tied, and certain fundamental decisions had to be made as to what could and could not be used to outfit the latest bombers. Tupolev decided to speak with the "key people," as he referred to the managers of organizations involved in defense issues. He began with Admiral-Engineer A. I. Berg, a major specialist in radar, which already had been introduced into aviation, the Navy, and air defense forces. In addition to everything else, the admiral belonged to the party; therefore, he understood the official ideology.

Their meeting was held at the air base in Zhukovskiy. The site was probably chosen intentionally, because our nonparty-member boss was afraid of listening devices, and didn't want to put the admiral in an awkward position. In the rooms in his rest area at the air base, where there were no telephones or wires except those for the ceiling light, there would scarcely have been time to install microphones.

Tupolev wanted to know if it was possible, using the electronics that existed in our country, to design and build a "cybernetic brain center" for the new No. 64 strategic bomber. This center would develop automatic navigation systems, which would allow our bombers to hit enemy targets even under the most "intolerable" weather conditions. The two most unlikely conspirators agreed that this was absolutely necessary, regardless of the consequences that such an action could have for both of them.

To avoid attracting the attention of the ideological overseers, who were always on the lookout for those who deviated from their one-and-only all-encompassing philosophy, it was necessary to give the cybernetic brain center an innocuous name that would keep those naive simpletons in the dark. Instead of a cybernetic brain center, they would be developing a "technical route calculator."

The second issue was more complicated. New types of specialists—software developers for onboard digital computers—would be needed, but our higher educational institutions, of course, offered no training in these critical areas. If we created new departments at the Moscow Higher Technical School or the Moscow Aviation Institute, all the social-activist teachers would ask why and for whom was all this?

This sort of sensitive situation required great caution. It was worth remembering that those who had dared to violate the "purity of the faith" had always been dealt with cruelly by the police.

Would our overall boss, First Deputy Minister Petr Vasil'yevich Dement'yev, an old Communist and member of the Central Committee, agree with the project we had conceived? After meeting with Berg, Tupolev went to see Dement'yev. Tupolev told us later, "Dement'yev heard me out with unusual attention, thought for awhile, then said, "If we shut down today—as has been ordered—all work on computer technology for all future aircraft, in a year or two these very same high-placed persons, forgetting about their orders today, will give the order to get rid of all these loafers [i.e., us] who have left the country without any combat aircraft. . . .

"So what do we do? This is what: only one special design bureau—Lanerdin's—is working in this area. Who do you have running this area? Kerber, isn't it? Let him go to Leningrad and tell Lanerdin that Dement'yev has declared aircraft 64 super-secret and of special importance. To prevent even the slightest leak on this topic, we'll set up special procedures for the order. Let the two of them put together the tactical-technical assignment for this airplane. You each can bring in two or three of your most capable and reliable men. You and I will approve it together, and we won't permit any outside coordination." At this point, Dement'yev was even amused: "Or else, you know, they will take it around for approval, and there won't be any place to sign it! They are always packed with signatures for coordination approval: several chief committees, all kinds of scientific research institutes, standards specialists, and God only knows who else. Is it any wonder that dozens of people know about these things right away? And when that is the case, you can put whatever classification markings you want on the documents, but somebody will still mess things up. Also, we will have less correspondence of all sorts on this topic. I think that Berg, you, and I can manage with conversations face to face. If it is extremely necessary, we can use telephones, but talking so that others won't understand.

"And don't worry, the one responsible for the 64 will be—who? It'll be you, as the Chief Designer, and I. So, you go ahead and set things up for yourself so that only two or three men—no more—are aware of all the details. And at the People's Commissariat, in addition to me, the only other person who knows will be Deputy People's Commissar A. I. Kuznetsov. If I'm not there, you can see him." And so we found a second influential ally who didn't tremble in the face of potential trouble. Our hopes rose for a favorable outcome of the matter.

With the help of V. I. Lanerdin in Leningrad, they determined that the "technical route calculator" would actually be a central onboard analog computer. For now, however, it would handle only control of the airplane's motion in the horizontal plane. We did not have the capability to work in the vertical plane, too. Even without this, our task (given our capabilities at that time) was daring. We agreed that we would plan for control in the vertical plane, but put it into effect later.

In the summer of 1946 Lanerdin invited Tupolev to inspect an operating model of the "technical route calculator." Lanerdin's special design bureau was housed in a former private residence on Skorokhodov Street. In one of the rooms in an apartment on the second floor of this old apartment, on a large dinner table, they showed us something that looked like a huge bookshelf laid on its side, crammed with numerous blinking radio tubes and ventilated by ten or so fans from various sides.

On instructions from the First Department of the design bureau, the windows in the room had been sealed so well that when the calculator was switched on the room temperature quickly reached near-tropical readings. Within two or three minutes, Tupolev begged them, "This is unbearable! Open the windows!" He said to the First Department representative who was about to protest, "All this was done for my airplane! I have to see better, to see whether there is anything secret here."

After looking over the apparatus, he glanced up at Lanerdin. "There is nothing of the sort here. Everything we see here is the distant past for the West." Lanerdin winced. Tupolev, paying no attention—or pretending to—said, "I'm interested in the reliability of your monster. How many hours for one failure? What has to be done to have reliable automation onboard our aircraft?"

The young engineer couldn't contain himself. Almost hysterically, he laid out for Tupolev all the troubles that had besieged his group. "Get rid of these dogmatic fools who have dared to give an entire realm of new knowledge the epithet 'pseudo-science.' Give us freedom of action. Organize the production of transistors and microcomponents. Train the programmers, and—I swear to you—we will do what you need within a year."

Lanerdin tried in vain to interrupt and calm him down, but the engineer didn't stop until he had told Tupolev everything that was overflowing the souls of all the young engineers.

"Reliability?" One of them repeated. "Why, there is no reliability at all!"

"I don't understand."

"The tubes fail unexpectedly! One won't make it through an hour, another one will hold on for several hours. There are at least a hundred here."

"Aha! And all this is in a room. So what will happen on an airplane, with its vibration and pressure and temperature changes?"

"Exactly!"

"So who supplies the tubes?"

Lanerdin pointed outside the window. "At the end of our street is Nevka, and on the other bank is the Svetlana Plant."

So we went to the plant, where we were met by the manager for experimental programs, a Corresponding Member of the Academy of Sciences, S. A. Vekshinskiy.

"Yes," he said, "our vacuum tubes are defective, but what would you have us do? We blow the tubes, and smelt the alloys for the anodes and filaments, and draw the filaments ourselves—finer than a human hair. Do you want to have a look?"

In a tiny foundry, next to a cupola furnace, in dirt and grime, two craftsmen in cloth coats were clip-clopping around in wooden shoes and sneezing.

"They are sneezing because of the drafts," explained Vekshinskiy. "They produce the alloy from the furnace. They keep the doors open in the fierce heat, and there you have it! A cold! They try to blow their noses right into the crucible, the scoundrels. If they hit it, poof! You lose 99.9 percent purity! So there is your reliability! I have pounded it into their heads, but they don't listen. I have given them handkerchiefs, but they turn around and give them as presents to their wives."

"And this is the St. Petersburg working class, our pride and joy?"

"You are mistaken. Their fathers and grandfathers were our pride and joy. They preserved their craftsmanship and increased it, and valued their positions at the plant. The ones today are drifters. The one arrived yesterday, and today he is getting his dismissal papers. I don't know if the fault lies in a lack of pride in one's work or in these awful conditions."

Vekshinskiy fell silent. Evidently he was afraid of attracting the attention of the wrong person, and continued only in the yard on the way to the passageway. "I can assure you, the tubes that we are developing currently are no worse than foreign ones in terms of their design. But I would be ashamed to deliver them to you. They are growing obsolete while the sluggish production end masters series production. Then the poor conditions in the shops, plus the workers' indifference, take their toll. What percentage do you think we scrap under these primitive conditions? As much as 60 to 70 percent."

A distressed Tupolev returned to Moscow. What he had witnessed in Leningrad proved far from what was necessary for automatic control of a modern bomber. Equipment of an immeasurably higher caliber was required. Quality control would require hundreds of qualified specialists, laboratories, a shop with air conditioning—in short, standards at a much higher level than what existed at that time.

The most tragic aspect, however, was the fact that it was impossible to change the situation merely with the minds and hands of enthusiastic supporters. We understood that our basic science—cybernetics—could no longer remain in the shadows as a "pseudo-science." Someone had to say this to the very highest leaders—perhaps even Stalin himself! But who would take up this cross?

Despondent, Tupolev decided to meet with Berg again. Meeting at the air base again suggested a certain cowardice. A more bold move was required. "Should I visit him at the institute myself or, perhaps, invite him to my dacha?" Tupolev asked.

As it turned out, the energetic Berg was also trying to solve this extraordinarily absurd problem. His assistant, an engineer named G. A. Uger, visited us along with two Czechoslovakian engineers, Staros and Berg (with the same surname as the admiral), who were lifelong Communists and had fled when Hitler occupied Czechoslovakia. In Russia, they continued to study semiconductors.

Recently, the American physicist W. Shockley had discovered semiconductor properties in thin coupled layers made from a group of rare metals. These "sandwiches" were called transistors and had the properties of several types of cathode tubes. Our visitors had achieved consistent transistor functioning in several radio circuits, and were now looking for an influential person capable of overcoming the inertia of our state bureaucrats. In their opinion, Tupolev was that person.

After seeing the miniature parts that would replace the unreliable and capricious electron tubes, Tupolev gave an immediate assessment of their work. "Magnificent," he declared on the telephone to Admiral Berg. "How can I help?"

The admiral recommended that Tupolev see A. I. Shokin, the Deputy Minister of the Radio Industry, who was responsible for industrial applications of transistors. Tupolev visited Shokin the next morning.

"Well, we have found another smart one in the government," he said with satisfaction upon returning. "They have at least some clever persons who work without fearing to be branded heretics. Our numbers are growing: There is hope that we can get over this obstacle."

Soon thereafter a decree was issued on the creation of an Academic Council on Cybernetics under the Presidium of the Academy of Sciences, and none other than Admiral-Engineer A. I. Berg was appointed its chairman.

Lanerdin's group never succeeded in achieving consistent operation of the "technical route calculator." However, his group made an enormous contribution in producing the aviation industry's first programmers and first circuit designers, who later created the system of external links necessary for operation of the onboard computer for our strategic bomber.

Meanwhile, we had lost at least three years fighting with the ideological dogmatists who had suppressed cybernetics and branded it a "pseudo-science"!

The Tu-4

The immediate post–World War II environment was difficult and unpredictable. Stalin interpreted the lack of new aircraft models to replace the "old-timers" that had grown obsolete during the four war years as criminal lack of foresight on the part of Air Marshal A. A. Novikov, the commander-in-chief of the Air Force, and A. I. Shakhurin, People's Commissar of the Aviation Industry. Both were removed from their posts and repressed. "Guilty parties," as always, were uncovered, but naturally this didn't help matters.

At the time of the war in the Pacific between Japan and the United States, there were instances—albeit rare—when the latest U.S. Superfortress strategic bombers were faced with a dilemma: drowning in the

Pacific Ocean, or landing at our airfields in Vladivostok. At President Roosevelt's request, Stalin granted them permission to do so. Since our country was not yet at war with Japan, according to international rules the crews of those airplanes were interned, and we provided security for the bombers. Thus, three nearly intact B-29s ended up on our territory, with the best functioning electronic equipment in the world. At the time, the B-29 Superfortress was the most advanced strategic bomber in the world.

With the Cold War taking shape, Stalin thought it made sense to reproduce the B-29 and all its equipment in order to re-equip our military aviation with airplanes that met current requirements for on-board equipment.

Confidential discussions were held with Tupolev, Il'yushin, and Myasishchev. Il'yushin turned down the assignment immediately, saying, "I have never worked on such a huge aircraft." Myasishchev and Tupolev agreed to it, and the lot fell to Tupolev. The decree from the party and government, signed by Stalin, obligated Tupolev to reproduce the B-29 promptly and in painstaking detail. The same decree ordered all people's commissars, ministries, scientific research institutes, experimental design bureaus, plants, and enterprises of the USSR, without exception, to reproduce upon our demand everything required for this high-priority project: materials, instruments, various components large and small, and so on. The only change permitted—and it was even directly prescribed—was replacing the American English system of measurement with the metric system.

This unusual decree ended with the following two paragraphs: Within a year, Tupolev was to complete all documentation necessary to build and operate the new bomber and, within a second year, the plant director V. A. Akulov was to have built twenty aircraft.

This decision forced enormous burdens upon us; nevertheless, it was evidently the only correct one at that time. Yet it possessed certain tragic dimensions. If the customary "But can we do this?" was considered, Tupolev was handed a working model and told, "Please make one just like this!" A simple goal, but an extraordinarily difficult task.

Our pilots in the Pacific, after changing the B-29's white stars to red ones, ferried the American airplanes from Vladivostok to Izmaylovskiy

Airfield in Moscow. First, we set out to get acquainted with the "marvels from overseas." Tupolev examined up close everything he could; he would have gone into the access duct between the pressurized compartments, but he turned out to be a bit chubby and got stuck. "Hmmm," he said, "that is for scraggly Americans; maybe they don't get enough to eat under capitalism." His first impressions: the bomber was truly outstanding.

Then he noted, "I have not found anything hitherto unknown to us in the aircraft's construction." But he immediately criticized the curved windshields in the pilots' cockpit: "Our pilots would get rid of them, since they distort the view of everything ahead of them too much. I would install flat angled windshields, like the ones on all our airplanes, but we are not permitted to change anything." And, in fact, all our pilots did curse the curved windshields.

Tupolev's initial assessment was that it would not be particularly difficult to build the airplane. "But what will we do about the armament and equipment? To tell the truth, Stalin didn't comprehend the complexity of the task before us. The gunner sits almost in the tail itself, but fires from the forward machine guns. Such a remote-control system must possess great reliability! Then again, what about all the radar sights, range-finders, computers, and automated plotters? Could our specialists reproduce them accurately? Let's assume they do, but look: They are connected by hundreds, even thousands"—here the Old Man began to get irritated—"millions of wires. How Kerber and Nadashkevich will figure these out, how they'll understand where even one of these"—here he tugged on a wire—"comes from and where it goes, I have no idea! What if the damned thing shorts out somewhere? Then what?"

He spit, became extremely agitated, and pointed to me and Nadashkevich. "Why are you keeping quiet? Answer me. How will you solve this puzzle?"

What could we answer? With all his shouting, he was right.

"Okay, let's go back to the special design bureau and get down to working on this Tower of Babel."

Not immediately, but gradually, our methodology for this unusual work of emulation began to take shape. To tell the truth, it was rather elegant. Tupolev appointed our chief production engineer, S. A. Vigdorchik, to be chief of repairs. We had to find space for carrying out the disassembly itself. At that time, in all of Moscow only one hangar could hold the B-29. The building had belonged to the Central Aero-Hydrodynamics Institute, and had been erected for Tupolev; however, while Tupolev was in prison they transferred the hangar as a production facility to another designer, V. N. Chelomey. Tupolev, who was by nature something of a property-owning kulak, couldn't stand it when they gave "his property" to someone, and so, taking advantage of the government decree, he returned the hangar to his design bureau.

Tupolev named his deputy, D. S. Markov, as lead designer for the Soviet version of the B-29 bomber. Before prison Markov had worked at Aviation Plant No. 1 as the chief designer for startup production of the American Vultee airplane, under a license obtained by the USSR, and had done a good job putting it into production.

Stalin ordered the first B-29 to be disassembled completely, so blueprints could be made for all parts. Various components of the B-29 were forwarded to specialized scientific research institutes, design bureaus, and plants for study and copying. The second B-29 served as a test aircraft, to emulate the flying characteristics of the B-29. The third would be kept untouched, to be used as a standard or template for our project.

The bomber was disassembled into structurally independent assemblies, each of which was assigned a lead designer with a team of engineers, technicians, and rigging specialists. Once removed, an assembly was placed on a specially prepared stand covered in felt; it was measured and photographed. Next, instrumentation and controls were removed, after first recording where hydraulic and pneumatic lines and wires led from each instrument, the numbers of these items, their cross-section and length, and markings. We photographed each component, which was then hung up, numbered, and entered in special registries. Every evening buses carried the designers and disassembled components to our design bureau. When a component or section of the aircraft was removed, it was taken to a plant, where it was disassembled down to the last rivet.

When we were working on the forward pressurized cockpit, we came across a crumpled plaque fastened next to the bombardier's seat, on the cabin wall: "At the request of the workers of the Boeing plant in Wichita, Kansas, this B-29 is named in honor of the visit to the plant by the Commander-in-Chief of the U.S. Air Force: *General H. H. Arnold.*"

There was an uproar in the party committee. Did this have something to do with the apparent solidarity of the imperialists with the workers? Such a sense of common purpose ran counter to our ideology. It was decided to destroy it and throw it away. Nevertheless, I preserved it. Clearly, the class struggle didn't interfere with their patriotism!

There was another complication: How would we know to which plant or which ministry we should send a given unit, instrument, electrical mechanism, radio, sight, piece of sheet metal, plastic, glass, fabric, rubber, a given wire, hydraulic line, or anything else we found? Was it our business to decide these issues, and were we capable of making such choices?

An anxious Tupolev went to A. I. Mikoyan, the Deputy Chairman of the Council of Ministers, to ask if they could appoint someone within each ministry involved with the effort to replicate the B-29, soon to be designated the Tu-4. Someone should be personally responsible for processing and delivering the new equipment. Tupolev said to Mikoyan, "It seems to me that without that we designers will drown in telephone calls, passes, and in waiting to be received by all sorts of 'management!'" Mikoyan agreed, and these authorities were appointed; most of them were deputy ministers or individuals in charge of key committees.

In my opinion, this was almost the first instance in our country of creating an informal, interministry body, which made it possible to introduce genuine direct links among all the participants in a period of absolute power wielded by the administrative system. Somewhat later, Korolyov successfully used Tupolev's experience to form his own famous Council of Chief Designers. Each chief designer was subordinate to his own ministry, but the council was subordinate to Korolyov and no one else (except, perhaps, the Council of Ministers and the Central Committee), and its decisions became mandatory for all industry. I am certain that this was the only reason that our missile specialists succeeded in those years. Without the council, no talents could have helped Korolyov, and the "management" would have tied him up in "valuable instructions."

I want to emphasize that Tupolev had an amazing ability to find within any multilayered system—it didn't matter what kind: a complex machine, an aircraft design, or joint efforts by several industrial enterprises—the very element capable of bringing down the whole structure, idea, or corporation. After finding it, Tupolev exposed it for all to see, and, not infrequently, to laugh at. It was subjected to the necessary measures and ceased to interfere with the work.

It was interesting that this association—which was unapproved anywhere administratively—existed on the principle of close "horizontal relationships" (as Tupolev called them) between our design bureau and industry in other ministries. This did not dissipate with the end of our work on the Tu-4, but remained our viable working situation. Whenever any problems occurred with the delivery of certain component instruments and assemblies for our new airplanes, we used "horizontal relationships" quickly to find the person who was responsible for the decision, reached this person, bypassing telephones, secretaries, and advisers, and effectively cut through the knots of contradictions that would have tied us up. That is how things were until the end of Tupolev's life.

The authorities from the ministries, together with their colleagues, began to visit our design bureau regularly, and quickly decided what to send, and where and to whom to send it. This new, very effective body of specialists was unusually successful in helping the dispatch of materials. These transmittals occurred when the representative from the plant received an assembly in person, its outline drawing, photographs, a weighing document, a diagram of wires or lines that connected the assembly to the airplane, and a short description. We determined the assembly code jointly; at this point, deadlines were set for delivering the first unit to us, and subsequent units to the series-production plant.

The number of units, instruments, and models sent off numbered in the thousands. To keep them from getting mixed up, Tupolev organized a special dispatching office under the supervision of the energetic engineer, I. M. Sklyanskiy. (He, too was a released prisoner who did time with us at the *sharaga,* while his brother E. M. Sklyanskiy was at one time the Deputy Chairman of Trotsky's Revolutionary Military Council. I don't know whether this troubled Tupolev at all, but it was Josef Markovich he appointed to a position of responsibility.) Once he had estimated the volume of documents that lay ahead of him, Sklyanskiy compiled a network timetable that filled an entire wall in one of the rooms at our facility. This was probably the first such timetable in the Soviet Union. Four cards were made out for everything that was sent to the cooperating elements: one for us, one each for our ministry and the cooperating ministry, and a fourth card for the manufacturer. With this type of inventory system, during all the years of this epic, not a single part was lost, and not a single instrument disappeared.

We gradually began to receive samples of materials, instruments, and assemblies for the future Tu-4. They arrived slowly and unevenly, and it was clear that industry was having difficulty restructuring. Tupolev decided to urge them forward. He arranged an exhibit showing who was supposed to deliver what, and when, and how it was being carried out. The exhibit attracted managers from our suppliers and even ministers. Word reached even Stalin, and he decided to visit.

Tupolev called in Markov, Minkner, and me, and ordered us to be prepared to provide explanations; he would inform us of the day later on, but we knew that it would be after midnight. The day turned out to be Sunday, when the plant was closed. After noon, our building was occupied by young men in civilian clothes, but with a military demeanor and ever-watchful eyes. After searching the building thoroughly, they locked all the doors, replaced plant security with their own men, and set up many extra guardposts.

They asked us, the "tour guides," into Arkhangel'skiy's office and gave us instructions: Comrade Stalin should be informed in brief, precise terms understandable by a nonspecialist. He did not like it when the briefer did not take his eyes off him, and reacted irritably if a speaker appeared suddenly from behind his back. We were not to keep our hands in our pockets, or leave the office until we were needed. Two majors would be present to escort us if we wanted to go to the toilet; if we wanted tea and sandwiches, they would bring them. Such procedures preceded any visit from Stalin.

From the office window we could see how some more young people, with the same military demeanor, were hanging around on the deserted nighttime streets, pretending to be normal people. The traffic light at the intersection had been turned off, and traffic was being directed by three traffic cops in blindingly white cuffed gloves and batons.

Time dragged slowly, we were bored and humiliated, always under lock and key. Whenever you have to kill time senselessly, you can't help remembering the past. Some of the radio equipment that I had to test spent time on the airplanes of the personal detachment of Ya. I. Alksnis. It was commanded by Brigade Commander Shakht, and one of the pilots was M. Grachev, a parachute jump instructor. In 1932, I had carried out a parachute jump from an airplane he was piloting, and received badge number 22137, and was proud of it.

Chance brought Grachev and me together over many years during vacations at Kislovodsk. One of his stories regarding Stalin is not without interest. It was well known that Stalin did not like to travel by air. It was much more reliable to travel by special train, where both the cars and crew had been checked out numerous times. He almost decided to travel even to the Teheran Conference by rail. However, when it turned out that the cars and personnel would be Iranian due to the narrower gauge, he reconsidered. Immediately, two Lend-Lease C-47s were ferried to Baku; their pilots were Air Marshal Ye. Golovanov and Lieutenant Colonel M. Grachev.

After greeting both of them, Stalin said, "So, who will we fly with? Perhaps it is better to go with the lieutenant colonel. Marshals do more sitting behind a desk than behind the controls. It will be safer that way."

After he had arrived in Teheran and was disembarking, Grachev shook Stalin's hand. Stalin then remarked simply, "Thank you, Colonel." Grachev reported that his next promotion came quickly out of turn.

A regal gesture, wouldn't you say?

About 2 A.M., Tupolev came in and informed us that the visit would not take place. The polite majors took us home in black Volgas.

As the final operation in all our work, the new Tu-4 was weighed "dry," that is, minus fuel, oil, and so on. The B-29 weighed 34,930 kilograms; the first one we produced weighed 35,270 kilograms. A difference of less than 1 percent: an excellent result.

"I'd be interested to know," Tupolev remarked once during those days, "how the Americans would assess our work. Although we'd be flattered to learn their opinion, I think they'd rate it highly if they knew what kind of hard-labor work it was!"

True, although the work was called copying, it was not easy, and we had numerous failures. For example, the military and industry demanded that all weapons be domestically produced. However, the decree had been prepared hurriedly, since they knew that Stalin did not tolerate delays, and so specifications for the intended armament were not attached to the documents.

When they looked into matters further, it turned out that the weapons for the Tu-4—bombs and machine guns—were dragging a rather long train behind them, such as locks for bomb racks, electric bomb releases, electric motors for traversing turrets, ammo boxes, and sleeves for ammo belts. Even different sights were required, since the ballistics of American bombs and our bombs and rounds were different.

Then there were arguments and demagoguery: "Well, what are we doing? Copying the Boeing or not?" For each trivial component, any participant could demand a piece of paper freeing him from responsibility.

Matters were coming to a dead end. It then became necessary to produce a new document, with a larger number of signatures, to record our work on the project. All this was decided at a glacial pace, with terrible worrying. Moreover, and most important, it slowed the production of blueprints. We were faced with the threat of breaking deadlines that had been set by Stalin himself.

And at this point a new disappointment appeared, this time involving the radio recognition system. This is also called the "identification, friend or foe" (IFF) system, abbreviated "S-Ch" (*Ty svoi ili chuzhoi*). One would have thought that it was obvious that the one used in our army should be installed. However, this wasn't specified in the text of the basic decree, and, with unexpected rapidity, our radio industry produced for us a device copied from the American one. This was installed in the first Tu-4 bombers, and no small amount of confusion resulted. Imagine our fighter operating with our own IFF device, encountering a Tu-4, which used a device copied from the Americans! The pilot of a Yak or a MiG would ask, "Are you a friend?" and, getting no response, would open fire.

Such a consequence could have been foreseen. But fear got the upper hand and even overcame common sense. Only one reality governed, the notion—based on fear—that those at the top can see things better!

American parachutes also presented problems for us. They are worn on the back, and serve as a soft back for the seat, while our packed parachute is a pillow on which the pilot sits. Therefore, the seat under our pilot is a square cup. As a consequence, we faced another dilemma with emulation. Did we have to work up new parachutes and special seats for the Tu-4? Everyone agreed that this was stupid, but they demanded papers with signatures, in other words, official permission to be smart.

From time immemorial, contacts in our plugs have been marked with numbers, while the Americans' plugs have letters. At first, out of fear, they repeated the American markings, but our installers had not been trained in the Latin alphabet. They confused things, and, as a result, valuable assemblies were ruined. What could we do? Since the alphabets weren't the same, one might think this would mean using numbers instead, but they were afraid to do that. They used the Russian alphabet. At this point, there was terminal confusion. Several assemblies burned up, and only then did they start using numbers. As a result, at one time the plants involved in producing the Tu-4 had three types of markings, and we had to develop special translation tables.

Nobody wanted to go to Kolyma, and everyone knew that, without a doubt, some scum would turn up and denounce us: "They are violating the instructions of Comrade Stalin himself." The specter of the camps haunted us.

In another instance, we took one of the skin plates off the B-29 and measured its thickness as 1/16 of an inch; when recalculated, the result in metric is 1.5875 millimeters. Not a single plant wanted to roll plates with that accuracy. Here's something for engineering psychologists to ponder: Why is it impossible to roll 1.5875mm sheets, but possible to do the same thing when it's called 1/16 inch? No matter how much we asked the metallurgists to be reasonable, the victory was theirs: We had to round off the thickness. However, if we rounded it upward, to our standard 1.75 millimeters, the airplane would become heavier, and its speed, altitude, and range would decrease. If we rounded down, to 1.5 millimeters, the strength specialists started to make a fuss, with good reason: The structure would be weakened intolerably. What a dilemma!

It would seem that these incidents were completely trivial and stupid, but each one slowed our work. The American B-29 came with stacks of manuals needing translation. We didn't have enough English specialists, so we brought in translators from outside. They knew the language well, but turned out to be unfamiliar with aviation terminology and technical terminology in general. When we sat down to edit their translation, we discovered that the Americans who wrote the instructions intended to communicate to their readers in a concise and visual manner, whereas our official language specialists proved to be quite verbose. There were some unusual and baffling items. In the American manual, for example, we read, "Start the putt-putt." What is a "putt-putt"?! We rummaged through all our dictionaries and encyclopedias but could not find the answer. It was only in Kazan', as they began to prepare the first Tu-4 for flight tests, that everything became clear. They started the emergency engine, and it began to throb, "putt-putt-putt-putt . . ." The manual contained a simple imitation of the sound of the engine exhaust. But we couldn't allow ourselves to do something so simple, and the brief "putt-putt" turned into "For this purpose, the above-mentioned unit, consisting of a small two-cylinder, two-stroke, air-cooled gasoline engine, should be started, setting in motion the four-pole DC generator with compound circuit, which serves to supply the airplane's electrical system when the main engines are inoperative."

During a little less than three years of rather close contact with many ministries and organizations because of the Tu-4, we recognized that the vast majority of their workers considered themselves beyond reproach. Bound by the decree to make "a copy," they preferred not to think about the problems that arose designing a precise copy, problems that were, at times, crucial for the success of the undertaking. They were severely limited by the premise, "The bosses know best." A second group, less numerous, understood everything very well, but their official zeal, generated mainly by fear, overcame their judgment and common sense. Even when they encountered a totally absurd situation, such as the IFF system, if it deviated from the prescribed "copying," they immediately informed the higher-ups. And that set things in motion! Inspectors, controllers, commissions, and meetings followed.

Fortunately, there was one more group of people—unfortunately, the smallest group—who were deeply convinced that even an order from Stalin should be carried out with wisdom. Some of these bold, wise people deserve recognition. At least they should not be forgotten. For example, the B-29 was equipped with high-frequency command radios that were already obsolete. The B-25s that were transferred to us under Lend-Lease had new, ultra-high-frequency (UHF) radios, with automatic tuning. We managed to obtain one such radio from the military. I told Zubovich, the Deputy Minister of the Radio Industry, about this, and proposed that we not copy the outdated SCR-274, but copy instead the new SCR-522 from the B-25. One way or another, I told him, we would be working with them soon, since the entire world was switching to UHF command communications.

Zubovich listened to me in a thoughtful manner. He answered, "I know my own close colleagues, and am not afraid of them; however, without a doubt there will be some son-of-a-bitch on the periphery who will write to the higher-ups that we are not carrying out the leader's precise instructions, and we will all do time, you can be sure. No, at this point we have to enlist the support of someone prominent and honest in the military, who can explain things to Stalin if we get called in."

And so we decided to talk to S. A. Danilin. He was the Chief of the Scientific Research Institute for Special Services, where technical specifications for outfitting aircraft were set. We had no doubts that Danilin would support us and would, if necessary, try to convince Stalin. In addition, it would be difficult to drag someone like him into court: He was a hero of the record ANT-25 flight from Moscow to America and enjoyed great popularity. (I am amazed at what simpletons we continued to be at that time, despite all that we had lived through; however, I am writing about things as they were.) Danilin agreed with us, and our "conspiratorial operation" to replace the radios went off without negative consequences.

The parachute story had a similar ending. When the absurdity of using American parachutes on the Tu-4 became clear to almost everyone, Tupolev sent me to General "X," the Chief of the Parachute Service of the Soviet Air Force. I told him everything. "Please," he replied, "tell Andrei Nikolayevich: naturally, use only ours! And tell him that if any difficulties arise, I will take full responsibility." Then, glancing at me, he smiled. "Do you need some paper? Where should I sign?"

Equally bold people turned up among specialists on engines and armament, but Minkner and Nadashkevich were involved with that equipment for us, and I do not know details in these areas. I can only relate one incident that showed me, convincingly and clearly, just what sort of bravery was required of the eminent figures who agreed to support us.

M. S. Gotsipridze, chief of one of the Scientific Research Institutes, and I were riding along Bol'shaya Dorogomilovskaya Street in Moscow. On the approach to the old, narrow Borodino Bridge, we heard behind us the typical twin-tone horn of a government vehicle, and we were passed by a ZIS-110 with guards. We quickly recognized Stalin in the black limousine.

At that instant, a truck carrying logs was climbing from the embankment road onto the bridge. Suddenly, the truck's engine died, and it stopped, blocking our way. A streetcar coming from the opposite direction stopped, too, and the bridge was now blocked at both ends. Stalin's ZIS-110 ended up right next to our vehicle. On a collapsible seat, behind General Vlasik, as Gotsipridze told me in a whisper, sat Stalin himself in a gray coat and a military service cap. We were separated only by the cars' windshields. I saw a very tired, old man, with pockmarks on his cheeks, with a greenish-tinged, smoke-stained mustache, and penetrating, evil eyes.

Somewhat anxiously we greeted the leader of the security detail and, touching our hands to our hats, bowed slightly. The look that settled on us seemed worried, at least to me: Stalin probably had no occasion to see unfamiliar people at such close range. However, he too bowed, and slowly raised his hand to his visor. At this point the road was cleared, the ZIS dashed ahead, and the apparition vanished.

This occurred in fall 1947. At the time I was struck most of all by Stalin's emaciated face, which was not at all like the carefully groomed, youngish facial image that looked down on us from thousands of portraits.

When I later recalled Stalin's look, which had flashed such mistrust, I thought: How must Zubovich, Danilin, and the parachute general feel under this gaze? For nothing, neither merit nor laws, could shield them from Stalin's fits of distrust.

Aircraft assembly for the Tu-4 soon began, and we moved to Kazan' to participate in this climactic phase of our project. The huge Kazan' plant existed for one thing only: producing Tu-4s on schedule. The pilots entrusted with test flights arrived from Moscow: N. S. Rybko, Mark L. Gallai, and A. G. Vasil'chenko. The day of the first flight had arrived. No matter how hard the security personnel tried, and no matter how many false rumors they spread about the date, they were unable to hide it. On that fateful day the shops were empty, and all roofs at the plant, all roads leading to it, and all clearings at the edges of the airfield were packed with crowds of people. When Rybko lifted the first Tu-4 into the air, thousands of people yelled, "Hurrah!" until they were hoarse. Everyone understood the enormity of our triumph.

After conducting several flights, Rybko ferried the airplane to Moscow for comprehensive tests. Soon thereafter, Gallai set out for Moscow on aircraft No. 2. Tupolev permitted me to fly with Gallai. Tupolev loved it when someone from his inner circle participated in such test flights, and he wanted to hear opinions on the airplane not just from pilots, but from his colleagues, too.

I have to admit, the flight didn't add any laurels to my biography. We took off normally, but the compartment soon became rather warm. The crew turned to the Deputy Chief Designer—me—to request that I turn down the air conditioner, which was forcing hot air into the cockpit. Alas, I was unable to handle this mission and, as we approached Moscow, the entire crew and all the passengers began to resemble tourists emerging from the Sochi beaches. The ceremonial reception was not held because we all had to change our clothes.

The long, tedious period of flight tests had begun. Day after day, in good weather, but more frequently in bad weather, twenty Tu-4s criss-crossed the sky. They flew day and night, to the north, west, and to Central Asia, at the highest and lowest altitudes, simulating failures of engines, control systems, radio communications, and navigation systems—in short, everything that could fail. In this way the pilots worked to give the combat units truly reliable planes suitable for combat.

Finally, the commission in charge of the flight tests was given a voluminous document containing all the trials and tribulations of the testing on a scale hitherto unseen. However, in vain did we think that everything would soon end with the necessary collection of signatures. As the document moved higher and higher within organizations, the debates grew more and more serious.

For example, one of the professors suddenly posed the question, "Well, does the aircraft, as constructed, meet our—that is, Soviet—strength norms rather than American norms? That is, it won't come apart in the air, will it?" At this point, the worm of doubt gnawed its way into certain other members of the commission. Tupolev tried to convince them. In Southeast Asia, where American B-29s saw combat for two years, the atmospheric conditions are much more severe than in Europe; however, the bombers had not been destroyed. He tried to prove that if there were some difference between American and our norms, then during the very detailed study of the structure, and during the transfer from the English system to the metric system, our requirements were taken into account. All of this was in vain, however. There was always a "doubter."

Finally, at one of the sessions, after exhausting all his polemical abilities, Tupolev picked up a Kremlin telephone and dialed a number. Everyone fell silent: What was this capricious person doing now? Tupolev, in his high voice, said, "This is Tupolev reporting, Comrade Stalin. Several people here think the strength of the Tu-4 is insufficient." After listening to the answer, he put down the handset and informed them, "Comrade Stalin doesn't share your opinion, and recommends that the completion of the document not be delayed."

Naturally, the document was signed rather quickly!

As Deputy Minister of the Aviation Industry, Dement'yev was a member of the commission. Many years later, as he sat with us at an air base whose administrative building he had nicknamed "Peterhof" (after the curiosities of Tupolev's artist Kondorskiy, who had painted the building with all sorts of unusual designs), Dement'yev told us how the document on the testing of the Tu-4 had been approved and how the name "Tu-4" had been created. Not particularly enchanted with the copying of another country's airplane, Tupolev had ordered that "B-4" be entered on its blueprints, to mean "four-engine bomber."

Here is Dement'yev's story:

One evening, Minister Khrunichev and I were summoned, with the document, to see Stalin at his dacha. We were met by the Chief of Security, General Vlasik. He led us to a terrace, asked us to wait, and asked the car to leave. About one half-hour went by; then, suddenly, the door opened silently and in walked Stalin, who nodded to us. He sat down at a table, without saying a word, and buried himself in the thick document. Then, filling his moustache with pipe smoke, he said, "It's exactly one year late!" And disappeared, taking the document with him.

We waited for quite a while—in a state of alarm, as you can understand. Finally, Vlasik came out. "Comrade Stalin has signed the document, you can return to Moscow."

Khrunichev was about to reach for the telephone to summon the car, but Vlasik stopped him. "The car is waiting for you."

And, in fact, a ZIS-110 was standing at the entrance, but it was not ours. Next to the driver was an armed officer with a package sealed in wax.

We set out for Moscow. We passed the Arbat, turned toward Hunter's Row, and began to go up to Lubyanka Square. Khrunichev and I did not say a word. Our reception had been so cool that we were just trying to guess who this officer was: some sort of aide or an escort? The car passed the square, turned onto Bol'shaya Lubyanka, but even there there were lots of NKVD buildings! Only after we had passed Sretenka Street, after the turn into Dayev Lane, toward the Ministry, did we breathe more freely.

At the entrance the officer silently handed the package to Mikhail Vasil'yevich. It was 3 A.M. We went up to the "minister's" floor, and locked ourselves in the office. Tired from the severe encounter, from the fact that Stalin had let us go, slighting us emphatically, and out of worried uncertainty as to what he had written in the document, both of us fell lifelessly into our chairs. Taking a deep breath, we opened the package and read, "Approved. President of the Council of Ministers and Minister of Defense of the USSR." And, in blue pencil, the date, the signature "I. Stalin," and the correction: "B-4" to "Tu-4."

Khrunichev stood up, wiped his brow, took some vodka out of his safe, and poured it into some glasses: "No, Petr Vasil'yevich, we won't last long this way!"

We drained the glasses without eating any snacks; it had turned out to be nothing, but our nerves were so tense that we weren't affected by the vodka.

Yes, it wasn't easy for our ministers under Stalin!

Television had appeared during those years. Complaints were coming in fast and furious to the higher-ups because at a distance of only 100 kilometers from the capital, you could barely see the Moscow TV center broadcasts. The signal troops proposed mounting a relay on the Tu-4, with a solid ten-meter antenna that would lower in flight. After receiving signals from the Shabolovo tower, the relay would amplify them and, from an altitude of 10,000 meters, would ensure long-range reception.

This proposal ended up at V. M. Molotov's commission, since the Ministry of Communications and our Ministry of the Aviation Industry could not agree. They convened a meeting. Since Tupolev was not in Moscow, A. A. Arkhangel'skiy and I attended. We met in the former Ryabushinskiy Bank building, opposite the former stock exchange on Il'inka. The session began with "Why is the Ministry of the Aviation Industry rejecting such a tempting, honorable task?"

Arkhangel'skiy proposed that they listen to a specialist, and Molotov nodded. I informed them that the relay required a lot of electric power, which an aircraft did not have. In addition, if the antenna did not retract before the landing, both airplane and crew would perish.

Molotov looked at me with irritation (our big bosses did not like to encounter technical difficulties), and said, "Get the power. As for a disaster, first of all, arrange things so that the antenna retracts. Second, make sure a reserve airplane is ready at all times." Then he dismissed us and never summoned us again. The idea faded.

It was interesting to meet the number 2 man in our government. At the time it seemed that he had an iron will and that doubts of any sort would never occur to him. He would have forced us to implement a foolish project, no matter what the consequences. If an accident occurred, then we would get the blame. I wondered how he felt in 1942 as he flew on another Tupolev airplane, the ANT-42 (TB-7, or Pe-8), over wartime Europe, hurrying to a meeting with Churchill and Roosevelt? Did he worry? After all, this airplane, too, was designed by "saboteurs." Or, right from the start, didn't he believe in the "machinations of the class enemies" who were unmasked with such talent by Vyshinskiy, Ul'rich, and K⁰ Company?[2]

The Tu-4 was linked with another page in our history. As everyone now knows, our first atomic bomb exploded in December 1955, after being dropped from this airplane. A. A. Arkhangel'skiy and A. V. Nadashkevich went to the test range from our design bureau. Our communications with them went over a high-quality telephone. On the day set for the drop, I was called to the telephone for a call from the test range. Arkhangel'skiy said in a worried voice, "There's a problem in the aircraft's electrical system. A safety fuse burnt out and the drop has been postponed." Beria, who had arrived at the test range, was furious. Arkhangel'skiy shouted over the phone, "Report immediately on what could happen and how to fix the problem!"

I had not left for my office yet when N. I. Bazenkov called (he was acting General Designer): "Get in here, and make it quick!"

I ran. In Bazenkov's office were three men in civilian clothes, but we could spot State Security men a mile away. One of them opened a folder lying in front of him, then asked, "Why did a 20-ampere safety fuse burn out on a Tu-4 aircraft?"

With emphatic calmness, I answered, "Several hundred Tu-4 aircraft have been produced, in several variants. They have thousands of fuses on them. Tell me the serial number of the airplane and where the fuse is located; we will find the diagram and try to answer."

Without moving a muscle in his face, this character repeated his question, word for word. I had to go up to my office, check the diagram, and figure out what was wrong, so, feigning indignation, I turned to Bazenkov. "If they hide something from me, there's nothing I can do here. Until I know the serial number, I can't answer the question." And I headed for the door. One of the three silently stood up and followed me. I went into the elevator—my State Security "screw" went, too. We arrived at my office, where they had already prepared the diagram and my electricians had gathered. The screw was at the doorway. His mission was to keep me from fleeing.

After figuring things out, we went downstairs, past the high-frequency cubbyhole, but they called me from there: "It's the test range for you again!" This time it was Lt. Col.-Engineer S. M. Kulikov calling, on behalf of General Finogenov, who was overseeing the tests for the Air Force. Kulikov informed me, "They figured it out: a 50-amp fuse for the bomb heater had been installed not directly into the main power circuit, but in series with a 20-amp fuse in the airplane's power circuit. When they turned on the heater, the 20-amp fuse blew. If you agree, send a high-frequency telegram. We will replace everything and fix it, and hook things up properly, but over your and Bazenkov's signatures."

Two hours later they called again and everything was fine. During those hours I tried to separate myself from the screw in order to call home to say that I could be delayed for some time, but I did not succeed. The screw unglued himself from me only after the test range informed us that everything had been checked out and that the experiment was set for the following day. When our guests had gone, Bazenkov and I recalled those not-too-distant years, and the *sharaga* within those very same walls, and we wondered what abyss our lives were teetering on.

Tupolev as Leader

In preparing the Tu-4 for the atomic experiment, we had to figure out how the airplane would be affected by the shock wave from the blast, and how the materials used to fabricate the airplane would behave in the thermal radiation field. Our strength experts solved the first problem quickly: calculations showed that the aircraft was strong enough, and no reinforcement was necessary.

The thermal radiation problem was more complicated. No one in Russia had worked on such studies, and there was no one from whom we could borrow results. We decided to conduct experiments. Tupolev asked the air defense troops for a powerful searchlight. We set it at one end of a corridor, and at the other end we placed samples of materials to be tested. Once we focused the searchlight beam on the test objects and turned it on, in an instant there would be a puff of smoke from the illuminated surface. These tests convinced us that the all dark areas on the belly of the airplane, all radar housings, and all antenna mounts needed to be painted with white nitropaint. In addition, we had to mount protective, fan-shaped metal shutters in front of all the crew members' faces.

To our great sorrow, soon after the atomic bomb test we lost Nadashkevich. In order to test the effectiveness of the new weapon, at the test range they built several types of buildings, sections of railroad lines, highways and subway lines, and set up trucks, tanks, planes, guns, rail cars, locomotives, and even small ships.

Nadashkevich, our "arms bearer," could not wait to assess the effectiveness of atomic weapons first-hand, and at the first opportunity set out in a Land-Lease Willys for the site of the burst, although it was protected by special fencing. It is not known whether he remained there longer than was permissible, or got out of the vehicle and went somewhere he shouldn't have, but within a year he came down with leukemia, and died two years later.

During our work on the Tu-4, numerous problems appeared in many different areas of science. We had to seek assistance from several Academy of Sciences institutes. Many academicians began to spend time with us. These meetings gave us insight into the relations among them, about which many stories were told.

Across from our design bureau was academician I. P. Bardin's Institute of Metallurgy. While it had no direct relation to aviation, there was a link through the steel used to make our engines. Our engines were heavier than foreign ones, in Tupolev's opinion because our steels were less heat-resistant and weaker. Our engine designers had to increase the thickness of cylinder walls, compressor blades, and so on, for reliability and durability. Tupolev also associated these shortcomings with Bardin's negligence. Meetings between the two on these topics grew more frequent, and they were fostered by one incident.

Bardin built new buildings energetically. He erected one where our polyclinic had previously stood in a rundown shack. Fearing that it would collapse, Bardin's construction supervisor ordered it hauled away. Tupolev then had to allocate space at his plant for the polyclinic. He would have had it moved, but God forbid he should accept the loss of his territory! Like a feudal prince, he considered it his forever. As they were completing one of the new buildings, Tupolev expressed to Bardin his desire that they have a common polyclinic, and that it would be wonderful if Bardin set aside one floor for it.

Bardin rejected this idea rather firmly. A conflict began, and V. V. Grishin came to make peace between the academicians. Upon seeing the new building from Tupolev's office, Grishin asked, "What is that they are building?"

"What do you mean, 'What'? It is a laboratory building."

"Doesn't Bardin know that industrial construction has been forbidden in Moscow? We will make you redesign it as housing."

"Nothing will come of it, Viktor Vasil'yevich: You would get a mangy, foul café out of a building like that."

The situation heated up when Tupolev proposed, "Let Bardin allocate some space for our joint polyclinic. Then there won't be any violation at all. The building will fall under social services, and I will use the land that is freed up to begin erecting the computer center that was assigned by the last decree of the Central Committee and the Council of Ministers."

Grishin liked this idea, and the Moscow Committee of the Communist Party of the Soviet Union issued a decree. Those present decided that now the academicians would be quarreling forever!

Several years went by. After the 20th Party Congress in 1956 they began to take down statues of the late General Secretary, Josef Stalin.[3] One of them stood in a niche in the facade of Bardin's institute on Koroviy Brod Street. One night the building manager threw a rope around its neck and—horror of horrors!—the plaster didn't hold, and the head was torn off. What was he to do? The institute didn't have any carpenter-model makers, or rigging specialists, or trailers. A worried Bardin called Tupolev late at night. Tupolev ordered his dispatcher to get the model makers and rigging specialists together. They draped the decapitated figure in a plywood cylinder, and the next night they lowered it onto a trailer and hauled it outside the city. The incredible disgrace that would have been obvious in the morning, when crowds of students and workers at our bureau and plant were hurrying to work, was avoided.

The institute of academician A. G. Iosifyan, which occupied the private residence of von Mekk near the Red Gate, designed the remote control for the Tu-4 weapons. When some sort of glitch occurred, Tupolev went to see Iosifyan. The ceiling in the director's office was decorated with very frivolous scenes. After looking around, Tupolev stated, "I understand why you have so many delays, Andronik Gevondovich: You probably spend more time contemplating the ceiling than delving into our system!" Everyone burst out laughing. Iosfyan, who had expected a somber conversation, was relieved, and the work-related discussion went smoothly.

A. P. Aleksandrov, a physicist, arrived because their atomic bomb was being mounted in the Tu-4. Difficulties had arisen. After exchanging greetings, Tupolev laughed, "What are you doing getting so shaggy, Anatoliy Petrovich?" I should add that Aleksandrov was bald, but would cover his pate inventively with what was left of his hair. Everyone has his own foibles. "In my opinion, you have hairs No. 6 and No. 11 positioned incorrectly." Everyone burst out laughing, and again the tension disappeared.

I have included these examples to show that Tupolev was a subtle psychologist and diplomat, and that stories that appear about his coarseness in such situations are fabrications. Moreover, these qualities were extremely helpful in calmly resolving the conflicts that inevitably arise in design work.

However, one aspect of his nature misled him at times—his mistakes in sizing up a person from a first impression. Corresponding Member of the Academy of Sciences A. A. Pistol'kors once visited us regarding the question of aerodynamic drag with the ever-increasing number of antennas placed on the airplane. As he walked into the office he caught on something, tripped, was embarrassed, and was delayed. Dressed in a modest suit, in old-fashioned glasses in a thin metal frame, he did not look like an academician at all. Tupolev did not restrain himself. Without standing up, he barked, "So come in, already, what did you get tangled up in out there?" This completely embarrassed Pistol'kors. Interceding on his behalf, I introduced Pistol'kors and said, "Aleksandr Aleksandrovich has been developing several antennas for us; they create no aerodynamic resistance." Tupolev was embarrassed and immediately changed from a prosecutor with an aggressive attitude into a gracious host. Peace reigned, and all issues were resolved quickly.

In another meeting, on a subject far-removed, an academician came to see us. He didn't look like a prince of science, either. Tupolev greeted him severely. "So you are concerned that radioactive phosphorescent materials used to coat dials and indicators on aircraft instruments have a pernicious effect on the potency of flight crews? I will have to disillusion you. My test pilots are reporting the greatest number of concerns precisely in this area. I am getting many complaints due to their overactive endeavors in this regard."

It would be possible to provide many examples of Tupolev's interaction and conversations with his Academy colleagues, all of which indicate one thing: there was no proper deference in their formal relations. They were the typical exchange of opinions on issues related to solving difficult problems that arise during the creation of new aircraft, and nothing more.

One of Tupolev's curious characteristics was his inner discipline, a relentless commitment to his work. Catching sight of a building under construction at the German Market, he decided, "We should stop by to see whether it would be suitable for our computer center, with full-scale test stands. We could have it turned over to us in a decree by the Central Committee and the Council of Ministers, then, poof! It burns down. True, it has an absurd look, but we could add on something and everything would be fine!" This was said in a convincing tone, as if the issue concerned some booth or newspaper stand.

On another occasion, as we drove past the horizontal gas tanks of a former gas works, he said, "You know, it would be good to build a supersonic wind tunnel here. You could pump lots of air into those tanks, then open the valve into the tunnel, and the air would rush in there at a tremendous velocity." After a short silence, he added, "So you say the experiment would be too short? Just look at how much empty space there is around here. There is an empty stadium—we will fill it with tanks, and we will put our compressor building on the banks of the Yauza."

These weren't abstract fantasies—they were more like dreams about how everything around him could be adapted to his favorite cause.

We used the Tu-4 as the basis for a seventy-seat passenger aircraft. As always, when he created an airliner Tupolev transferred everything possible from the most recent variant of the military airplane (in this case, the Tu-4) to the new airplane, without any modifications: the wing, engine mounts, landing gear, tail, and instrumentation. The only thing new was a pressurized fuselage. This was the first pressurized cabin of that size in our country, and, perhaps, in foreign experience, too. The fuselage was a cylinder about 30 meters long and 2.7 meters in diameter, with numerous lights, doors, and hatches, and it had to maintain passenger comfort up to an altitude of 10,000 meters. We solved this problem, and in fall 1948 F. F. Opadchiy lifted the Tu-70 into the air.

Tupolev and the leaders whom he had invited from the Ministry of the Aviation Industry and Aeroflot participated in one of its demonstration flights. Everything went splendidly, but gradually the temperature in the cabin began to rise, as had happened in the Tu-4 when I had been there. The guests asked for some cool air. Tupolev called over M. N. Petrov, who was responsible for pressurization and air conditioning for us, and ordered him to turn down the heat. When he returned from the cockpit, Petrov reported that the instrumentation failed.

Tupolev became furious. "I will order Opadchiy to stop the airplane right now and put you off!"

Naturally, the air conditioning was fixed subsequently. The Tu-70 was ahead of its time. Aeroflot had many Li-2s and Il-14s, which could handle air transport at that time. But our Tu-70 would carry all sorts of delegations and commissions, at least until Gen. Vaso Stalin began to use it to transport his soccer team, "VVS" [Air Force]. Until 1953 the Tu-70 carried this team the length and breadth of the Soviet Union, from match to match. Then Tupolev returned the airplane to his OKB, where it lived out its service life.

The Tu-85 Bomber

The Tu-4 proved to be an outstanding bomber, but its range did not suit the Air Force. At first they thought of attempting to solve the problem with "minimum fuss," that is, by making substantial modifications to the existing airplane. The results turned out to be more than modest: the Tu-80's range increased by 13 to 15 percent; however, it was necessary to double the range. It was clear that a new bomber had to be developed.

A new configuration of powers had occurred in the world. The United States had become our new probable enemy, and it was 6,500 kilometers away. Accordingly, the new bomber had to be capable of reaching America, destroying the major targets, and then returning home; in other words, it had to have a range of 13,000 kilometers. After discussions at appropriate levels, it was decided to entrust Tupolev, V. A. Dobrynin, and A. D. Shvetsov with the new project. Dobrynin and Shvetsov were assigned the task of designing the engines.

Analysis showed that the new bomber's weight would be more than 100 tons, and its wingspan 56 meters. One could not help remembering how they had reproached Tupolev for designing the doors on the assembly area at his experimental plant to be 60 meters wide, while they mockingly asked him what planes he intended to roll out through them.

The engine designers were faced with an equally complex problem. Getting such a giant off the ground required engines with a takeoff power of 40,000 horsepower. To fly 13,000 kilometers, they had to have low fuel consumption while cruising to assure optimal range.

Designing the aircraft and engines also required solving endless problems that, at times, seemed contradictory. We should give the two designers their due because they managed to work out these problems. They utilized such innovations as skin panels up to eight meters long for variable-cross-section wings. In accordance with notions of weight savings, Tupolev ordered them to be shifted in the direction of the outer wing section. Our metallurgist, I. L. Golovin, and our production engineer, S. A. Vigdorchik, spent many days correcting the panels' stubborn tendency to buckle when cooled.

New types of alloys for the landing gear and engines were introduced, causing some problems. The engine designers also had to work toward unheard-of—as they put it, ironically—"locomotive" power.

How much the introduction of liquid oxygen cost us as well. When Tupolev saw a cluster of 64 pressurized steel tanks in the mockup, he literally exploded. "What kind of monstrosity is that?! Thirty years ago, when I was in high school, our physics teacher told us about liquid oxygen and Dewar flasks for storing it. Please order some and use it!"

"But no one in the world uses liquid oxygen in aviation."

"That is outstanding. Won't you be flattered to be the first?"

When this issue became more involved, Tupolev stepped in, calling academician Petr L. Kapitsa. For some reason Tupolev considered him the general designer for compressing all gases. Kapitsa answered, "I haven't worked with oxygen in a long time, but I will try to help you. There's a certain Chief Directorate for the Oxygen Industry; its chief is Gamov, a progressive man. I think he will help." And, in fact, Gamov did help. Soon we were visited by two engineers from the All-Union Scientific Research and Design Institute for Chemical Machinebuilding, K. S. Butkevich and Yu. Miroslavskaya.

The parts for the KPZh-30 oxygen device were manufactured rather quickly, but an unforeseen obstacle appeared. Both halves of the KPZh had to be tinned, inside and out, but the institute didn't have a tin-plater.

Tupolev called Gamov again. "You can't imagine what kind of stupid things we are running into. You know, I remember quite well how the gypsies used to go around the courtyards yelling, 'Who needs tinning or soldering?'"

"I have one like that," replied Gamov, "at the plant in Karacharovo. But is it wise to involve him in defense work, and how will the organs view this?"

"Well, I will take responsibility for that. Would it be possible to send him to me?"

I was ordered to meet the gypsy. He was a real Aleko in wide, velveteen trousers, wearing a vest with a chain with charms, and with an earring in his ear. When I took him to Tupolev, those in the reception area were amazed: What else had our irrepressible Tupolev thought up, if they are bringing gypsies in to see him?

Our meeting was short and successful. At first "Aleko" refused, but when Tupolev promised to take his whole family for an airplane ride, he agreed.

The KPZh was assembled, tested by the military, and put into series production. However, everything new had to undergo severe real-life testing. The ubiquitous quests for insidious enemies—prompted by the leader's enigmatic statement, "The closer to socialism, the more aggressive are the class contradictions"—had not yet died down. They didn't overlook the KPZh, either.

We were preparing the 85 for its next flight. That evening, lead engineer N. V. Lashkevich put his seal on the tarped airplane and went off to sleep. Late at night the guard heard suspicious noises inside the airplane. The vigilant watchmen raised the alarm: the enemy had gotten into the airplane and was working his black deeds! When the sleepy Lashkevich opened the airplane, it turned out that the release valves on the KPZh were bleeding off excess pressure.

Later, when one of the provincial plants began to produce the Tu-16, the oxygen had to be transported from Moscow. The first tanker arrived half-empty, the second did likewise. There was immediately talk of malicious intent to disrupt the production of combat planes! The seeds of suspicion fell on fertile soil. They began to look for guilty parties. They threw the guard escorting the tanker into prison. No one thought about the fact that it was summer, the tanker had no thermal insulation, and the driver drove it for several days. What was the oxygen (which evaporates actively even at a temperature of $-183°C$) to do but evaporate? It never occurred to them to seek the answer not in stupid politics, but in sober engineering calculations.

Finally, everything necessary was ready, and they began to couple rail cars with liquid-oxygen tanks to high-speed trains.

Meanwhile, we prepared our Tu-85 for its most serious examination: the range flight test. The route had been selected: Moscow to Kamchatka, the release of cement-filled bombs there, and a return flight to our air base. All of us whose assistance would be necessary in an emergency situation—who would give advice to the crew via radio—had been at the base since evening. Tupolev and Dobrynin arrived the next morning, at dawn.

The crew took their positions. The engines roared to life, and the overloaded airplane dissolved against the background of the sky growing light in the east.

Everyone gathered in Tupolev's office. On his desk was a huge map of the country, and on it was a tiny model of the Tu-85. Clocks were ticking evenly. From time to time Ye. K. Stoman would go to the radio for a report. He moved the model further and further toward the east. How were our comrades onboard doing?

At midnight they reported that the bombs had been dropped and they were now plotting the return course. Tensions eased around the office, but our superstitious host was vigilant: "There are still 12 hours of flight time ahead. Half a day!"

We wanted to sleep, but no one closed his eyes. The devoted Polya, Tupolev's office assistant, brought coffee to everyone. It was already growing light when we received the report: "We have passed the dam at Kostroma; touchdown at 0832."

Donning our raincoats, we set out for the runway. Those with the best vision looked east: "There they are!" The descending 85, its engines rumbling, turned and came in for a landing. The unheard of flight of 24 hours' duration and 13,000 km in space was over.

Our friends were hugged and kissed as they deplaned. Tupolev gave the order, "Everyone into my office, quickly, and it will be time for a rest. After all, you have flown a whole 24 hours." The young smiling navigator, Kostya Malkhasyan, corrected him: "Not 24 hours, Andrei Nikolayevich: 23 hours, 42 minutes, and 19 seconds!"

In his office Tupolev laughed out loud and, like a fussy host, said, "Polya! Polya! Now where has she gotten to? Better bring everybody a shot of cognac and a cup of coffee. If there aren't enough shot glasses, pour it into regular glasses, mugs, or even saucers if necessary."

Laughing and joyful, everyone drank out of whatever he had. Only an embarrassed Vladimir Alekseyevich Dobrynin said to our host, "I don't drink alcohol, Andrei Nikolayevich." However, caught up in the general enthusiasm, he, too, moistened his lips.

Yet another important page in the history of our aviation had been turned. The tired crew went home. Tupolev used the HF radio to call the Minister and L. P. Smirnov, chairman of the Military-Industrial Commission of the Central Committee, and reported the success.

After resting, we again took up our usual, joyful, and favorite work, which contains so much that is interesting and eternally new.

The Tu-14 and Tu-16 Badger

Even before the war, in 1941 our fellow countryman B. S. Stechkin was working on the theory of jet engines. Unlike piston engines, they could operate at high altitudes without losing power, had even simpler designs, and had a great deal of power. The design teams of A. M. Lyul'ka and V. Ya. Klimov set about to develop them. The former succeeded in creating industrial models at the Putilov Plant in Leningrad. When the Germans drew close to the city, the models were disassembled and buried in the ground.

After 1945 the work was begun again. Tupolev knew about them, foresaw their potential, and chose his own method of getting acquainted with them. After redesigning the Tu-2 somewhat, he mounted on it two license-produced Rolls-Royce Nene engines that had been acquired by Klimov. The modified aircraft, designated the Tu-12, enabled our group quickly to become familiar with the operating characteristics of jet bombers. Single-engine fighters had already been studied by the Lavochkin, Mikoyan, and Yakovlev OKBs.

After acquiring this experience, we began to design the Tu-14 combat aircraft ordered by the Air Force. We lost the competition with S. V. Il'yushin's special design bureau, which created what subsequently became known as the famous Il-28. The Il'yushin airplane went into large-scale series production, while the Tu-14 was built as small-scale production. Because it had a larger bomb bay and better navigation equipment, it was used by sailors as a torpedo airplane, and was better suited for flights over seas and oceans.

After the straight wing Tu-14 had been created, we needed to flight-test the aerodynamics of a twin-engine swept-wing bomber. Our greatest concerns related to the junction points for the engine nacelles and wing, and the large cutouts in the wing, at the same points, for the retracted landing gear. These Soviet designs were unique; no test data existed for similar designs in the West. Academician V. Struminskiy was a great help in solving this difficult problem. Aircraft No. 82, piloted by A. D. Perelet, finished the entire sequence of flight tests. There were no surprises with our swept-wing configuration. I clearly recall Tupolev saying to his closest assistants, shortly before his sixtieth birthday, "Only now do I feel that we are ready to work on a heavy, long-range, strategic jet bomber."

For Tupolev's 60th birthday, in 1948, the chief designer for aircraft engines, A. A. Mikulin, presented as a gift a model of a heavy jet aircraft. Plexiglas gas jets streamed out its engines, serving as a base for the model. It was as if Mikulin were saying, "Take my engine, Andrei Nikolayevich, and you will have a fantastic aircraft!"

Work was already underway on a jet bomber to replace the piston-engined Tu-4. However, the existing Klimov VK-1 jet engine produced a thrust of only about 3 tons, whereas Mikulin's AM-3 had 7.5 tons of thrust—an unheard-of figure at that time. Naturally, this changed the entire configuration of the aircraft substantially.

The dimensions of the AM-3 engines were also unprecedented, and we had to figure out how to fit two engines into the airplane. If we suspended them under the wing, they would have reached the ground. As alternatives, we considered mounting them on the sides of the fuselage, at the tail, as the French later did with their Caravelle and Il'yushin did with the Il-62, but at that time there was still a lack of theoretical studies for such a bold step.

Irritated by the delay, Tupolev gave the order to transfer Pavel G. Butkevich, the chief overall design specialist for the engines, into his rear office. Several times a day Tupolev would call on him, demanding a solution. God only knows how long this would have continued if a photo of the English Comet aircraft hadn't fallen into Butkevich's hands. It had two small engines in the root portion of the wing.

Butkevich began to draw Mikulin's engine in the same position; however, its diameter was about four times larger, and cut across both wing spars. When Tupolev noticed the sketch, he scrapped the design completely. Moreover, growing angry, he ordered two more overall

design specialists, V. P. Sazarov and I. B. Babin, to work with Butkevich and ordered their boss, S. M. Yeger, to drop everything and work
only on the AM-3.

"You are absolutely forbidden to abandon your work for any reason.
I will order four cots to be set up for you, and you can sleep on them
until you find a solution."

They were already getting close to one. The design slowly began to
take shape. The wing root portions of the wing spars developed into
circular ribbed frames that were relatively light even though they were
cast. Channels that fed air to the engines were designed to run through
them. The frames were cast and strength-tested. Later, during series
production, they were replaced with simpler stamped versions. The
very complex calculations were confirmed so thoroughly that Tupolev
held a meeting and summed things up: "Enough! Put all doubts aside,
and on with the work!" For the first time, it seems, he betrayed his
partially superstitious custom of not rushing any final conclusions until
an airplane had been test-flown.

Of course, work on the Tu-16 did not consist only of positioning the
engines and improving the aircraft's structure. The Tu-16, with a top
speed that surpassed anything at the time, required new instrumentation, radar, sights, autopilot, and much more. Our colleagues from the
other branches of the defense industry, recalling the epic of the Tu-14
and the "horizontal relationships" that developed then, responded in
businesslike fashion to our requirements, and they managed to accommodate our orders for components rather quickly.

By early 1952 the Tu-16 was assembled and at the airfield. On a
frosty, sunny day, the prototype Tu-16 lifted into the sky for the first
time. We prepared for this flight carefully. No one anywhere had tested
jet aircraft of such size and weight, with swept wings. True, the Americans had produced the Boeing B-47 somewhat earlier, but it was 15
tons lighter; more important, we worked on the Tu-16 without test
flight data on the American aircraft. Test pilot N. S. Rybko, who was
exceptionally well experienced and had gone through fire and water,
was even more careful than usual on this occasion.

The first flight brought only joy. The Tu-16 proved to be stable,
highly responsive, and a forgiving aircraft. Aviators in general, and
pilots in particular, are inclined to anthropomorphize their airplanes.

Listening to their conversations at airfields or the design bureau, you can overhear airplanes referred to as being nervous, calm, cantankerous, obliging, stubborn, obedient, an angel or a devil. I should note that an airplane's flying characteristics stemmed from a technical concept.

Soon we received more good news. During its maximum-range flight, with a payload, the Tu-16 reliably maintained a speed of 1,000 km/h. Stalin immediately gave the order to put the Tu-16 into series production, without waiting for the end of the testing.

Then we received information, first in the form of a question posed by Stalin to our aviation minister, M. V. Khrunichev: "Wouldn't it be possible, by mounting two more engines and increasing its size somewhat, to obtain an intercontinental bomber capable of flying to America and back?"

It was impossible, according to calculations, Khrunichev answered. Then Stalin wanted to have a talk with Tupolev in person. Tupolev used to tell us about this meeting as follows:

Stalin was gloomy. "Why, Comrade Tupolev, are you refusing to carry out the government's assignment and build an intercontinental jet bomber which we need greatly?" Stalin asked.

I explained that, according to our very careful calculations, with existing engines it was impossible to do this, since fuel consumption was too great.

Stalin approached the table, opened a folder, removed a sheet of paper, looked it over, and put it back. "Well, there is another designer who is trying to create such an airplane. So why is it working for him, Comrade Tupolev, but not for you? That is strange!"

After waiting a bit, evidently to assess my reaction, although I was silent, he continued: "I think that we are capable of creating working conditions that are no worse for this designer than for you. That is what we will do, in all probability."

And he dismissed me with a nod of his head. I understood that he had remained extremely dissatisfied.

As a result of that conversation, within a few days we received an order to transfer to Vladimir Myasishchev's new design bureau about 200 of our engineers and designers, and also a portion of our equipment and machine tools. His bureau was developing its own prototype bomber.

It was a very good bomber, but it fell short of Stalin's requirement: it lacked the range to fly to America.

Tupolev's visit with Stalin took place in 1952, and within a year Stalin was dead. N. S. Khrushchev, the new General Secretary, ordered several Tu-16s to be transferred to China, at that time an ally, and then, after the revolution in Egypt, to the Egyptians. Finally, so they said, several ships and airplanes, including the Tu-16, were sold to Indonesia as a part of a growing friendship with then-President Sukarno.

After returning from an official trip to lecture to officers in the Egyptian Air Force, our chief overall designer, S. M. Yeger, told us, "I don't know what they think of us in Egypt's ruling circles. The officers in their air force, however, are actually pro-English. This is no great surprise, since they all studied in England. I suspect that the detailed information on the Tu-16 published in foreign technical journals can be traced to Egyptian sources."

The Americans gave the Tu-16 the code-name "Badger" for the aggressive animal that doesn't like its neighbors. This was accurate, since the Tu-16, with its potent defensive armament, can permit itself "not to like" someone, particularly enemy fighters. The Tu-16's payload was so great that the magazine *Aeroplane* informed its readers that with this airplane A. N. Tupolev had reached the zenith of his design work.

It is good that this aviation magazine did not know about our difficulties with the Tu-16. Our prototype Tu-16 turned out to be too heavy, a characteristic that drastically reduced its range. During series production, we managed to reduce structural weight of the aircraft substantially and quickly. Tupolev gathered all the specialists on aircraft structures. The squabbles began right away: who had exceeded the limit most, why, and so on. After listening for awhile, Tupolev asked them to calm down.

"You are princes, by God—real feudal princes! We have a common cause, remember? That is enough explaining who is more to blame! Give me your suggestions!"

D. S. Markov broke the silence that had descended. "I believe," he said, "that we can give series-production people the necessary blueprints, but only after we have reworked them to solve the weight problem."

Someone stuttered, "What about the static testing, then? Will it happen that we have tested one airplane, and the plant will build a different one?"

"Yes, that is exactly right, and I will assume all responsibility. Reworking all the blueprints isn't necessary, since the vast majority of the parts will remain unchanged. On many of them the changes will be so insignificant that new static tests won't be necessary. However, with the largest and heaviest items, where we can save considerable weight, we will fabricate and test them as rapidly as possible. Not only that, right after we have tested them, we will fabricate them ourselves for the series-production plant, for ten airplanes. For complete static tests, however, we will take the airplane from the series plant and transfer it to the TsAGI; we will let them break it out of turn. I have already agreed on this with A. I. Makarevskiy, the chief over there."

Markov's plan proved to be successful, and the Tu-16 reached the Air Force on schedule.

I would like to share some other episodes in the "life" of the Tu-16. Many such stories are told, as with other Tupolev airplanes, but these stories are often exaggerated. What can you do? Tupolev is part of aviation folklore. I will relate the authentic ones.

A group of our designers were invited to visit the Navy, to assist them in problem-solving. Naval tradition has developed a strong attachment to wide bell-bottoms, caps with long ribbons, and the sailor's affinity for the color blue. I could see that the radomes on all the radars were painted blue. I casually asked them how things were going with their target location range and their bombing accuracy. They replied, "Within norms . . . Well, it is like it is within norms . . . almost within norms . . . or, perhaps, just a little bit less than in the manual."

I asked them to bring me the manual. It read, "The exterior surfaces of radomes are to be painted only with paint specially developed for that purpose. . . . The use of other paints, particularly zinc and lead whitewash paints, leads to errors in aiming and to a reduction in range of visibility due to losses of electromagnetic energy in passing through the radome." Need I mention that blue paint is simply whitewash with the addition of blue?

The sailors were embarrassed, and the next morning they were cleaning off the blue paint. Here tradition was sacrificed to requirements of modern technology.

We increased the range of the Tu-16, but via another method. Instead of remaking the airplane, we employed in-flight refueling. The tanker, which was another Tu-16, and the airplane being fueled approached each other in the air, with the wing of the airplane to be fueled slightly below the tanker's wing. The tanker lowered a long hose from its wing, which the airplane to be fueled caught with its wing. The hose slid along the leading edge along the automatic coupling assembly. Then the airplanes carefully separated, and the slack in the hose was taken up until the coupling at the end of the hose engaged the receptacle on the airplane to be fueled. The line was now ready for fueling and fuel transfer began.

It would appear that the operation was not complicated, but it was. Two 70-ton aircraft were cruising at high speed in the turbulent atmosphere performing a stunt as though they were in a circus. It took almost a year to perfect this procedure, but we succeeded, and the long-range Tu-16 bombers became *very* long-range indeed.

Naturally, the security "organs" struggled to prevent any information leaks. Driving away from our air base I was stopped near Pekhorka by a police major, who asked me to drop him off at Lyubertsy. At the fork where the road from Zhukovskiy enters the main highway, I asked the major, "Why is there always a reinforced detail with motorcycles on duty here?"

"We are protecting you," he answered. "As soon as a car with foreign license plates appears from the direction of Moscow, the detail stops it and has it turned around with, 'The road to Zhukovskiy is closed.'"

"But don't you think that a foreigner from an embassy, who speaks Russian, could take a cab to Lyubertsy and easily show up in Zhukovskiy?"

By the way, the "organs" protected the students at the Moscow Aviation Institute against information that was too new as well. The institute graduates would come to us amazingly uninformed both about the latest achievements of our design bureaus and about worldwide aviation. Their senior theses had been completed on old equipment. It took a considerable amount of time to educate them on our current technology and its capabilities.

There were heroes who, as the saying goes, lived among us. On 28 October 1954, pilot Molochanov and his crew took a Tu-16 from the series-production plant on a checkout flight. At an altitude of 9,000 meters the Tu-16 suddenly went into a flat spin and would not pull out of it. Speed and loads began to increase, and the airplane dropped 4,000 meters. Molochanov gave the order to bail out, and then ejected. Air mixed with snow began to rush into the cockpit through the open hatch, and in the dizzying blizzard the copilot A. I. Kazanov noticed the disoriented navigator crawling toward the open hatch without a parachute. Kazanov then decided to try to save the airplane and the man. The Tu-16 continued to descend in a huge, gently sloping death spiral, and true airspeed reached 1,000 km/h. The added G-forces soon reached the nominal strength limit of the aircraft's structure. This caused the landing-gear locks to break under the strain, and the huge four-wheel undercarriage assemblies tumbled out of the nacelles. Air resistance increased abruptly, speed began to drop, and finally the airplane began to respond to the controls. By landing safely, Kazanov saved both the airplane and its navigator. For this feat, he justifiably received the highest award, a Hero of the Soviet Union.

Rumors persisted that the Tu-16 was subject to flat spins. It was only through serious studies that it was proven that accidental circumstances caused the accident: a nonstandard load, an error in centering, and an autopilot malfunction.

By chance, Nikita S. Khrushchev, on an official visit to Belgrade, needed certain documents left in Moscow. He needed them immediately. If they were delivered on an Li-2 or Il-12, there would be no "immediate," so the Tu-16 was pressed into service. After arriving at Belgrade in two hours, the bomber landed at the airport and, on the tower's instructions, taxied to a hardstand. The tower was not familiar with the power of our engines, and positioned the airplane next to other parked aircraft. Behind the airplanes, on a pad in front of the air terminal, stood kiosks for newspapers and souvenirs.

Later, when the airplane was ready for takeoff, many people, including military attachés from almost all the embassies, gathered to see the spectacle of the Tu-16's departure. First the mild hiss of the starters was heard, and then the bass roar of the powerful AM-3 engine. The commander released the brake and accelerated. When the dust cleared, no kiosks were left on the pad. They lay on the lawn off in the distance, blown away by the jet stream.

The Tu-16 ended up involved in politics (this time internal politics) on another occasion. Arriving at the air base in Zhukovskiy, I saw careful double-checking of passes, and several parked government ZIS limousines. Many KGB officers and nameless civilians were there. On the hardstand were about 10 military Tu-16s.

Another Tu-16 taxied over, and before its engines had completely stopped a ZIS drove up to it and a reception group followed on foot. The hatches opened, and the crew descended, but no one exited the top gunner's hatch. Finally, feet wearing low boots with galoshes appeared, then rumpled pants, a jam-packed briefcase, and then the entire figure.

The reception group freed the man from his parachute straps. Before us stood a civilian, very disoriented by the flight experience. They almost carried him to the car and helped him into his seat. Next to the driver sat an armed officer. The car left through the gates and headed for Moscow.

What did all this mean?

"I don't know myself," answered M. N. Korneyev, the chief of our air base. "I came to work and the duty officer informed me that he had received a call on the radio, an order from the minister, to receive military Tu-16s and to send the arriving passengers on to Moscow without delay."

At the design bureau, too, no one knew anything, but the episode prompted alarm. It was only that evening (and not without the assistance of the BBC) that things became clear. At the Presidium of the Central Committee, Malenkov, Molotov, Kaganovich, Saburov, and Shepilov, among others, had challenged Khrushchev, and demanded that he step down from all his posts. Khrushchev calmly stated that he had been elected by a plenary session of the Central Committee and was subordinate only to the plenary session. Convening a plenary session involved collecting all the secretaries of the central committees of the party and republics, and the regional and oblast' committees from all over the country. Our country is so large, and passenger airplanes were so slow (the Tu-104 did not yet exist) that waiting for the session participants to gather in Moscow was risky in itself; who knew what could happen?

Khrushchev found the answer. He ordered that all participants in the plenary session from eastern Siberia and Central Asia be delivered using the Tu-16 jet bombers. The only problem was where to put the airplanes. They couldn't land at Vnukovo, Domodedovo, or Sheremet'yevo—public places—because there would be witnesses to their arrival. In addition, the Tu-16 was still a highly secret aircraft, and the military had kept it away from unauthorized eyes. Tupolev's air base was private, however, being situated at the airfield of a flight-research institute like the bases of other experimental design bureaus. All members of the Central Committee were delivered to Moscow in time, and at the plenary session the vast majority of them supported Khrushchev in his successful campaign to remain in power.

Over time the Tu-16's many modifications kept the airplane from becoming obsolete. Moreover, variants of the Tu-16 included reconnaissance aircraft, missile-carriers, and torpedo bombers. In Tupolev's opinion, the Tu-16 was noted for its great reliability, and excellent flying performance. During its operational life, the Tu-16 suffered few accidents, given its sound performance characteristics. Among the troops, the Tu-16 acquired a reputation for safety.

The Tu-104 Passenger Aircraft

After the Tu-16 had flown thousands of hours in military service, Tupolev began to develop its passenger version, designated the Tu-104. To begin, our design bureau had to solve many interesting and difficult problems. The first hurdle, however, was the sluggishness and stagnation of the bureaucrats at all stages. As we encountered numerous fundamental changes necessary for the successful addition of jet propulsion to passenger aviation, we were amazed at the amount of arguments against these changes. Customarily, the arguments began with the statement that no one, anywhere in the world, was doing this. Next we heard that the passengers would refuse to fly, that it was too fast and too high, and that some sort of solar radiation would make it too dangerous. Finally, it would be necessary to build concrete runways, new air terminals, repair bases, and hangars, to switch from gasoline to kerosene, and to retrain crews—what huge expenses! We also had to prove that our Aeroflot wouldn't be ruined.

Tupolev put forward the decisive arguments at one of the high-level planning meetings:

Military aviation in all advanced countries, including ours, have been using jet aircraft for a long time, and nothing happens to the crews.

Large, four-engine airliners with ordinary engines all require concrete runways for takeoffs. They have been built in such tiny countries as Denmark, Belgium, and Holland. The same is true of air terminals. Is it really possible that a nation like ours can't manage this? In an age when air transport is growing by leaps and bounds, we cannot keep up using the slow-moving Li-2 and Il-12.

As for economics, let's study our books a bit. I believe that not many people understand the issues. The Tu-104 is five times faster. In one day our airplane can fly to Sochi and return. This means that one Tu-104 will replace at least twenty Il-12s.

So, calculate the salaries of nineteen extra crews, maintenance personnel, towing vehicles, hardstands, fuel, and so on.

The English helped us out on the economics. There were rumors from England that the de Havilland Company was working on a "Comet," the first passenger jet airliner.

The press fell silent, until the journal *Aeroplane* carried an article about the loss of two Comets over the Mediterranean Sea, the first on 10 January 1954, the second on 8 April 1954. The persistent British raised several sections of fuselage from the bottom and determined the cause of the disaster. The pressurized cabin had two square recesses for the radiocompass antennas. Whether it was due to careless work or to the concentration of stresses in the corners of the recesses, hairline cracks developed. During large drops in pressure inside or outside the cabin when flying at high altitudes, enough cracks opened that caused the skin to burst. The subsequent decompression destroyed the entire cabin.

We had managed to avoid this danger in our airplanes. When the scientists at the TsAGI, for example, tested the pressurized cabin on the Tu-70, which was about the same size, they detected a similar phenomenon as early as 1947, and recommended to the designers that all recesses be made round. Successful operation of the Tu-70, Tu-4, and Tu-16 confirmed the accuracy of these recommendations.

England might have been ahead of us in building passenger jets, but we were the ones to begin regularly scheduled passenger flights in such airplanes.

On 15 September 1956, after comprehensive testing, the experimental Tu-104 set off to Irkutsk on its first trip with passengers. On board were the relatives of the employees of the design bureau and the Aviation Ministry. Ye. M. Beletskiy, the Deputy Chief of the Civil Aviation Fleet, also flew with us.

The flight went splendidly, and all participants became enthusiastic advocates of such airliners. A minor comical incident did take place, however. On the return trip, after we had passed Omsk, we received a radiogram: "The flight to Moscow is forbidden; return to Omsk." Beletskiy managed to find out the reason. Khrushchev was returning to Moscow from Warsaw in an Il-12. Fearing a crash (we were flying at an altitude of 10 kilometers from the east, and he was at 2,000 meters from the west!), the NKVD had decided that our flight be turned away, as an extra measure for safety.

After consulting with Beletskiy, we nevertheless decided to continue our flight, rightly believing that no one would notice our arrival. Experience showed that we were right. Many years later, a student from the Federal Republic of Germany, one Matthias Rust, flew a small aircraft to Moscow, then landed not just anywhere but on Red Square.

The time came when the Tu-104 made its first overseas trip. Pilot A. Starikov carried State Security General Serov and his retinue to London to discuss an upcoming meeting between the leaders of both states. The English press wrote, "Russia amazed the Western world by showing off the Tu-104, which was better than all airplanes we have seen" (*Daily*

Mail). In a television interview, a British air marshal stated, "The Russians have far outstripped us in building such airplanes, and we have no jet engines of comparable size."

Finally, we, too, deserved the honor of seeing Europe! Before a first flight outside the country, to Paris in our case, it was customary to instruct someone on how a Soviet person should act abroad. A not overly educated and bad-mannered comrade from the regional party committee lectured us: "Keep in mind the intrigues of the capitalists, the White emigrés, and our foes. Beware of provocations, shun persons who speak Russian, and avoid all types of suspicious places, such as cheap cafés and cafeterias, markets and stores, and third-rate movie houses. Don't walk around alone, try to go in groups."

Ironically, the embassy rented us rooms in the most plebeian of hotels, which sometimes lacked even basic conveniences (equipped with a shower at the end of the hall instead of a bathroom), putting two or three persons to a room. They paid us so little that we could eat only in the cheapest of cafeterias.

No excesses or provocations occurred, although the Parisians stared at us, dressed Khrushchev-style in baggy sportcoats, wide trousers, and felt hats pulled down over our ears. We traveled in groups and, since we were afraid of losing one another, unintentionally pushed aside people coming in the opposite direction, interfering with pedestrian traffic on the sidewalks.

Talking loudly or, at times, very loudly, we did not melt into the crowd, but instead attracted attention, especially from those whom we were supposed to avoid according to Moscow's admonitions. A taxi driver asked, "So where are you from? Russia? So tell me, how are things there? I would like to go home, but I am afraid: I am a Cossack. And the papers say that all of them were sent to Siberia, or else shot."

A sales clerk in a kiosk next to our hotel said, "So how are salaries over there? About 200 rubles? Not bad [he was thinking of prerevolutionary prices]. But why do you stay in such God-forsaken places? Saving your money? For what? Take a look at yourselves! You are dressed like clowns! For that kind of money you could look a little better!"

A woman selling things at a "flea market" (a cheap second-hand market, which the embassy told us to avoid) commented, "When I saw you, I thought you looked like tourists, but they all hurry off to the Samaritan and Richelieu stores, while you try to find something on the cheap. Maybe it is the same as under the Tsar, you have only enough money for a single loaf of bread?"

A casual passer-by, well-dressed, with a briefcase said, "You are from Russia? No, I won't go there, the Jews and commissars are in charge there [he was clearly one himself]. L. Smirnov sends him to . . ., and he disappears immediately."

When we displayed the cabin of the Tu-104 for the public, a man and woman walked in. He was middle-aged, with an officer's bearing; she wore an embroidered Ukrainian blouse. After glancing around, the little lady, clearly trying to start a scene, asked, "If I stow away here, will they shoot me right away in Moscow?"

"What do you mean, ma'am?" I said. "They will send you back to Paris on the first flight."

The man said, "Katrin, what are you being so discourteous?"

"Oh, Papa, stop it! Can't you see? They are all from the GPU [State Political Directorate], and their airplane isn't a passenger airplane at all, it is a camouflaged military spy airplane!"

These were the conversations we had with our former countrymen.

The Tu-104's triumphal parade around the world continued. This airplane opened a new era in aviation, and we were proud. The Tu-104 even made a small contribution to astronautics. S. P. Korolyov, then the head of our space program, complained to Tupolev, "We are in great need of trainers to teach future cosmonauts how to behave in weightlessness before they fly into space. No matter where I turn, everyone turns me down; they say you can't create something like that on the ground."

Tupolev replied, "Perhaps I can propose something. If we take a Tu-104, fly it to an altitude of 10,000 meters, and then begin to follow a sine curve as we descend, the airplane will experience a negative load in each upper portion of the curve."

"How long will it last?"

"About a minute."

"That is too little. We a need a bit more."

"So fly about twenty sine curves, and you have twenty minutes."

All seats were removed from the cabin, the inside was covered in soft porolon, a foam-like material used for insulation, and the problem was solved.

Korolyov always recalled Tupolev with gratitude, because his idea helped greatly in the instruction and training of future cosmonauts under real conditions.

The Tu-95 and a Tragedy

Stalin's absurd (from an engineering point of view) idea of turning the twin-engine Tu-16 into a four-engine intercontinental jet bomber nevertheless made an impact on Tupolev. He always considered himself morally responsible for long-range combat aircraft. However, jet engines that were economical enough for such flights did not yet exist and were not foreseen in the near future. The international situation did not permit waiting. Based on Tupolev's proposal, the decision was made to build a heavy bomber with turboprop engines, which, although slower than desired, would have greater range.

This heavy bomber required four engines, each producing at least 10,000 hp. The only ones available were N. D. Kuznetsov's series-produced NK-6, which had been increased to "only" 6,000 hp each. Moreover, calculations indicated that a propeller for a 10,000-hp engine would have an inordinately large diameter, and the velocities of its blade tips under certain flight conditions would exceed the speed of sound, which resulted in a sharp drop in the propeller's efficiency.

The government entrusted the creation of a new powerful engine to Kuznetsov's aeropropulsion design bureau, and the creation of a coaxial eight-blade propeller for the new engine to V. I. Zhdanov's bureau. Unfortunately, they felt it would take two or three years to accomplish these tasks.

The aircraft designers, together with the scientists at TsAGI and TsIAM, the Central Scientific Research Institute for Aircraft Engine Construction, found a faster solution. Two NK-6 engines would be installed in each of the four engine nacelles on the first experimental airplane. Each engine in the pair would use common reduction gearing to drive not a single common propeller, but two coaxial, counter-rotating propellers with a slightly smaller diameter.

It was an exceptional design. Stalin signed the decree on the new bomber, noting that "Tupolev finally changed his mind and got serious about the right effort!" Only one issue had yet to be resolved. Although our plant had prepared the huge plant for transport to the airfield, its landing gear was 12 meters wide. The road we would have to tow it on, from the turnoff on the Kuybyshev highway to Zhukovskiy, was narrower than that. Many leaders of all sorts had promised Tupolev that it would be widened by now. As usual, everything began and ended with words.

Tupolev resorted to his "horizontal connections" and telephoned war hero and Marshal G. K. Zhukov to ask for his help. "Georgiy Konstantinovich, we have built a certain weapon for you over here, and we can't fire a shot with it!" "Why's that?" "The road's too narrow to get it to the test range." "Andrei Nikolayevich, I am not a highway engineer, after all . . ." "That may be so, but you have combat engineers, and I have been out there, took a look, and think that they could handle the job within two or three months. Could you possibly help out?" Soon the combat engineer chiefs arrived. Tupolev took them to the plant to see the airplane, then to see the road, after which the engineers went back to report.

Within a week or two the work was in full swing. Tent cities with field kitchens popped up along the road. The soldiers got down to work. Tupolev visited them frequently, acquainting himself with the commanders and many privates. They nicknamed the new road "Tupolev strasse." The name stuck, and can be heard even today from the older local residents: "It's over there, behind Tupolev strasse." The road was ready on schedule, and "airplane No. 95" was transported safely to the test airfield.

Flight tests went smoothly. The crew carried out more than ten flights. There were no complaints about the operation of the paired engines, and both test pilots (A. D. Perelet and V. P. Morgunov), as well as the entire crew, were satisfied with the airplane's performance under highly varied conditions: day and night, high altitude and low levels, and normal and adverse weather conditions.

On 11 May 1953 the "95" was scheduled for its next flight—unremarkable in all aspects and observed only by those who were authorized at the air base.

The radio operators raised the alarm unexpectedly; they had been amazed by how calm Perelet's voice was: "I am near Noginsk. There's a fire in engine No. 3. Clear the runway. I will come in for a landing straight from our route."

Several minutes later, Perelet reported, "The fire is spreading, the engine nacelle and wing are on fire. I have given the crew the order to bail out," followed by silence. Our radiomen sought the Tu-95 in vain, but the loudspeaker only hissed and crackled. Holy Mother of God, what had happened? We rushed to our vehicles and headed for Noginsk.

When we were still a distance away, we saw a column of smoke, its top drifting slightly, off to our right, over a small, swampy patch of forest. We had to go in on foot to reach the crash site.

We came upon a horrible sight. The remains of what had been just a few hours before our pride and joy were burning in a huge crater. The tires on the huge landing gear were smoking. Not far away a second, smaller crater, created by an engine that fell separately, was slowly being filled by rivulets of water from the swampy soil. Frogs that had already recovered from the accident were frolicking. There was not a trace of the crew, eleven of our friends.

Spreading out into a line, everyone combed the swampy woods. With us was Tupolev, his face ashen, his coat and pants muddy. He had lost his service cap, probably catching it on a branch, so his gray hair was out of place.

Bazenkov shouted, "Over here, over here!" Among broken branches lay our badly smashed navigator, S. S. Kirichenko, wrapped in white parachute fabric. When he ejected, the parachute canopy had opened improperly and wrapped around him, enshrouding him, and so he fell to earth like a stone.

Residents in the nearby village helped us determine what had happened. "The airplane was coming from Noginsk, from over there, northeast. Something huge broke off and fell to the ground. We saw parachutes in the sky [an argument broke out as to whether there had been six or seven]. Then the airplane tipped toward one wing, went down, hit the ground, and exploded. But the crew are all in the village."

All of them? We sent men into the village to find out.

They returned, bringing copilot V. P. Morunov and the lead test engineer, N. V. Lashkevich. It was terrible to look at him—the harness straps of the open parachute had left bloody traces across his face.

Three men remained in the village: engineer A. M. Ter-Akopyan, assistant flight engineer L. I. Bazenov, and flight electrician I. V. Komissarov. Later we found engineer K. I. Vayman and radioman I. F. Mayorov. Bazenkov sent them to the hospital in Zhukovskiy on a U-2 that had arrived at the accident site.

We were missing three men: command test pilot A. D. Perelet, flight engineer A. F. Chernov, and technician A. M. Bol'shakov. Lashkevich recalled that Bol'shakov had stood, unconcerned, over the open hatch, making no attempt to jump. Commander Perelet and his good friend, flight engineer Chernov, had tried in vain to pull the heavy aircraft out of its roll until the very end. God only knows whether the efforts of three men would have been enough to straighten it out if the copilot had not hurried to bail out.

Nearby, soldiers from the Monino airfield dug up the engine that had broken off. Our chief engine specialist, K. V. Minkner, urged them to "please be as careful as possible. Any fragment, any damaged part, will help us explain the cause of the accident."

The ubiquitous representatives of the NKVD arrived from Moscow, as well as real specialists from our Ministry's Institute of Aircraft Engine Construction. The latter included a thin, middle-aged Georgian. One of the soldiers showed him a fragment he had found in the dirt of some part similar to a pinion gear. After examining it on all sides, the Georgian placed it in his pocket. This did not please me at all. Finding K. V. Minkner, I nodded at the suspicious Georgian.

Minkner, smiling completely inappropriately, turned to the Georgian and said, "Robert Semenovich, Kerber suspects that you are hiding something."

It turned out that this Georgian was Professor Kinosashvili, a leading authority on metallurgy. After examining the fragment of the reduction-gearing pinion, which transmitted the output of both engines to the propellers, he wondered whether it had been made of the proper metal. The surface of the break indicated a fatigue failure of the pinion, although the total operating time of the reduction gearing had been only minimal.

Kinosashvili's suspicions were correct. It was subsequently found that the wrong metal had been used at the plant that made the reduction gear.

Subsequently, logic, strengthened by analyses, made it possible to envision how events unfolded. When the pinion gear broke, its fragments penetrated the reduction-gear housing. The oil it contained splashed out onto hot areas of both engines and ignited. Carried by the onrushing air, the flames crept along the wing, engulfing the lower engine mount, which connected the two engines. The mount began to deform and, after losing its rigidity, broke, and the engines with the propellers tore away.

We all gathered in A. A. Arkhangel'skiy's office at the design bureau building the evening after the disaster. No one wanted to talk; we were all sitting silently when the telephone rang. It was the Noginsk NKVD: They had found the remains of Perelet and Chernov. Looking around at us, Tupolev said, "Kerber, you are the youngest one here. Please go and see what is going on."

With my team, I arrived at the site at about 2 A.M. The soldiers on duty were making tea over a campfire; next to them, on a bed of fir branches, lay a small baby elk. Its mother had probably abandoned it when she was frightened by the noise, fire, and smoke, and now the soldiers were trying to feed it. It was a very quiet night, like in a fairy tale, if not for the plywood box nearby with the hideous pieces of the bodies of Perelet and Chernov. We wanted to return immediately, but the soldiers asked us to stay a little longer. They cut off some fir branches, brought some fragments of red calico from a flag (from God knows where), and covered the box with them. We set out on our return.

As we approached Moscow, the driver asked where he should take us. We naturally replied "the Bauman morgue," since it was next to our design bureau. The appearance, at night, of people with lacerated fragments of corpses put the coroner on guard. My not-entirely-comprehensible account of the disaster (we did not have the right to reveal what kind of disaster or what kind of aircraft) disturbed him even more. He called the police department, which was across the street, and the proper authorities appeared at once. They clearly wanted to put us behind bars, with criminals and drunks.

I had with me the home telephone number of Colonel of State Security Pil'shchikov, our Deputy Director for Cadres. After the call they let us go.

It was about 5 A.M. We were exhausted and went to the design bureau to rest a bit.

The following day, Lashkevich appeared after lunch, with his head bandaged. He had come from Beria himself, and told us about his talk with the leader's best friend.

"Help us figure this out," Beria ordered Lashkevich. "A strange bunch has gathered around Tupolev. The surnames alone tell you something: Yeger, Stoman, Minkner, Kerber, Val'ter, Saukke. . . . A nest of Germans and Jews. And notice: not a single Communist. Perhaps that is the explanation for the disaster? Perhaps they are at work here? You, an old member of the party, keep a closer eye on them. I won't object if you involve several more solid Communists to help you; we will cooperate with you. Write down everything for us. If necessary, I will invite you again."

Thank God for Lashkevich! He ignored Beria's ideas. But all this meant that we were again under suspicion, which was vile, insulting, and terrible. Faster than you could turn around it would all start anew: "Why did such-and-such happen? Why did that occur? Answer, don't try to evade the main issue." It was two months before the presumed cause of the disaster was finally "approved" and the design bureau could return to devoting all its efforts to work.

It was necessary to build a duplicate of the 95. The 12,000-hp NK-12 engines had been manufactured and finished at N. D. Kuznetsov's design bureau. We put all the old blueprints into production, and reworked only those for the engine nacelles. We worked with a certain fury, as if exacting vengeance for the deaths of the airplane and our friends.

Finally the new prototype was at the airfield. Whom would Tupolev choose to take the first flight? M. A. Nyukhtikov, the same pilot who had once not been afraid to take up the Tu-2 dive bomber built by the imprisoned saboteurs, was chosen. (The "organs" had asserted to him that he was surrounded by saboteurs, spies, and counterrevolutionaries who had sold out to the enemy.) After the war, Tupolev, deeply convinced of Nyukhtikov's abilities and persistent about having him transferred to us, was finally successful.

The Tu-114

As the test flights were coming to an end, Tupolev began to think about a passenger version of the 95, the Tu-114. At first, he worked only with B. M. Kondorskiy, our artist, and a specialist in interiors. At last he brought us in. We saw the plan for the passenger cabin, divided into compartments. Written in Tupolev's rough scrawl were "first-class cabin," "special cabins," "bar-restaurant," "second-class cabin." What sort of whim was this? Could we manage this configuration?

The model shop had already begun work on the 114. We were busy there for a while, with everything except the passenger cabin, where carpenters, decorators, and painters worked in private. Finally, after looking it over for the last time with his wife and Kondorskiy, Tupolev met with us, his deputies, in his office one morning.

On the table in the bar-restaurant was a bottle of cognac, a dish with candies, and shot glasses. In an expansive, clever mood, our boss poured drinks for everyone. "Gentlemen, I called you together to tell you some very unpleasant news. When we began to configure the 114, I thought, 'Will the head of our government really sail on a ship for an entire week, as previously, when he travels overseas? And this is when other presidents and premiers fly across continents and oceans on their Douglases, Boeings, and Lancasters in a matter of hours!' No, we, too, need an airplane for this, and not one with an ordinary configuration, but a hybrid variant, in which the government won't be embarrassed to fly and the people will be happy to fly. So let's drink to our hybrid variant!"

The second Tupolev passenger jet appeared on the air routes (the first had been the Tu-104), but before this occurred, he tickled our nerves. During one of the test flights of the 114, the right landing gear would not lower. Aleksei P. Yakimov had only about one-half hour of fuel left. Tupolev summoned me. "The electrical mechanism in the landing gear is yours. We have to calm them down and figure things out." At the command post I took the microphone and began to speak as calmly as possible: "Aleksei Petrovich, it's me, Leonid Kerber; Tupolev is right here. Let's figure this out."

The flight controllers took away the microphone in irritation. "You are talking like you are on the phone with a friend. You can't do that here, it has to be by the book: 'Watermelon-1, Watermelon-1, this is Apricot-5. Come in, commander of item 114 . . .'"

Tupolev told them, irritably, "I thought Yakimov needed to be calmed down and told that we have found the reason that the landing gear has lowered, that we will tell him immediately, and that he can land normally. Your rules are foolish! Some ass put them together! What are you afraid of? That the CIA might hear that they call Yakimov? Well, then, write this down: 'Crash a 100-ton airplane and it is your responsibility.' Kerber and I will help Yakimov land it while violating your rules—that is our contribution."

We really had found the problem. I won't go into details except to say that to eliminate the problem the crew had only to switch off all sources of electricity on the airplane for several seconds. This meant being deprived of radio communications with the ground during those very same seconds, and having dead silence. We had to convince them, but once they decided to go ahead, there were seconds of silence followed by a burst of exultant voices: "It worked! It worked! It worked!" The gear lowered out of the nacelle and Yakimov landed the airplane confidently.

After the Tu-114 had accumulated the required flight hours, the long-distance flights began. These also involved the first series-produced airplanes from the Kuybyshev plant. For the first flight of the series Tu-114, Tupolev went to Kuybyshev, taking us with him. He was taken to the oblast' party committee dacha, while we stayed in the plant's hotel. Tupolev asked us to get him in the morning. When we arrived, he was still in bed. Calling in the duty officer, he asked, "Is it too late for them to have breakfast?"

In the dazzlingly clean dining area, with its starched tablecloths, china from the DDR, silverware from Czechoslovakia, goblets of Bohemian crystal, the pretty young maids in aprons and hats offered us our choice of fried sturgeon, pork chops, mutton kabobs, cognacs, vodkas, wines, beer from the socialist countries, and a light snack consisting of crab salad, flank of beluga, ham, vegetables, and fruit.

In the dining area Tupolev drank a shot of vodka, snacked on whatever came to hand, and, after wiping his hands, said, "They don't live too badly!" And he was exactly right. The workers' dining areas had very modest lunches, and even in our Moscow director's dining area we never thought of anything similar. You could treat yourself like that only at our most fashionable restaurants, and then only for lots of money. But here everything was free.

Regular routes began for series-produced Tu-114s from Moscow to Khabarovsk in eastern Siberia. It became possible to open regular service abroad later. First was Cuba. Next was Delhi. The Japanese expressed interest in the Tu-114, since they had to fly a long, expensive, roundabout route to Europe, with stops in Calcutta and Teheran. Our government agreed to the opening of a Tokyo-Moscow route, across Siberia.

Before beginning joint routes, the Japanese company JAL wanted to study the Tu-114 in greater depth and detail, so a Tu-114 flew to Tokyo. It was interesting and instructive to see how the Japanese specialists studied both the airplane and its manuals meticulously. After convincing themselves that the airplane fully met all technical and performance standards, they opened the route. Next to our flag on the wings of the Tu-114 appeared the white crane emblem of JAL.

With the start of operations by the large-capacity Tu-114, passengers began to complain. Although they landed safely, finding a speedy way to downtown Moscow was a real problem. Ye. F. Loginov, the Minister of Civil Aviation, convened a meeting. Helicopter designer N. I. Kamov proposed, "Transfer them to my Ka-model helicopters, and they'll be in Moscow within ten minutes."

Tupolev was terribly amused. "A Tu-114 lands and disembarks 100 passengers. Your Ka can hold four. So, 25 helicopters lift off! God help them as far as noise goes, but the helicopters will collide. It would be time to open a new cemetery. No, we have to connect Sheremet'yevo and Vnukovo airports with the city via surface subway lines. That is also an effort, but twenty kilometers of rail line—that can be done in a year."

Many years have gone by, but everything is still as it was then.

A Flight to America

Our head of state, Khrushchev, was preparing to visit the United States in 1959 for a UN conference. Tupolev proposed—some felt too boldly—that he fly on a Tu-114. Aeroflot still did not carry "simple" passengers extensively, and here Tupolev was suggesting that they take the head of state across the ocean!

Formally speaking, it *was* impudence, but Tupolev never thought in formal terms. The prototype for the Tu-114 was the military Tu-95 aircraft, which had been in series production for several years. These planes had been flying often and successfully over oceans and continents, to the farthest reaches of the globe. The Tu-114 differed from the Tu-95 only in its fuselage, which carried passengers. Everything else—wing, crew cockpit, engines, tail surfaces, landing gear, and equipment—was exactly the same. However, the structure of a huge pressured fuselage had been tested on the Tu-70 for hundreds of hours. Thus, no new risks were involved.

As Tupolev recalled it, Khrushchev not only was enamored of the idea of flying, but also decided to take his entire, large family with him.

The flight required convincing the top leadership, obtaining the concurrence of the Minister of the Aviation Industry, Aeroflot, the KGB, and, naturally, the CPSU Politburo. Tupolev had many obstacles to hurdle, so many that catching Tupolev in his office became almost impossible.

While he worked on global issues in the upper echelons, we were involved with practical issues, of which there turned out be more than enough. First and foremost, the "organs" gave us problems. Even in ordinary times they didn't leave us alone, but when . . . well, you simply couldn't take a step without their representatives. They were most interested, of course, in the safety of the passengers. For three-fifths of its journey the airplane would fly over ocean. What kind of rescue devices did we have in case of an emergency landing at sea, and how would they be used? How should one jump from the airplane into the water? How should inflatable life jackets be handled? How should one climb into a life raft?

At the KGB's insistence, we made a full-scale mockup of the Tu-114 fuselage, including the door, in our model shop and transported it to the swimming pool at the fashionable government village in Lenin Hills, which the people have nicknamed "Il'ich's testament." V. I. Bodganov, the lead engineer, was entrusted with this work. He recalled,

The mockup of the fuselage section was set up in the swimming pool. I instructed our future passengers how to put on life jackets, and began rehearsals. I don't recall that Khrushchev and Nina Petrovna, his wife, jumped, but their children and the rest of the multitudinous relatives and other members of the delegation trained very actively. Wearing swimming suits and their life jackets, they threw themselves from the doorway into the water, swam to the life rafts, and clambered aboard them. True, this was done somewhat in good fun, with happy yells, in a swimming pool with heated water, but our future passengers nonetheless acquired the necessary skills.

Another issue was navigation support. Numerous military Tu-95s, with their equipment identical to that of the Tu-114, would fly for days and nights over all the oceans. They would always end up at their assigned positions confidently, and would return to their home airfields no less confidently. However, the specialists at the KGB proposed the addition of Soviet ships in the ocean every 200 miles along the entire flight path. Their radios would serve as radio beacons for our navigators, and in the event of a landing at sea the two nearest ships would pick up the passengers. This wise proposal was approved.

A third concern was the flight manifest. Such a manifest, filled out for every flight of any aircraft, records the names of the crew and the mission and is signed by the flight chief. On this occasion, however, they decided "upstairs" that the manifest would be signed by the minister and Tupolev, and stamped by all the chief designers associated with the Tu-114.

Tupolev invited about fifty such designers to our base, familiarized them with the "proposed long-range flight by one of our leaders" and asked what equipment checkout they would like to undertake on such an occasion. The first to speak was the autopilot designer, I. A. Mikhalev. "I consider it necessary," he said, "to remove the autopilot from the airplane and check it out on test stands." The majority of the designers agreed with Mikhalev.

At this point Tupolev burst out, "So, in your opinion it is necessary to disassemble the entire airplane for the sake of reliability? You are ordering me to remove the engines, landing gear, and controls? Why your autopilot alone, Ivan Aleksandrovich, is solid cables, about a hundred or so, and over the entire aircraft there are probably about 2,000 of them! Right now the crew has no complaints about the operation of our equipment. So we will unscrew plugs, and then we will tighten them again in a week, and will they become any more reliable after this? Nonsense! That is right when your equipment will start to malfunction. No, spare me! I won't give permission to undo even a single one. Get in the airplane, fly around as much as you have to, check out your equipment in flight as meticulously as you want, and convince yourselves that it is functioning properly, then enough! Sign the manifest!"

That is what they did. All the signatures were collected gradually, and only the two main ones (I don't want to mention them) "disappeared" and sent their deputies instead. After all, it is easier that way! And these two weren't designers, but commissars, public figures placed in their positions to provide ideological leadership.

Earlier, in June 1956 the Tu-114 airplane returned from the Paris Air Show. It was rolled into a hangar and the engines were changed. They performed a checkout flight, then two or three main route flights, and then the airplane was ferried to Vnukovo. On 26 June, with Politburo member F. R. Kozlov, who was traveling to New York to open a Soviet exhibit, onboard, together with his entourage, the creators of the

Tu-114—Tupolev, A. A. Arkhangel'skiy, and S. M. Yeger—the airplane took off for America.

In essence, this was a general checkout of the Tu-114 before Khrushchev's visit. Everything went without a hitch. The agencies were convinced of the reliability of our equipment, and the resolution of the Central Committee Politburo on Khrushchev's flight was passed. Two Tu-114s were prepared for the long flight: a primary, commanded by A. P. Yakimov, and a reserve, commanded by I. M. Sukhomlin. N. I. Bazenkov, Tupolev's deputy for the aircraft, and I, his deputy for armament, were to fly on the primary airplane with Khrushchev and his closest entourage. Our friends laughed and called us hostages (with some element of truth).

The delegation, which was referred to as a government delegation, had something of a family flavor, I would say. It included Khrushchev's wife, Nina Petrovna; her sister, Anna Petrovna, with her husband M. A. Sholokhov; their daughter Rado, with her husband A. Adzhubei; their daughter Yuliya and her husband N. Tikhonov (who, at that time, was President of the Dnepropetrovsk Council on the National Economy); Khrushchev's son, Sergei, with his wife, and numerous others. Those who were not family members included A. Gromyko, Minister V. Yelyutin, corresponding member of the Academy of Sciences V. Yemel'yanov, and medical doctor Prof. A. Markov. From the daily press there were Yu. Zhukov, L. Il'ichev, P. Satyukov, O. Troyanovskiy, and, once again, A. Adzhubei. There were also about ten bodyguards, several stenographers and typists, chambermaids and valets, a cook, and waiters. Together with the crew, two foreign pilots, and us—the "hostages"—there were about 70 to 75 people onboard, so there was plenty of room in the cabins. The Khrushchev family occupied four sleeping compartments, while close relatives had the rear cabin. I occupied the front cabin with the crew and foreign pilots. The central cabin was the restaurant.

I should note that throughout the flight, Khrushchev turned out to be (at least with us) truly democratic. Strolling around the airplane, he would stop to chat, make jokes and laugh. Nina Petrovna, too, was plain, kind, and exuded feminine warmth toward those around her. All of us, particularly those for whom this was a first, observed Khrushchev with great interest and, it seemed to me, great affection. Even today I feel sincere respect for him. Just think about what it took, at that time, to reject Stalin's "iron" course! Beginning reforms, breaking up what was powerful resistance that was both open and hidden. Freeing from the camps millions of innocent people (who, by the way, had not been hidden away there without Khrushchev's concurrence). Changing housing construction to assembly-line production of houses and apartments, which, although not the most comfortable, were urgently needed. Beginning the conquest of the southeastern deserts and oil deposits. And, after honorably and bravely admitting that he had gone too far, without fearing for his own authority, removing the missiles from Cuba.

It is true that he did not fix the nation's agriculture. But was this only his fault or problem? We are too inclined to accuse someone of our own failings and misfortunes, although it was said long ago and then taught in school, "If your neighbor thinks it is a good idea to work on something, wouldn't it be good if you turned it to your advantage, too?"

Turning, if not to myself in particular, but to the aviation industry, in which we weren't exactly bringing up the rear, I recall Khrushchev's restructuring of the entire system of Soviet industry, the elimination of ministries, and the creation of councils of the national economy. This destructive and ill-conceived idea coincided with the creation of two long-range bombers, Myasishchev's M-4 and the Tu-95 in our design bureau. A worried P. V. Dement'yev, our minister, asked us for the locations of all our suppliers and related facilities. He plotted them, then drew lines from Moscow to them. The nondelivery of any part from even a single city could stop our aircraft production. Our minister was hoping that the complexity of organizing this network through new people would make an impression on Khrushchev, and would shake his confidence in innovation.

Nothing came of this, however. The next morning they brought the diagram back from the Ministry, and Dement'yev told Tupolev on the phone that Khrushchev firmly held to his position about how national security should be defined; he did not want to hear any opposing viewpoint. We all fell silent, except, perhaps, for expressing concern among ourselves.

Let me return to the flight.

Bazenkov and I lived in the same building and agreed that the car should pick us up at 3 A.M. We went to bed early, but could not sleep. The weight of our responsibility kept us awake. It was clear that Khrushchev's visit could become a watershed in our relations with America.

At dawn we arrived at Vnukovo. The two Tu-114s sat by the terminal, each with a guard posted. It was quiet, broken only by the chirping of awakening birds. Both the flying field and the planes were covered with a thick dew.

We wanted to have another look around the airplane. After checking Bazenkov's documents, the guard saluted and let him in. I was not on the list, however. It looked like I would be staying on the ground. I could only guess at who was playing jokes.

At about 5 A.M. the passengers and those coming to see them off began to arrive. When Tupolev arrived, I told him what had happened. He listened, became clearly irritated, and went into the terminal. He returned in one-half hour with the Deputy Chairman of the KGB. (I don't remember his surname; it was something like "Ivanovskiy.") The general set everything straight in an instant. Nothing but glitches. . . . Tupolev smiled. After we had returned from America, he told me what he had said. "I told him outright: 'What? You mean you are getting rid of my specialists, who are providing flight safety?'"

At 6:45 A.M. Khrushchev, after kissing all the members of the Politburo, ascended the ladder, waved his hat, and entered his cabin. The door slammed shut. The engines could be started. Flight engineer L. A. Zabaluyev switched on engine No. 2. The propellers turned, but the engine did not start. He tried again, with the same results. You can understand what we were feeling. Through the porthole I could see Tupolev and Minkner, our Deputy for Engines, growing concerned as they inspected the aircraft.

Khrushchev walked into the front cabin and up to Bazenkov and me. "What is going on? Why can't they start the engines?"

Out of pure inspiration I answered, "They are performing so-called cold rotation of the engine rotor; it is necessary before starting," clearly recognizing that, in the event of failure, the "cold rotation" could turn out to be very hot for both us and the crew!

Next to us sat Sukhomlin's reserve Tu-114. Could it be possible that the head of our state would have to transfer to it in front of the entire diplomatic corps? Disgraceful!

The door into the cockpit was open, and we could see the pilots' distress. But once Zabaluyev started engine no. 3 easily, on the first try, then no. 2 started up as well.

The airplane taxied out to the takeoff runway, and engines 1 and 4 were started en route. Yakimov began his takeoff. One minute, two, three, and right on seven, as set, we were in the air.

The incident involving engine no. 2 turned out to be very interesting from an engineering standpoint. At the airport in Washington, a person looking exactly like one of us was waiting for us with a radiogram from Moscow: "Report immediately: Why didn't the engine start?" The fact that we had flown across the ocean from one hemisphere to the other, and had delivered the head of state to Washington on time and in good health, was of no importance to Moscow. "Why didn't the engine start?"

Why indeed?

I asked crew member O. S. Arkhangel'skiy, an engineer from our department in the design bureau, to look into the problem. It turned out that a single plug, containing a single wire which fed the starter's signal to squirt kerosene into the engine during starting, had been poorly soldered. When they attempted to start engine No. 2 the shaking caused this wire to lose contact. But when Zabaluyev started engine no. 3, and the airplane shook entirely differently, the ill-fated wire again made contact.

Two conclusions present themselves. First, the reputation of a great nation may suffer due solely to an improperly soldered wire. And second, how wise Tupolev was when he didn't permit them to disassemble thousands of plugs, for how many of them could have had damaged contacts after disassembly and reassembly?

And so we were airborne, and came onto our course. I had been ordered to note the airplane's position on a map of the Northern Hemisphere (a very basic map from a school atlas) every half hour, and show it to Khrushchev. Khrushchev sometimes let me go at once. At other times he would ask questions about communications with Moscow, about whether we were running late or on-time, or about how the crew was feeling.

After returning to my seat, I placed the map on the seat next to me. Once we passed Yarmouth, we did not have far to fly. As I reported to Khrushchev for the last time, I grew bold and asked him to autograph the map. He smiled and in the midst of the deep ocean, he wrote, "Onboard the Tu-114. A splendid airplane, thanks to its creators and builders. We are approaching Washington. 17 hrs 26 m. 15 September 1956. N. Khrushchev."

However, after the landing and the confusion that ensued, I noticed that the map had disappeared. A truly historic document was gone. I won't recall how Tupolev reviled me for this loss. It is not worth remembering. . . . All that remained for us was a photograph in which Khrushchev was seated at a table, I was next to him, and the ill-fated map was on the table.

However, all this still lay ahead, and we were passing Velikiye Luki, Riga, Stockholm, and Bergen. Now the ocean lay under us. At 10:18 we passed the destroyer *Smelyy,* which was waiting for us 200 miles from Bergen; at 10:58 we passed the trawler *Dobrolyubov;* at 11:18, another trawler, the *Zavolzhsk;* and at 12:35, the destroyer *Stremitel'nyy.* Looking out the porthole with strong binoculars, I could see how the ocean was tossing the ships from side to side. Then came more trawlers named after authors: the *Lev Tolstoi,* the *Belinskiy,* and the *Novikov-Priboi.* At 13:43 we passed over Gander, where we could see the American continent to our right.

It was lunchtime. The sound of forks, spoons, and goblets, and the banging of plates could be heard in the salon, while pleasant smells wafted from the kitchen. I showed some initiative and called Bazenkov for lunch. He was doubtful: Would it be proper?

"Lord have mercy, I had a look and the stenographers and secretaries are sitting in there. So are we, the designers of this airplane, really unworthy of this great honor?"

Wavering, Bazenkov agreed. We took our places next to a table with four women, and tried to involve them in ordinary table conversation, in vain, for they were silent as mice. Evidently, they had to be silent toward those whom they did not know. Instructions . . . service. . . . Well, to hell with that and to hell with them!

At the first table along the starboard side were Nina and Anna Petrovna with their husbands Khrushchev and Sholokhov. The ignorant waiter had seated them near the propellers, where vibration was higher than normal, and the full, heavy cut glasses kept trying to slide onto the floor. The Kremlin cooking was excellent—appetizers, soup, steaks, cognacs and wines, all of high quality. Bazenkov and I decided not to drink, since we were more or less working. The delegates, however, gave the drinks their due. Dessert was coffee with ice cream, after which, according to an old Russian custom, the delegation went off to have a nap.

But then we received an insult intended to remind us who we were. About 15 minutes after our meal, someone in black came up to us and said, "Please refrain from visiting the restaurant; we will supply you with dry rations." Naturally, we did not go there again, and also left the dry rations for the "maitre d'hôtel." On the trip home, we ate what we had bought in Washington. But giving the foreign pilots dry rations was enlightening—they must have gotten an interesting impression of our democracy.

At 18:13 we passed over Boston, at 18:45, New York, and at 19:22 we came in for a landing at Andrews Air Base, 30 kilometers from Washington. The brakes squealed, and the airplane stopped next to a red carpeted path. Below us we could see President Eisenhower and three aides.

The delegates saw the American business-like attitude right away. As they walked to the podium set up for the greeting, they noticed footprints drawn on the concrete, with "N. Khrushchev," "R. Adzhubei," "N. Tikhonov," and so on, next to them. Without secretaries and without escorts, they all ended up in the right spot instantly.

The official greeting ended, and the delegates moved to their cars. Suddenly there was a hellish roar, like machine-gun fire, from the thirty escort motorcyclists who had all started their Harleys at once. The frightened guards ran to us. Once their fears were laid to rest, the cavalcade of luxurious Cadillacs carried Khrushchev and his entourage into Washington, D.C. We would not see them again for two weeks; meanwhile, we had our own agenda.

We were put in the care of American Air Force Major Miller, who knew German; I knew some, too, and this made our interaction much easier. Miller wanted to know "who's who" immediately. Then they gave us two buses, and a stationwagon for Bazenkov and me. Both the buses and the car had the inscription "United States Air Force." The Tu-114 was towed to a hardstand, surrounded with a light fence with the inscription "Police Zone," and had red flashing lights set around it. Now it was under two sets of security, Soviet security personnel inside, and American soldiers with carbines outside. At the entrance through the fence was a table at which sat a duty officer; he had a radio, a telephone, and a crew list in English. We received badges labeled "Kerber," "Bazenkov," and "Nuchtikov." You couldn't get through the layers of security without one.

Major Miller led us into a cafeteria that served all personnel at the base, both military and civilian, white and black. In the line at the counter were soldiers, workers, officers, generals, secretaries, and drivers. . . . Everything was done quickly and politely. The food was tasty, varied, and cheap. We were amazed when a Negro sergeant calmly approached a table where people were sitting, and without asking took a sugar bowl or bottle of ketchup. That is not considered polite where we are from. Of course, we understood that all this was purely democracy on the surface but, to tell the truth, it wouldn't have been bad if we had had the same sort of superficial democracy. . . .

Near the Tu-114 hardstand were several other Soviet aircraft: two Tu-114 weather reconnaissance airplanes, which flew about 100 kilometers ahead of us, and another Tu-114, which carried the correspondents and administrative personnel. Finally, there was the Il-18 of Khrushchev's chief pilot Tsybin (due to his age, he was not permitted to fly passenger jets), who delivered the sturgeons, smoked sturgeon, caviar, wines, vodkas, and champagne for receptions and banquets. There were nearly ten Soviet airplanes. Being inclined to mockery, our brothers-in-flight renamed this airfield near Washington "Nashington" ["Nash" in Russian means "Our"].

Another small anecdote. Before we flew home, Bazenkov and I wanted somehow to thank Miller for his help and his touching attention to our entire group. For gratitude, what could be better than Russian vodka and caviar? But there wasn't any of either left aboard the Tu-114, since the tourists had cleaned it out. I decided to go to Tsybin on the Il-18; our people said that he still had a whole grocery store onboard.

Tsybin greeted me coldly. "Unfortunately, I can't help you at all." What did he do? He took it all back to Moscow, across the ocean, all the vodka, cognacs, crab, sturgeon, and other victuals. We owe thanks to his crew, though, for helping us out with their own larder. Major Miller was touched.

Once everything was taken care of, theoretically, we could go into the city. We could, but didn't go, because someone was missing. They sent someone after him, but the ones they sent disappeared, too. Gradually, in searching for one another everyone spread out, and the bus was empty. Once everyone gathered again, and we could set out, it turned out that Ter-Akopyan had forgotten his bag somewhere.

Miller was nervous and asked sarcastically, "Maybe I should always have the bus arrive an hour later?" This was sad, particularly against the background of the businesslike, yet obliging American approach. During our two-week stay, not once did we leave the airport on time.

From this point on I will describe my impressions of America selectively. The impressions are strong ones, but the same ones that all Soviets get when they are there for the first time.

They put us up in a Hilton Hotel. Bazenkov and I could not manage to pay for a room with two beds on our high (by Soviet standards) salaries, so we asked them to put a folding cot in our room, and L. A. Smirnov moved in. The three of us somehow managed to pay for it.

I went down to the reception area, bought stamps, and dropped my letters home in the box with the "Air Mail" sign. Naively, I hoped that the letter would reach home the next day or the day after, but I received it myself, two weeks after I had returned home.

To be in America and not visit New York makes no sense. During the two weeks that the delegation was getting acquainted with America, we—per the schedule established by Moscow—were to go to Andrews every day and "be present." I decided to talk to our ambassador, A. D.

Men'shikov, and request that he permit us to visit New York. I was helped by our Air Force attaché, N. M. Kostyuk, with whom I had collaborated for a long time when he had been deputy chief of the Air Force Scientific Research Institute.

We set out early one morning, just as it was growing light. From the city, we got onto a turnpike, a highway that didn't have a single intersection all the way to New York. The signs read, "Minimum Speed, 60 mph." Accordingly, we didn't have the right to drive slower than 108 km/h! We weren't used to such a high-speed road (and still aren't, thirty years later).

Cities, farms, and fields flashed by. The farms were very rich, the fields were well cared for, enclosed with light gridwork fences. Clean, solid-colored cows wandered over lush grass. The farmers' houses usually had two stories, with a garage next to them. Under overhangs were tractors, plows, and farm equipment, washed and shining with bright colors, like children's toys. But then, all this was accessible to us in the Soviet Union, too, if there was a desire to work properly! In the yards, healthy, beautifully dressed children, some white, some black, frolicked under fruit trees.

At about noon it was time to rest and have a snack. We stopped at a gas station with a curious gas-pump sign: "Gasoline, 28 cents/gallon [generally speaking, this was expensive]. Cost price: 10 cents; profit: 4 cents; federal highway construction tax, 14 cents." So that is where these lovely turnpikes come from!

We traveled further. The factories were concrete palaces with clean, shining windows. It seemed that all the big enterprises had clean windows; what savings in electricity! With us, if they wash them at all, it is once or (when forced to) twice per year. I heard somewhere that several Dnieper hydroelectric power plants work just to make up for our dirty factory windows.

There were parking lots filled with countless numbers of cars and grandiose self-service stores with a huge selection of goods. Moreover, this wasn't Broadway or Fifth Avenue, but in areas where there are only farmers and workers from neighboring enterprises.

As we got closer to New York, we saw docks, with hundreds of cranes, piers, warehouses, and airfields. The cranes were constantly in motion.

No airplanes stood idle at airfields: they would land, be fueled immediately, be loaded, and take off again. It was obvious that the usage factor for people and equipment was several times higher here than with us.

We descended into a tunnel under the Hudson and soon appeared in the wide world, already on the island of Manhattan, that symbol of the United States. We decided to devote the day to seeing the city, and the following day to museums and shopping.

We drove around the city for an entire day. The skyscrapers were magnificent! Majestic, and made of steel, aluminum, glass, and plastics. There is something impudent about them, and they illustrate the capabilities of architects and engineers splendidly.

Twilight was falling. Overwhelmed by the experience, and since it was dark, we decided to get some rest. This wasn't difficult, since there were as many hotels as you could want. It was much more difficult to park the car. The streets were packed solid with vehicles. Selecting the 25-story Hudson Hotel, we circled it for a long time, waiting for some American to get the idea to free up a spot for us. He finally did. Our Cadillac was parked.

In the hotel, two clerks sat in a small booth. No questions were asked; they silently handed each of us a cardboard card. We had to fill it out with last name, first name, point of origin, and how long we expected to stay. We received the keys to our rooms immediately. A Negro boy led us to the elevator, which went up to the 23rd floor. We had a room for two, two bathrooms, two bathtubs. We were amazed that there weren't two TVs.

We decided to have dinner in our room. We went downstairs, found a tiny shop whose owner was an old Russian Jew, an emigré. We bought Smirnoff vodka, herring, pickles, and black bread. We spent the evening sitting around comfortably; the vodka, however, despite the two-headed eagle, the blue, white, and red flag, and the inscription "Supplier to the Court of His Royal Majesty," was utterly rotten.

The next morning we went to the Guggenheim Museum. It was a concrete tower with numerous elevators, and had splendid lighting that was well thought out in engineering terms. One of the walls of the tower was solid glass; paintings were hung on an inner wall. And they were beautiful, with works by Flemish, Dutch, Spanish, Italian, French, English, German, and Russian artists.

There were several canvases by Repin, Levitan, Korovin, Yuon, and Kustodiyev. We recalled a consignment shop on the Arbat. When he returned from the plant in Fili, Bazenkov liked to drop by there to see whether they had anything worthwhile, but reasonably priced. They always had something worthwhile, but reasonably priced—almost never. The worthwhile things were always purchased by "them," that is, the foreigners, and then exhibited there. Signs on the molding read, "Personal property of Smith, Boll, or someone else; declared value is ____ thousand dollars."

In the morning we bought wall hangings for the crew, at their request. The embassy personnel took us to the shop of Ya. G. Birkenshteyn, a Russian refugee, where they take everyone who arrives from the USSR regardless of rank and position. Birkenshteyn was a curious fellow who had preserved the language and customs of the turn of the century. He had fled Kishinev in the face of the Jewish pogroms of 1905.

"Mrs. Ulanova is a 'very practiced lady,' a demanding buyer, but don't trust her!" he told us. On Igor' Moiseyev, he said, "Now that is what I call a Jew! He doesn't know what *kashir* food is! They say he dances over there with you. I don't know how he is at dancing, but he knows what is what in textiles! I selected the material for his suit myself. A polite man: he thanked me."

Birkenshteyn walked out to the car with us and saw us off with, "Of course, all that is fine, communes, kolkhozes, state factories. But believe me, without private initiative you will flop." Prophetic words!

We didn't feel comfortable as we crammed tens of rugs into the trunk of our Cadillac with diplomatic plates. What did they think, the curious people gathering on Orchard Street? The most interesting thing to me was that the rugs with the "Shishkin bears cutting wood" were labeled "Made in Italy," had come to the United States, and were now being taken to the USSR.

When we said goodbye, Birkenshteyn whispered to me, "I, dear sir, give Russians a two-percent discount." What a complicated world we live in.

We set out on our return trip. A farewell loop along the Roosevelt embankment, around Manhattan, and then across one of the splendid suspension bridges, the George Washington Bridge, and we were headed for home on the turnpike across New Jersey. We could see far from the bridge, and I cast a final glance at the island of skyscrapers. They are like the snow-capped peaks of distant mountains, and they sparkle in the rays of the evening sun. Would I ever see you again? A stupid question: of course not, never. And that is too bad.

The minimum speed limit helped us to reach the Stratford-Hilton on 15th Street in Washington in only six hours.

After we had landed in America, before Khrushchev set out on his trip around the country, he instructed Bazenkov and me, at President Eisenhower's request, to familiarize U.S. generals with the Tu-114. There was good reason for the generals to assume that the Tu-114 was the brother of the Tu-95 intercontinental bomber and missile launcher.

We carried out his order. The generals came with their families, and, to our surprise, the wife was usually behind the wheel, with the general beside her and the back seat full of kids. The first to arrive was Nathan Twining, head of the Joint Chiefs of Staff. I showed him the airplane. His examination was of a superficial nature, however. A "banquet" consisting of vodka, Tsinandali, petit fours, and chocolates was set up hastily in the airplane's restaurant. There was no caviar, crab, or sturgeon; it had all been eaten.

After the Chief of Staff came the Chief of the U.S. Air Force, also with his wife and children, and everything repeated itself. His questions were very shallow. We had already begun to think, "Well, thank God, we are off the hook!" when the next general turned out to be a genuine specialist, and precisely in navigation and special equipment; in other words, he was a colleague of mine. "Behave!" I told myself, and right then I was confused by our navigator, K. I. Malkhasyan. He decided, from his experience in dealing with the first two chiefs, that this general was like a character from a Chekhov short story and just for show, too. Accordingly, when the general asked what the periscopic sextant was, Malkhasyan answered cheerfully, "An astronomical autopilot sensor." The general became unusually interested, asking, "So how do you fly off only one star?"

We had to mumble something about an inaccurate translation. When we ran low on generals, we heaved a sigh of relief. Our main comforting conclusion was that, as specialists, their leaders are by no means better than our combined-arms generals.

We saved some time to familiarize ourselves with the sights of the capital of the U.S.A. Our first trip was to an American national shrine, George Washington's plantation at Mount Vernon. It is not far, about 15 kilometers, on the banks of the Potomac. The two-story house reminded us of our old landowner estates. It contained antique furniture and everyday items—a plow Washington used in the field, a shovel he used for digging, a sickle his wife used for harvesting, and a spinning-wheel on which she spun wool. Everything was about the same as we have. And the doubts are the very same, too: Did he plough, did he dig, did she harvest, did she spin?

The most interesting trip was to the Smithsonian national museum and its aviation section. The whole history of American aviation is there: the Wrights' flyer; Lindbergh's *Spirit of St. Louis,* which crossed the Atlantic Ocean; Admiral Byrd's aircraft, which reached the South Pole; the airplanes of Wiley Post and Amelia Earhart; and military aircraft from the First and Second World Wars. To tell the truth, we felt a bit sad, since we do not have such a museum, and we are more indifferent to our history.

The Tu-114 had been standing for a long time without being operated, so we had to check it out in flight. We decided to fly for about two hours, from Andrews across the Chesapeake Bay to Cape May, then about 500 kilometers into the ocean and back "home," via Richmond, to Andrews.

Richmond, in the state of Virginia, was the capital of the Southerners during the Civil War, and is only about 200 kilometers from Washington, the Northerners' capital. We covered the distance dividing the two cities, which were once at odds with each other, in several minutes. Won't flights from Moscow to Washington be the same in 100 years?

Yashin and Kostyuk, the air attachés at our embassy, flew with us. They wanted to familiarize themselves with the latest Soviet aircraft and we had questions for them. It was a calm, businesslike discussion. This was a rare stroke of luck for them, since they had no direct contact with Soviet design bureaus. Everything learned in the United States had been sent through the Ministry of Defense, where someone filtered out the useful information. As a result, most of it grows old in safes and on shelves, without any benefit to anyone.

The engines, equipment, and controls operated impeccably during our checkout flight. The airplane was ready to return to its motherland.

On the last evening before our departure, Khrushchev arranged a huge reception at our embassy. Bazenkov and I were invited, too. Of course, we didn't have any tuxedos or calling cards with us. So what were we to wear? Fortunately, Khrushchev himself attended in an ordinary suit, and this took care of our concerns.

Precisely at 7 P.M., our premier took his place as host on the upper landing of the main staircase in the private residence, to shake hands with all invitees. Good God, Bazenkov and I could not get our fill of looking at all the stars! There were ministers, the secretary of state, admirals and generals, financial aces, industrial figures, ambassadors and envoys, famous actors, literary figures, pianists, and saxophone players . . .

The reception was standing only. The numerous tables held cognacs, vodkas, Georgian wines, different types of sturgeon, red and black caviar, veal and suckling pig, grouse and quail, Astrakhan herring and crabs, Siberian butter, Nezhin low-salt pickles, marinated mushrooms, in other words, everything that the pilot Tsybin could deliver from Moscow in his Il-18.

Bazenkov and I stood modestly near the wall, without interfering with the beau monde. At a table directly in front of us, an ambassador from one of the African nations, dressed in a white burnoose from head to toe, and the wife of the ambassador, naked down her back to her waist, and possibly somewhat lower, were hard at work. They drank, I can say, downed some real Russian vodka, and tried the caviar (with spoons). That was apparently the custom in Africa. We didn't try to make our way to the tables, and didn't want to. However, we didn't refuse the wine that the servants brought around. Van Cliburn, who had been invited by Khrushchev, performed one of our classics and then "Moscow Nights." Khrushchev, who was well versed in music, was touched. This was the crowning event of the day.

The entire event was a big success. Most of the newspapers gave it high marks, and did not forget to mention the hellish stuffiness, since there was no air conditioning in the embassy's old building.

As this splendid undertaking came to an end, Khrushchev again took his place on the landing, and again reshook several hundred hands.

Our hotel was not far from the embassy. We returned to our room, nearly starving. Fortunately, we had set aside a bottle of Stolichnaya and a mixture of food—bread, pickles in sealed, transparent packages, and smoked sausage. Reviving ourselves after being in the high-society atmosphere, we consumed the victuals after proclaiming a toast to Columbus.

Our flight departed at 10 P.M., but we began to pack that morning. I slipped a copy of *Doctor Zhivago* I had bought into my suitcase.[4] Our things were already in the trunk of the car when two gentlemen walked up to us and, turning over their lapels, showed us police badges. Unload everything! Bazenkov was on the verge of clarifying our treatment, but the embassy employee who had arrived just in time explained that there was information that a bomb had been placed in someone's luggage. Accordingly, FBI agents would be checking our baggage. As if I didn't have enough problems with my copy of *Zhivago!*

Everything was searched, however, and the agents were not interested in my banned novel. Our things were packed again, and we set out for Andrews through the city in the evening. In the airplane we took up our previous seats in the first cabin.

There was the darkness of a southern night, which was unusual for us. It was warm and stars were shining, film and TV crews were bustling around, getting ready to film the farewell ceremony for government figures from the two superpowers.

In the distance we could see the lights of a motorcade. Dozens of floodlights flashed on. It became bright as day. Our delegation was being escorted by the vice-president, the secretary of state, the head of the Joint Chiefs, and many other high-level people. There were farewell handshakes, short speeches, the whirring of movie cameras, and dozens of microphones extended toward Khrushchev.

Finally we boarded the airplane. These were our last minutes in America. Yakimov requested permission for takeoff. The engines were started. As during the departure in Moscow, Khrushchev came into the front cabin and sat ahead of us. The airplane began its takeoff roll.

Here things were done differently than they were at home. No other departures were postponed because of Khrushchev's departure, and we could see red flashes here and there through the porthole. According to a law passed in the United States, a powerful red light flashes continuously on every airplane at night, an "anticollision beacon." Khrushchev turned around to us and asked, "What sort of flashes are those?" I answered him. "And do we have those?" When I replied that we don't, he said, "Well, we should get some! They are useful, after all!" After we returned, I told Tupolev about this red light. He assessed the proposal at once and, within five days, we had designed a device in our design bureau, manufactured it, and tested it. As always, things bogged down when it came to actually introducing it. It was only when they learned at the Committee on the National Economy that this was Khrushchev's idea that everything was resolved with lightning speed.

We were climbing, the glow of Washington began to fade, the lights of other cities passed beneath us, and soon we were alone in the dark, bottomless abyss. Late at night, the delegates were sleeping in their compartments. The escorts were asleep in armchairs. Ministers, guards, advisers, and secretaries were all asleep. After checking that everything was going smoothly, Bazenkov and I began to doze, too.

Awakening to the touch of a hand, I was called into the crew cockpit. It was 3 A.M., so we were somewhere near Greenland. As I passed through the cabin, I caught myself on a metal handrail and felt a jolt of electric current. Where was such a noticeable electric potential coming from?

I understand as soon as I entered the cockpit. Tongues of blue flame streamed along the windshield. It was as if the propeller spinners and tips of the blades were on fire, and trains of fire flowed off them into space. All protruding parts of the aircraft—antennas, the turret of the astronomical sextant, wing fairings, air scoops—shone with cold, blue fire. Radio communications were not working, and the pointers on the magnetic and radio compasses moved about their dials arbitrarily.

On the horizon, in pitch darkness, the Northern Lights shifted about and then went out. These were the forces of nature clearly reminding us,

"Who are you, and what is your miracle of creating the Tu-114 and its important passengers? Nothing! Blades of grass, and nothing more . . ."

The crew was worried, and they should have been! M. A. Matyukhin bent over toward me. "Isn't it dangerous? Shouldn't we decrease our altitude?"

"Has this been going on for a long time?" I asked.

"About 10 or 15 minutes."

Calming him and myself, I said, "I think that, if it had been dangerous, something would have happened to us already. We have evidently ended up in a very powerful magnetic storm. The air is maximally electrified, and St. Elmo's fire, or smoldering cold discharges, has appeared on all pointed areas of the airplane. But since nothing has happened, then, most likely nothing else will occur. We are not in danger. However, I beg of you, keep a close eye on the course! It is not worth getting off course, and if we take the head of our nation to the North Pole or Africa, it will be quite embarrassing."

Nyukhtikov reassured me. "As soon as this show began, we switched to piloting by gyroscopic instruments, the aircraft artificial horizon and Pioneer." (Pioneer was an instrument that showed course deviation, in addition to other variables.)

The sight was sobering and majestic. I went into the tail cabin to look out the porthole at the wings and engine nacelles. The lights had been dimmed in there, and it was semi-dark. Everyone was sleeping soundly, thank God, or else some nervous passenger might have awakened and easily caused panic onboard. Only in the last row was a worried guard whose face had gone white; he was glued to the window.

"I want to wake the colonel!"

I asked him not to awaken him.

Through the porthole, long tails of cold, bluish flame could still clearly be seen flowing, clustering and flickering, off the wing cantilevers, the antistatic devices, and the tails of the engine nacelles into space.

After explaining to the guard what this was, in layman's terms, I returned to the pilots. Everything remained unchanged. The fireworks continued for half an hour, during which time we had flown about 400 kilometers. At about 4 A.M., it began to grow light in the east.

Dawn was approaching, and the St. Elmo's fire faded as the magnetic storm quieted down. Radio communications with Moscow were re-established. During the entire time, Moscow had been wondering whether a bomb had exploded onboard and the airplane had been lost.

The compass needles settled down, and the sense of alarm (why hide it?) gradually disappeared. Although we understood the phenomenon theoretically, it was still quite majestic and mysterious.

How lucky it was that the distinguished passengers had been resting! At about 6 A.M., Khrushchev came out of his compartment, drank a glass of tea, and began to dictate his Moscow speech to the stenographers.

The other passengers woke up, too, and the typical morning traveler's bustle ensued. Breakfast was served. At about 8 A.M. we passed Stockholm, then crossed the Baltic Sea, and after passing Velikiye Luki, we slowly began to descend toward Moscow. Ruza and Zvenigorod flashed by under the wing.

Taking advantage of the fact that the axis of the landing runway at Vnukovo nearly coincided with our course, Yakimov didn't circle, but came in for a straight-in landing. The brakes screeched, gangways were rolled up, and the doors opened.

Khrushchev entered the cockpit, thanked the crew, shook everyone's hand, went out onto the gangway, and descended to the Moscow ground. He was met by the radiant Brezhnev, Suslov, and the other members of the Politburo. There were smiling faces and friendly embraces. This was understandable, if the overseas visit of the head of our nation and the atmosphere of warmth during our stay there planted optimistic hopes. Didn't this mean that a new, friendly era had begun on our troubled planet?

Intelligence Gathering and the Tu-22

After the past few tales, the reader might be thinking Tupolev's design bureau had switched to passenger aircraft entirely. This would be in error, however. At the time, the Soviet press wrote extensively and eagerly about every new passenger airliner, but never a word about military aircraft. What was amazing was that "their" press managed to obtain such materials somewhere. Where did they come from?

The legions of security specialists and enforcers held general trade-union meetings at which they would urge us, "Be more vigilant. To a spy, a chatterbox is a godsend." After a leak, they would seal the windows of the bureau even more tightly than before, they would increase the number of 24-hour guardposts, eavesdrop on conversations among colleagues, and look under the smocks of our draftsmen for information leaks. But all this was in vain, for the spies snuck away and the publications continued.

But the sources were obvious. The pilots who participated in "fly-bys" would tell us, "We fly to Moscow along the Leningrad Highway, and the forest clearings are full of embassy cars. Their passengers are decked out in photo and movie equipment and photograph us continuously, at all headings. Radio-reconnaissance equipment is nearby on tripods, and its operators track us with their antennas."

Even the great Leonardo da Vinci noted that there is nothing more constant than the human head. By recording the pilot's head through the cockpit windows, a photograph has a unit of measure. It is then easy to reproduce an airplane's geometry from photos, with certain minor errors. Then, the angle of sweep of a wing provides an idea of the upper limit of the airplane's speed. The area of the wing, multiplied by the load per square meter used at the time, yields the weight of the airplane. Knowing the speed and weight, we know engine power. When a photo shows a bomb bay and a number of defensive gun positions, the airplane's function and even its crew composition can be determined. How easily the secret treasure chest opened!

Journalists from the socialist countries were another channel for leaks that was overlooked. Permission was given for photography even during static displays of equipment. These photos would usually appear in these countries' papers, and there were probably fewer restrictions on transmitting these photos elsewhere, such as to the West.

During the Cold War era another path appeared. The Americans would launch air balloons carrying automatic cameras from Japanese territory. After reaching the troposphere, they would be carried by air currents and drift westward across our territory. Our fighters downed some of them. The photos that were developed had recorded mainly forests and fields, but there were successful intelligence photos of important state facilities as well. For example, we were shown a photo of one of our airfields, in which Il-28 bombers and MiG-15 fighters could be recognized. The efficiency of this information-collection system was monstrously low.

Today spy satellites operate. One of the Western magazines contained a photo of the Voronezh airfield. It was easy to make out the words "No Smoking" on the front of the hangar.

After the Domodedovo air show in 1967, numerous photos of Tupolev's Tu-22 appeared in the Western press. Tupolev put this supersonic airplane in the category of his "unfortunate creations." An experimental airplane, for reasons still not finally determined, broke up and test pilot Yu. T. Alasheyev and his comrades perished.

Although V. F. Kovalev continued and completed the tests, there were incidents here as well. On a test flight in the winter of 1961, after noticing a drop in oil pressure in the right engine, Kovalev killed the engine. The heavy airplane began to lose altitude rapidly. Convinced that he wouldn't reach an airfield, he made a belly landing near the Pekhorka River. When the airplane hit the frozen ground (there had been severe frosts at the time), the cockpit tore off, slid several meters, and stopped. All hatches and windows were jammed due to the frame bending on impact, and the crew was trapped. The tail section with the engines was burning and the pilots could see the reflections of flames on the snow.

Kovalev told navigator V. S. Pasportnikov via telephone, "You can do what you want, but I am not going to burn alive: I am ejecting." On the ground, this is suicide. Fortunately, vehicles from the Flight Research Institute dashed up and freed the captives. The accident commission dug around for the initial cause of the accident. The oil line from the engine to the pressure gauge had broken right at the boundary between the engine nacelle and the airplane. This was a truly unfortunate location, because it gave rise to arguments over who was guilty—the engine specialists or the aircraft specialists. In such situations the predicament sometimes heats up and those that are not entirely at fault can be named.

In this case, things happened otherwise. The chief designer for the VD-17 engine, V. A. Dobrynin, a typical representative of the old Russian intelligentsia, put everyone in his place, saying, "Let's stop squabbling and search together for the truth and a solution to exclude such incidents once and for all. Is that correct, Andrei Nikolayevich?"

"Exactly so, Vladimir Alekseyevich."

I mention this example but there were other incidents, too. Another chief designer, from the younger generation, behaved aggressively in a comparable situation and ordered his subordinates to search the correspondence of his engine department for any "compromising" materials.

After the accidents, subsequent tests were passed and the supersonic Tu-22 bomber served the Air Force faithfully. It is true that many improvements were made to its structure, and it is even possible to say that it became a modification of the original concept.

The Tu-126 "Mushroom" (Soviet AWACS Aircraft)

The next project, which was assigned to us in the early 1960s, turned out to be quite interesting, but it lay in an area with which we had not been previously involved. Our country's northern borders were not sufficiently covered against penetration across the Arctic by foreign bombers. The ground-based air defense radars worked in wavelengths that propagated according to the laws of optics. Due to the curvature of the earth's surface, these radars could detect aircraft in flight at ranges no greater than 250 to 300 kilometers.

Taking into account the speed of the approaching aircraft, there was insufficient time to bring air defense to combat readiness. Alternative solutions were considered, such as putting the radars on the ice, closer to the North Pole, or elevating the antennas on high towers. The first idea was rejected due to Papanin's experiment: the ice fields of the central Arctic were drifting toward the Atlantic. The second raised concerns as to whether it would be possible to build tens of Eiffel Towers along the coastline.

The third alternative was placing air defense radars on a huge aircraft. The most difficult task involved in this would be mounting an 11-meter-wide revolving antenna on an aircraft. After prolonged and difficult searches, our configuration specialists found a solution in the form of a very strong pylon over the fuselage, on top of which was placed a flat oval radome. This solution, in turn, brought up two basic questions. First, would the eddies flowing off the housing ruin the efficiency of the tail rudder? And second, how would the gyroscopic movement from the mass of the huge rotating antenna affect the ability to keep the airplane on course?

Before agreeing to the project, we had to create aerodynamic models and subject them to studies in the Central Aero-Hydrodynamics Institute's wind tunnels. The wind-tunnel tests showed that it would be necessary to mount an additional dorsal fin with a larger area. With regard to the gyroscopic moment from the antenna, the autopilot specialists worked on compensating for it. After solving these basic problems, Tupolev and M. V. Keldysh reported to the government and the construction of the airplane, nicknamed "Mushroom," was assigned to the design bureau.

As usual, it was decided to preserve unchanged as large a number of assemblies from the Tu-114 as possible. Many innovations were made to the fuselage. At the mounting position for the pylon, which carried the wave guides for the antenna and a ladder for access, we added several heavy frames connected by additional reinforcing stringers. Almost all the portholes were eliminated, and a large, ventral fence-fin was installed.

Tupolev assigned management of work relating to Mushroom to Bazenkov and B. D. Khasanov, the chief of our affiliate at the plant where the airplane was being built. Tupolev retained technical management of the specialized subunits at the affiliate and construction of the test stand for ground tests of the aircraft radar at the Moscow design bureau.

Selection of a site for building the test stand was not simple. The KGB required that it be at least 800 kilometers from the state border and any seas, and that the emissions be directed only eastward. "Interesting," said our boss. "So how do England, France, and Germany, who have less territory, work on such things?" We selected the test-stand site jointly with Metel'skiy, the Chief Designer for Radars, using an An-2 airplane, since, in addition to the above requirements, it was necessary to have an airfield adjacent to the site. We were forbidden to transport components of the radar on trucks. After agonizing over this, we found an isolated site east of Moscow, with no cities, large plants, or dacha villages nearby, but with an airfield not far away.

As required, they built a full-scale mockup of Mushroom at our affiliate, then convened a commission made up of representatives of the Air Force and Air Defense Forces, who examined the mockup and approved it.

I will not detail the endless number of technical issues we had to resolve, but I must mention three of them. When we went to the ball-bearing plant and asked them to build an antenna ball bearing with a diameter of about one meter, that would have to operate under flying conditions with enormous changes in temperature, pressure, and vibration, they ridiculed us. Only when we showed them the mockup were they convinced that this wasn't nonsense. The chemical specialists refused to make a honeycomb housing 11.5 meters wide. We had to fabricate it in our laboratory, from sections joined by plastic bolts.

During their inspection of the mockup, the medical specialists stated that the crew in the forward cockpit would contract leukemia and die from the several hours' exposure to the radar beams, and that the measures we had adopted to screen them were ineffective. Of course, this could not have occurred, for the lower limit of the emissions diagram was about three meters above the pilots' heads.

However, time put everything in its proper place. Mushroom was driven to the airfield. Our test pilot, I. M. Sukhomlin, arrived and walked around the airplane, hemmed and hawed, chattered away a bit, and climbed up the ladder into the cockpit. He tried the engines, then taxied around the airfield, where half the plant had gathered to get a look at the unheard-of marvel. Laughing, he reported, "Okay, what the heck, we will try it out! And maybe, God willing, we will even return alive." Of course, the latter was said in jest.

It was decided to carry out the first flight on a day off because since the plant would not be operating, there would be several thousand fewer observers, and the housing project where most of them lived was far away. It was "Hurry up and wait." Bazenkov was worried, justifiably, since he had a bean-shaped object 12 meters in diameter and almost 1.5 meters thick in the center, sitting above the fuselage on a pylon 2 meters high. They started the engines. Sukhomlin made runs here and there on the taxiway and then set out on his journey. Soon he had the airplane about 3 meters off the ground, kept it straight and level, and began to gain altitude. Everything went absolutely fine. After reaching about 500 meters, he flew two circles, then flew over us twice at an altitude of about 100 meters, and then came in for a landing. One more Tupolev airplane, this one with somewhat fantastic lines, had been born safely.

That evening, as usual, they washed the newborn and went off to sleep. Before I went to bed, I noticed a copy of *Izvestiya* with a photograph entitled, "Over the Motherland's Wide Open Spaces," which showed the Tu-112 flying over a birch forest. With a very sharp pencil, I added a dotted Mushroom to the aircraft and placed it on Bazenkov's nightstand. The next morning dawned noisily. Yells could be heard coming from Bazenkov's room. "The swine! Scoundrels! What are they doing? Who gave them permission? What will happen now that they have declassified our airplane for the entire world to see!"

Seeing that he was serious, I erased my drawing. A little bit of strong "medicine" put him in a good mood. The morning was truly festive. Our protracted labors were finally complete, and we had won a huge victory.

The Tu-28

Let's jump back to the predecessor of the Tu-22, the experimental No. 98 aircraft, a frontal supersonic bomber. Its favorable test flights with pilot D. V. Zyuzin were completed at an unfortunate time because Khrushchev had just uttered his sacramental phrase, "We can now destroy a fly in space with a missile." This turned out to be enough to halt all work on aircraft numbers 98 and 91.

This was understandable in the case of the Tu-98, despite our desires, since, after all, the Army had a certain choice of surface-to-surface missiles. In the case of consigning the Tu-91 to oblivion, however, even the most contorted mind couldn't come up with any arguments.

As Tupolev read the press reports on the course of battles in Southeast Asia, he concluded that a special aircraft was required to support infantry detachments operating split-up in the thick of the forest. It had to fly low and slow, be capable of being mass-produced, and carry very diverse types of weapons. Experience from World War II had shown that aggressive air defense artillery would bring down low-flying airplanes quite often, at a high rate. We lost several thousand even of the fully armored Il-2 ground-attack aircraft. This meant that introducing armor into the structure was not the answer, since it was heavy and labor-intensive. However, it was possible to attempt to shield the crew against small-arms fire using the propellers in front (two coaxial propellers) and the engine to the rear. The shaft connecting the two would divide the cockpit in half. On the left would be the pilot, on the right, the navigator. We would limit ourselves to a speed of 500 km/h, and an altitude no greater than 5,000 meters. This meant a straight wing and no need for a pressurized cabin. All weapons would be suspended from three pylons: one under the fuselage and two under the wings.

The Navy supported the idea, as did the Air Force. Through combined efforts they developed the tactical and technical requirements; the government approved them; and our design bureau began work. P. O. Sukhoi was designated lead designer; he was subsequently replaced by V. A. Chizhevskiy.

The final product was the Tu-91. After undergoing all sorts of tests—it launched rockets, laid naval mines, fired from cannons, launched torpedoes at ships—now, it had been abandoned. Khrushchev's illusory idea of replacing aircraft with missiles killed it, as it did the Tu-98. After suffering this affront with the Tu-91, Tupolev was still unable to live with the fact that an advanced airplane such as the Tu-98, which had been the vehicle for solving numerous aerodynamic problems and for creating our first twin-engine airplane with a speed of about 2,000 km/h, had been rejected.

Tupolev always reacted to such incidents bitterly and had trouble accepting them. "They spent a bundle of the people's money, and all for nought. This can't be. If we have created a truly outstanding airplane, it means that we must find it an application." And he did!

After prolonged searching, consideration, and consultations with the military, radar specialists, and air-to-air missile designers, Tupolev formulated a plan. "We created the 98 bomber; let's turn it into a long-range fighter-interceptor. These airplanes will shift our air border far to the north. This is possible if we install a radar sight with sufficiently long range instead of the navigator's compartment in the nose, fill the bomb bay with fuel tanks, and put missile pylons under the wings. Calculations have shown that, at economical speeds, this type of fighter can fly defensive patrols for several hours. We will ask the government to order Chief Designer F. F. Volkov to create a sight, and Chief Designer M. R. Bisnovat to build missiles with the required range."

This proposal was approved, and the airplane was designated the Tu-28. Pilot M. V. Kozlov conducted the flight tests. (He would subsequently perish in a Tu-144 accident during an air show at Le Bourget. Upon seeing an unknown aircraft crossing his course and fearing a collision, Kozlov was forced to use a rough maneuver to veer off abruptly. One version of the accident holds that it was due to overloads exceeding nominal ratings.)

Somewhere in the deserts of Kazakhstan, the Tu-28 launched its missiles against target simulators. After completing all tests, we transferred the airplane to the military. Many years of experience in cooperating with the bomber department at the Scientific Research Institute enabled us to anticipate their desires and requirements. Now we were working on fighters, which had what was considered a traditional cockpit layout. Maj. G. T. Beregovoi was the designated test pilot for the Tu-28. We agreed to implement numerous requirements from him and from department engineers regarding improvements to the cockpit interior and ease of using instruments and control; however, one requirement gave us much difficulty.

As we wound up our work on the Tu-28, we began to formulate the document papers to investigate state testing. Beregovoi demanded that we add a button for "return to horizontal" in the Tu-28 autopilot. When a pilot in an ordinary fighter has been distracted by a dogfight and has just executed numerous acrobatic maneuvers while chasing the enemy, and then enters clouds, he is incapable of immediately recognizing what is sky, what is ground, and the pitch of the airplane. He would have only to press this button, and the autopilot would automatically return the airplane to horizontal. We tried to convince Beregovoi—in vain— that the Tu-28 was not an air-to-air combat fighter, but an entirely new class of aircraft: a flying platform for launching long-range air-to-air missiles. The missiles were launched when the target was still tens of kilometers away. We also added that the Tu-28 had no cannon or machine guns for firing at the enemy and, finally, that the Tu-28 was not intended for the overloads typical of ordinary fighters. Our arguments fell on deaf ears because "that is the way things are done on fighters."

Whether this involved rational ideas or "democratic centralism," fortunately, Senior Chief A. A. Manucharov eliminated the requirement for the button. In general, however, this infamous button opened up a Pandora's box. We had to install a second autopilot on the airplane, which meant a full cycle of new tests of the airplane in all possible modes, in sum, nearly six months' worth. And all that, you understand, cost a great deal. However, money was no object at that time.

After passing the entire cycle of complicated tests, the Tu-28 went into series production, and we were rightfully proud of the foresight of our general designer, who had been able to give an abandoned airplane a second life. It is unfortunate that the Monino Air Museum portrays this interesting airplane as "not fully operational."

We displayed our last airplane associated with air defense to leaders, headed by Khrushchev, in 1962, at one of the southern test ranges. I should note that there are various approaches to such displays, probably related to the individual traits of leaders. Stalin never visited the aviation design bureaus or flight bases. They said he was afraid of terrorists. Khrushchev did this eagerly, but always unexpectedly. Brezhnev introduced a specific ritual. All the chief designers and their deputies, plus military leaders, ministers, and their assistants would be summoned to the test range ahead of time. At the appointed hour a motorcade would bring Brezhnev and the Politburo members to the first group, which had been formed up along the runway. The guests made their way unhurriedly along the groups. Explanations were provided by the chief or general designers, but the guests did not pay any particular attention. We had the impression that what they were told in detail did not interest them. This looked more like some event, or unusual show.

On one of Khrushchev's official visits, we demonstrated the Tu-16, equipped with means for jamming radars. Khrushchev became aware of the complaints of Tu-16 pilots flying near the border. From across the border they would make the Tu-16's onboard radars become distorted, incapable of proper detection. An alarmed Khrushchev asked that the navigator-radar operators and specialists from Air Defense to gather at one of our air bases. Marshal P. F. Batitskiy also attended. After examining the airplane and listening to the specialists, Khrushchev asked to see everything in action, to see how an air defense site operates.

Our Minister, P. V. Dement'yev, asked to combine this demonstration with the one under preparation on innovations in aviation technology. Khrushchev agreed, and added, after thinking a bit, "At the end of August or early September."

The first day at the range was dedicated to jamming. A flight of Tu-16s, among which was our airplane with jamming equipment, approached the test range from the northwest. When the flight was 200 to 250 kilometers away, the display screens for the ground-based radars showed blips for all three airplanes. When the order was given to commence jamming, the screens on the ground-based radars were immediately covered with bright dots that nearly merged into a solid spot. No matter how hard the air defense operators tried to find the blips for the approaching airplanes, they were unsuccessful. The experiment was repeated several times, until Khrushchev said, "Enough!"

He turned to Batitskiy. "Marshal, do you recall what you were telling me in Moscow? You said, 'This is tendentious information from interested parties. On the test range our operators will show you how easy it is to get rid of any jamming.' Is this incompetence or the desire to reassure me? This is more valuable to me. Although it is a bitter fact to face, it is true."

The marshal, a large man with the face of a grand prince from Russian paintings, obviously began to get nervous. Khrushchev continued to attack him, and those around them were worried. But P. S. Pleshakov, the Deputy Minister of the Radio Industry, defused the situation.

"You know," he said, "our institutes have developed a method that makes it possible to see moving targets, even in jamming."

At first Khrushchev growled, "You are doing the same thing, reassuring me."

"No, it's true," insisted Pleshakov, and the conflict died down.

Life Changes and the Tu-124

In spring 1962, Yuliya Nikolayevna passed away suddenly and unexpectedly. They had lived together in harmony for 40 years, and during all those years I do not believe there were ever any serious disagreements between them. Her death was very hard on Tupolev.

Their children were settled down to their own married lives. They had their own children, their own lives, friends, and interests. Some of his old comrades (dating back to the early days with N. Ye. Zhukovskiy) had become chief designers; some, like V. M. Petlyakov, had perished in aircraft accidents; and still others had distanced themselves from him and showed no desire to communicate with him. Isolation set in unexpectedly, the friends that had been ready to support him at difficult times were no longer there, and the only soul who deeply understood him had passed away.

Slowly and without comment, he began to withdraw. Even with his only friend from his student years, Sasha Arkhangel'skiy, the intimacy appeared to have melted, like snow in springtime. "And this was by no means because I wanted it," confided Tupolev sadly. The inexorable solitude of old age was slowly but surely drawing him in.

People feared for his health; after all, Tupolev was already 74. The doctors insisted that Tupolev spend the winter at the Academy of Science's "Uzkoye" sanitarium. We tried not to alarm him, and said that, if necessary, we could consult with him there, but only when especially critical matters piled up. I had occasion to go there to consult with him.

It was a marvelous day, windless, sunny, and frosty. After the quiet hour, Tupolev had gone out for a walk in the park, wearing a winter coat, his *papakha* fur cap, and thin felt boots. Upon seeing me he was overjoyed, took me by the hand, and peppered me with questions. It seemed even at Uzkoye his tireless thirst for activity and creativity continued.

Noticing a squirrel among the branches, he slipped his hand into his pocket, brought out a handful of pine nuts, and started to feed him. Looking at the nuts, he mumbled, "Lyalya brings them," in a tone expressing both joy that his daughter cared about him, and grief that his wife was gone. When it came time to leave, he gave me a heap of instructions.

We were involved in the rapid creation of the Tu-124. Aeroflot had enough large, high-speed aircraft with the Tu-104, Il-18, and An-10. What they urgently needed was a high-speed jet for medium-distance routes on which the slow-moving Li-2 and Il-14 still predominated. Settling down to his new assignment, Tupolev decided to use the Tu-104 as the structural basis for the new design. Since the Tu-124 had half the number of passengers of the Tu-104, it could use less powerful engines, and, naturally, the airplane's size could be reduced. Very similar in flight to its older brother (the Tu-104), the small Tu-124 gave rise to the legend that Tupolev had taken the plans for the Tu-104, reduced their scale by one-third, and that was all there was to it. When the legend reached his ears, Tupolev smiled. "It is a nice thought, but how in the hell can you reduce people proportionally, too? It is really too bad, but they don't make them from our blueprints!"

The Tu-124 became quite popular, in part due to the following incident. During takeoff a passenger version of the Tu-124 flying out of Tallinn lost an unpinned bolt in one of the links in the front landing gear. The bolt didn't carry any sort of load, and landing without it would have been safe. Nonetheless, to lighten the load on the front gear during landing, the controller at the Leningrad airport ordered the pilots to circle to use up fuel. Nervous about the number of aircraft in the airport traffic zone, and fearing a possible collision, the crew apparently failed to notice that they were about to run out of fuel. When the engines died, the crew had to land, but the closest level site was the surface of the Neva River. Raising fountains of water, the Tu-124 went into the water just past the Okhta Bridge. Launches quickly evacuated the passengers and crew. After about an hour and a half the Tu-124 settled to the bottom. Two days later, the Tu-124 was raised by crane, dried out, and repaired.

For the aircraft's designers, nothing supernatural had occurred. When it ordered a passenger airplane, Aeroflot had foreseen the possibility of forced landings on water, and had instructed the designers to ensure an hour of buoyancy in such cases. For passengers, the incident involving the Tu-124 was a sensation, and a saying made the rounds of the air terminals: "Tupolev's airplanes don't go down, even in water!"

In fall 1964 Tupolev was vacationing in Miskhor but he phoned Moscow frequently to find out how things were going. When some complications arose with the Tu-124, Tupolev asked D. S. Markov, K. V. Minkner, and me to come see him. At about 11 A.M., loaded down with blueprints and calculations, we landed in Simferopol' with our own test aircraft. Within an hour, we reached Tupolev's residence.

The issues were resolved rather quickly, and Tupolev insisted that we stay for lunch. We all sat down on the terrace; the mountains shone blue in the distance. After looking around and winking at us, our boss took out a bottle of cognac. "So there! The doctors and my daughter don't let me have any. They are suppressing me, but good." We all smiled together, imagining how it could be possible to "suppress" him. "Let's say that you brought the cognac from Moscow," he continued, conspiratorially, "and we understood that his vacation had been good for him, and he was 'in shape.'"

That evening we returned to Moscow. Since we had been creating airplanes all our lives we really didn't realize how they had changed everyday life for business people. And after flying to the Crimea in one day, resolving several urgent issues, and returning home immediately thereafter, just as if we had gone to Zagorsk or Klin, we were even a bit proud.

Not long after this, the Tupolev family settled in Kostyanskiy Lane, and things became crowded again. Tupolev's son Aleksei had a son, Andrei, and soon thereafter, a daughter, Tatyana. Life was such that the family was expanding exponentially, and, willy-nilly, required more and more of the infamous "living space." For the umpteenth time it became necessary to find more "living space," and the Moscow City Council granted the Tupolevs two apartments on Stanislavskiy Street. Lyalya took over her father's care. Tupolev's health had declined, and who, if not she, constantly watching over him, could have been a better doctor?

Tupolev, his daughter and her husband, and his first and (as happens in such cases) favorite granddaughter, Yuliya, lived in one apartment. In the other were Aleksei Andreyevich and his wife, Maya Aleksandrovna, son Andrei and daughter Tatyana. When she was born, Tupolev joked, "How about having the engineers figure out when my descendants will fill a whole district in Moscow? After all, I started out with one cot in a dormitory at the Moscow Higher Technical School on Koroviy Brod, and now I need 10 beds for the family. That is after 50 years. What will happen in the next 50?"

The Tu-134

The only problem with the Tu-124 was its passenger load capacity. When it reduced traveling time to one-third of what it had been on the same routes, the number of people who wanted to take advantage of Aeroflot's services leapt upward again. A passenger jet was needed for local routes, but one with a much greater capacity. At about this time, designer P. A. Solov'yev produced a more powerful engine, and the proper favorable conditions had been created.

As he considered how best to carry out his assigned mission, Tupolev proposed retaining the fuselage structure but lengthening it by two meters, and shifting the engines toward the tail. This would increase wing quality, and make the Tu-134 airliner easier to control.

These changes permitted the plant to shift to production of the new model. During testing, however, the lead aircraft unexpectedly went into a steep dive and crashed on a routine flight.

Tupolev investigated the accident himself. The auxiliary control surfaces (trim tabs) on the Tu-134 were controlled by small electric motors. Tupolev wondered, "What would happen to the airplane if there was spontaneous shorting in the wires, which caused a false signal to be generated, and the elevator unexpectedly and rapidly shifted to its extreme position?" A wind-tunnel experiment provided the answer: a prolonged, steep dive, which is apparently what happened.

Then Tupolev formulated the solution, a scheme that would prevent any similar situation from being repeated under any circumstances. He ordered the developers (our engineers V. Ivanov and V. Venikov) to keep him informed about their work on a regular basis, and consulted with them constantly.

The problem was a difficult one, and the solution was elegant. The developers proposed mounting a device in the rudder that would prevent the trim tab from deviating more than 2 degrees from its set position. This ensured that the airplane would never go into an impermissible operating mode. Pilot A. Kalina carried out the flight testing, which was a great success. The overload limiter units were installed on all our airplanes, and subsequently on others.

Traditionally, an order for a new instrument went to the autopilot specialists, which would require several months. In this case, after spending weeks analyzing the causes of the crash, we began the development within 10 days, and within one month the overload limiter units were already flying. In similar cases, Tupolev would depart from tradition, take the responsibility himself, but achieve the necessary results within very short deadlines. Not many people followed this path.

During his early years, as far back as his work on the ANT-1, Tupolev frequently looked in on the laboratory, and he conducted most of the aerodynamic and strength calculations. He made sketches for the draftsmen of almost all assemblies in the airplane's structure.

While working on with the ANT-4, it became clear that one man could not handle the volume of work. We therefore created teams for the wing, fuselage, and tail led by Petlyakov, Arkhangel'skiy, and Nekrasov. Specialists such as these could readily be entrusted with much. Our workload became easier to manage.

As the war approached, however, and especially after it, during work on the Tu-4, even this rigid an organization interfered with our work. Tupolev was getting bogged down in details, so we organized specialized departments headed by deputies. Things improved, but parochial (or, perhaps, even feudal) emotions appeared among us. Today I can recognize that matters had improved, but endless amounts of time were spent reconciling our arguments.

Tupolev liked to cite Aleksei K. Tolstoi:

> Our passions were direct.
> Order? Not a bit!
> We know that without authority,
> You can't get very far. . . .

And so Arkhangel'skiy and I used our authority! One of its manifestations was my personal team (or, more precisely, an overall design department). When we gave the order to transfer there one or two good specialists in each area, the resolution of thorny structural questions occurred much more quickly and—more important—without unnecessary feelings of anxiety. All fundamental decisions were always made in my presence.

Several successful years passed in this fashion. When scientific and technical progress were advancing rapidly, however, and aircraft design (particularly the outfitting of airplanes) had grown immeasurably more complex, life forced us to switch to the current organizational structure.

In 1952, during the celebration of our design bureau's thirtieth anniversary, I stated, "Progress in aviation is fostered by the collective labor of people." Today, 20 years later, I would add only a few words to this statement: ". . . the collective labor of extremely highly qualified specialists in various fields, at all levels from workers to masters, engineers, programmers, physicists, and mathematicians."

However, that is the nature of discussions. But in practical terms, I am trying meticulously to understand what caused this lamentable incident, and how to prevent them, once and for all. I even realized [he laughed] how your overload limiter was constructed. By the way, no matter what anyone says, it was a clever design!

The only person who makes no mistakes is the one who does nothing, and so Tupolev, who did very many things, also made mistakes at times. The following incident took place when the T-134 was being created. Several colleagues had tried to convince Tupolev that engines might not start using electric starters powered by the standard storage batteries. The rotors of the new, more powerful engines had too much inertia. He did not agree, however. They were correct, however, and it was necessary to install an additional storage battery.

At this point, Tupolev flared up: "How much do three of your damned batteries weigh?"

"By themselves, 360 kilograms, and with the cable and relays, they will hit half a ton at least!"

"So why the hell didn't you insist on installing a single auxiliary engine instead of them?"

"Just a minute: we were the ones trying to insist on it!"

"You were trying to insist on it, but you didn't insist on it, and now, of course, I am the guilty one. You should have convinced me otherwise; you see, you should have, and now we have the devil-only-knows what!"

We had to install a small auxiliary engine. Naturally, this caused lots of reworking, and Tupolev reproached his colleagues: "You buckled under to my authority, that is what. You think that authorities never make mistakes? Instead of that, it would have been better if you argued and defended your point of view, and didn't agree with my opinion. And now I am upset with you precisely for that reason. Authorities make mistakes, but the truth prevails! Understand?"

We understood very well, but his personality was such that you couldn't always insist on your own, possibly wiser, decision. He didn't particularly like to admit his own mistakes, however. This didn't prevent him from recognizing his subordinates' right to make mistakes. And if a mistake wasn't too bad, he didn't torment the one who made it, but reproved him gently.

One way or another, the Tu-134s that had been put into series production went first to Aeroflot, and then onto the international market. They were purchased quite eagerly, of course, since the new airplane met the highest international standards. It could transport 80 passengers at a speed of 900 km/h over a distance of 3,000 kilometers. If that wasn't enough, if clouds had socked in the airfield when the Tu-134 reached its destination, the T-134 could make an automatic landing approach.

Tupolev had been interested in blind landings for a long time; however, true to his principle of unconditional safety for civilian aircraft, he did not rush things. When people who were "too progressive" (and they could be found in aviation, too) would bother him with, "Let's introduce it right away!" he would reply: "Automation is good, there's no doubt about that. But as long as it can cause even one accident, it is best to wait a while. When you can prove to me that there is only one failure in 100,000 landings, come see me, I will be glad. But for now, the honor was all mine . . ." and he would extend his hand to bid farewell.

When several tens of T-134s had been produced by the series plant, automatic devices reached precisely that level. After visiting the designers of the devices and being convinced, Tupolev invited the equipment developers, Ye. V. Ol'man and G. M. Gel'dman, to visit him.

"The results of testing on your equipment for the Li-2 have made an impression. So I have a request. My pilots are outstanding—you know that, don't you?"

Both designers eagerly agreed.

"Great! In that case, dear comrades, let's you and me and my assistants all make a couple of dozen automatic landings on the Tu-134. It will look good for you, and I will be calmer for it."

Later he admitted that this was a purely psychological test for them: Would they agree at once, or, expressing doubts, would they prefer to decline politely? The designers did not flinch, and the automatic landings were carried out. Then they performed a thorough checkout at several airfields here and overseas. Only after these tests did Tupolev give his approval for the use of this automation on passenger airplanes. In 1972, Tupolev was awarded the State Prize for this lovely airplane.

The Tu-154

Tupolev's lively public activity did not let up, either. He attended the Pugwash Conference of Independent Scientists of the world as a delegate representing Soviet scientists and delivered a spirited speech. He was elected president of the Soviet-Bulgarian Friendship Society. In these activities, as with anything else in his life, he carried out his responsibilities conscientiously. So it was in this case. He traveled to Bulgaria, became friends with Tsola Dragoycheva, the president of this society, spoke at its meetings, and did much to help the Bulgarian airline "Tabso" to introduce the Tu-134. They loved and respected him in

Bulgaria, evidenced by the fact that the first biography of our chief designer appeared here before any other Socialist nation.

Only great artists are granted the fortune to create immortal works. Aircraft engineers have been deprived of this possibility. Their creations are not immortal. No matter how excellent they were, the Il-18, An-10, and Tu-104 became obsolete. After serving the motherland honorably, the Tu-104 went on permanent display at Vnukovo Airport. One year after that a Tu-114 was placed on a pedestal at Domodedovo.

To replace them it was necessary to create an improved aircraft. The government entrusted this to Tupolev, decreeing that the new airliner, the Tu-154, have the speed of the Tu-104, the range of the Il-18, and the economical operation of the An-10. The Tu-154 had to have instrumentation for automatic landings in fog, be absolutely independent of airfield maintenance equipment, and be able to compete with the best airliners of Western nations. These proposed requirements resulted in an unusually large volume of theoretical calculations.

"As I was finding space for our computer center and had to crowd you"—Tupolev was speaking to his assistants—"you raged and reviled me. You think I don't know that you were muttering, 'Tupolev's losing his mind.' However, it is only due to this center that we succeeded in completing the plans for the Tu-154 within such short deadlines. And what would have happened after that, when the volume of calculations and modeling is growing ever larger?"

He was proud of the computer center, and eagerly showed it to visitors. The center illustrated his passion for advanced methods in engineering. We were remiss when we didn't award him with an honorary gold engineer's badge, the same type that was given to his teacher N. Ye. Zhukovskiy many years ago. It was in Zhukovskiy's design bureau, long, long ago, that the loft-template design method was first introduced. It made work easier for the designers and increased the accuracy of aircraft manufacturing. He was one of the first to introduce, at his design bureau, precision casting, deep forming, the manufacturing of parts via chemical etching, magnesium casting, electrographic reproduction of blueprints, and a host of other progressive methods in the design and construction of aircraft.

And, typically, Tupolev really would work to achieve these things himself. Everyone knew that when one of these innovations was being introduced there was no sense in stopping by his office. You wouldn't find him there, for he would be "on the field of battle," in the department, shop, or laboratory where it was being introduced.

So how were the numerous and, at times, contradictory requirements for the Tu-154 to be met? Together with D. S. Markov, the lead designer for the airplane, and S. M. Yeger, his right-hand man in matters of overall configuration, Tupolev sought the basic designs for the first Soviet airbus.

First we need a few more doors, and a fuselage that is as close as possible to the airfield surface. That way, the unproductive time to load and unload the airplane is reduced greatly. You say large wheels interfere with lowering the fuselage? Well, replace the four large ones with six smaller wheels. Over on the Tu-144, where there is a very thin wing, Aleksei Andreyevich [his son] and I are figuring on a landing gear with twelve-axle bogies.

Second, to improve stability and ease of control, particularly during takeoff and landing, we will make the wing clean. We will take the engines off the wing and place them behind it. We will choose engines with thrust reversers for braking the airplane on the ground. By cleaning up the wing we can equip it with mechanized systems across the entire wingspan (this is at the request of pilots, and makes it possible to increase or reduce lift).

Third, we will achieve full autonomous operation of the Tu-154 on the ground. We will put in a small auxiliary engine that will condition the air in the cabin on the ground, illuminate the cabin, start the main engines, and, if necessary, lower the landing gear and the landing flaps in flight.

Suddenly he became furious.

We were flying with the military from Astrakhan in a Tu-104. It was devilishly hot, and then they delayed our takeoff. Well, you know what the military is like: they have a lot of endurance. But I looked around, and generals, and then all the others, were taking off their jackets, undoing neckties, and pulling off sweatsuits. I was interested to see what would happen next. Was it going to be like a sauna, where they would undress all the way? And how is it, then, on normal flights, when there are women and children onboard? Are they supposed to get undressed, too? The devil only knows how you and I torment people!

In fall 1967 they finished the assembly of the airbus, and pilot Yu. Sukhov took the Tu-154 for its first flight. On 23 July 1971, *Izvestiya* wrote, "And then one July morning the announcer at Vnukovo Airport announced: 'Passengers on the Tu-154 are requested to proceed for departure.'" One more "Tu" began to serve the people.

The Tu-104, Tu-114, Tu-124, Tu-134, and Tu-154! What a magnificent constellation of beautiful passenger airplanes! While it may appear that Tupolev only worked on civilian airplanes for all those years beginning in 1952, he developed and produced outstanding military airplanes, too. But it is not yet time to talk about them.

Although Tupolev was in his seventies when he worked on the Tu-154, he remained, as before, a tireless worker. All his life, exactly at 8:30 A.M., Tupolev was already at the plant. Where he would set foot only God knew! It could be a shop, the garage, a construction site for the next building or residential building, the design bureau, the computer center, one of the many laboratories, the warehouse, the supply department, the medical station, or the dining area. This wasn't a parade review. He would delve into all the details, make note of problems, berate the lazy, praise the diligent, take note of needs, and—always—provide help. His words never deviated from the task at hand, and everyone trusted him. His demeanor would fill his colleagues with renewed energy. Once he had made the rounds of his huge enterprise he would go to his office. His first task was always to review the "synodics," as he called the summaries of work on all experimental and series-production airplanes. Only then would he immerse himself in his holy of holies—solving engineering problems.

Tupolev had a well-developed sense of humor, and he knew how to respond in comedic situations. At times, during long drawn-out quests for a solution, when those gathered were tired and couldn't reach an unambiguous decision, and everything seemed hopeless, Tupolev's clever word, some unexpected, apt comparison, or his loud, rolling laughter would immediately ease the tension and inject new energy. On one occasion, after a prolonged discussion of a complex matter related to aircraft structure, when two opposing designs had been formulated and no compromise had been found, he suddenly asked, "What tediousness! Wake up: How should we build our bridge, along the river or across the river?" Everyone laughed, and things were back on track. His ample supply of proverbs, puns, and anecdotes provided an important reservoir of wisdom to guide us. To a colleague who was offended that he had been given an insufficient award, he said, "Money and gifts blind the eyes of the wise." To a designer whose part had broken and failed a test, "Put your trust in God, but do your counting yourself." On a meeting held for organizational reasons, "Awfully splendid, but awfully untalented." After listening to a concert at the Scholars' Center, given by one of the middle-aged minstrels, "Frankincense and moths." After watching a thundering serial film, "For rural landowners." At a review, "For long ones, you have to come with a picnic basket and a cot." He was a wellspring of such unexpected comments.

Tupolev possessed an outstanding memory. His voice was often heard over the telephone (it saved time and he never called people for trivial reasons): "You promised me on such-and-such date to decide such-and-such. The deadline was yesterday."

His thriftiness showed up in everything he did. At one time they started to use textolite everywhere, for everything. After he had a look at it, he said, "Why, that is just cambric! Just think how many ladies' blouses you could make from that! I beseech you most humbly, wherever possible, replace it with plywood."

On another occasion, as he was walking by a shop, he noticed that a scrap box had a ferrous metal shaving mixed in with nonferrous metal. The shop chief got a good dressing-down!

At the same time he could be highly sensitive and humane, especially toward older workers and office workers who had worked with him for a long time. A middle-aged carpenter/model maker was no longer able to work on the scaffolding around an airplane. Noticing this, Tupolev ordered them to find some easier work for him. Someone carelessly recommended retiring him. "The man's given us 30 years of his life! Do you understand? 30 years! And just like that, you'd put him on a pension? Make his work a bit easier and he will still give us a lot of help."

In our professional literature people appear flat, as one dimension, mere outlines that offer little insight into their character. As to their professional lives, they are portrayed in a stylized way: They chose it in early youth, passed through life without a shadow of a doubt concerning their choice, and noticed for nothing other than their official duties. When they die they regret not that they are leaving life, but that they didn't give instructions to their successors on how to act. Tupolev cannot be contained in that category. He loved life in all its diverse manifestations. He also loved the arts: painting, the theater, and literature. His love was active and specific, and resistant to being classified easily. He read a lot, despite being inordinately busy, but he read selectively.

"I can't stand all kinds of 'isms.' The names aren't important—it is the talent that counts. Take such works as *The Death of Ivan the Terrible* [Tolstoi], *The Living Corpse* [Tolstoi], *Three Sisters* [Chekhov], or *Vassa Zheleznova* [Maksim Gorkiy]. You sit there, and you are afraid to say anything. It is the same in painting. When I stand in front of Serov's *Girl with Peaches,* or I look at Repin's *Zaporozh'ye Cossacks,* or Surikov's *Morning of the Execution of a Streltsy,* or I see Levitan's *Above Eternal Peace,* the force of the artistry is humbling."

A New Department and Dismissals

Department "R," a new department linked to cruise missile technology, appeared in our system. Since a cruise missile was an ordinary, miniature aerodynamic system, there were no difficulties of any kind with its organization at first. The design teams for wings, wing center sections, fuselages, tails, and landing gear allocated some of their designers from their own staff. They were headed by the most capable of the young designers, and so the new department was formed.

Organizational difficulties began with the two sections (engine and instrumentation) of the new department. In the engine section, automatic control was needed over engine operation, air intakes, and exit nozzles during various flight phases. Some type of electronic-command device was supposed to operate all of these, according to the altitude and speed, condition of the atmosphere, angles of attack of the system, and several other variables. The existing engine department of the design bureau had no such specialists. Developing such a device would have been awkward, and ordering it from outside required the compilation of detailed, well-supported specifications. There weren't enough technical specialists in the engine department to compile such specifications there.

After discussing the situation, Tupolev and his Deputy for Engine Matters reached a decision not to create an independent engine team in Department "R." All design and theoretical issues would remain under the existing engine department, which had accumulated impressive experience in adapting engines to existing aircraft that flew at speeds close to the speeds of the planned missile. All fundamental issues that arose regarding the missile's engine would be resolved by the Deputy Chief Designer of the OKB for Engine Affairs together with the chief of the new department.

The attempt to find an acceptable model for testing the instrumentation was more difficult. Instrumentation, by its very nature, permeated all systems in the missile—flight control, balance, navigation, execution of commands received via radio from the command post, control of engine mode—performing the functions of the circulatory and nervous systems of a living organism. Creating an active team in Department "R" under these conditions made no sense. The existing teams could work on these areas, but they would be commanded by the missile deputy in the area of missile issues. I did not agree with this.

Before a decision was made, Tupolev called me, his deputy for instrumentation, to see him. For the first time in my life, I experienced his anger, and it was terrible. Everything began on a rather peaceful note. "How will we build the bridge: along the river or across the river?" he asked, smiling cunningly.

After hearing me out, Tupolev said, unexpectedly, "This is how it will be. There are lots of issues in your area, and most of them are not too complicated. Assigning them to the missile designer is probably not correct; you will solve them, and he will solve all basic issues. Thus, responsibility for proper functioning of all instrumentation is yours, while he will make decisions on basic issues involving instrumentation. Do you understand?"

"No, Andrei Nikolayevich."

"What is the problem?"

"The issue is, either I am responsible and give the orders, or he is. But for him to decide while I am responsible cannot be."

"That is exactly how it will be," shouted Tupolev, furiously. He jumped up and with terrible force banged his fist on the desk. The six-millimeter-thick glass that covered the desk cracked into a pile of fragments. His face turned white. I was terrified that he might have a heart attack. "Get out of here!" he yelled.

I went toward my office. As I was going up the stairs, the bell rang. "Come to my office."

When I returned, he said, "As of tomorrow, you are fired. Hand over your files to Sklyanskiy, and you can go to hell, for all I care."

Vera Petrovna [Tupolev's secretary] had probably told people about the shouting coming from his office, because several deputies gathered in my office to find out what sort of unheard-of scene had taken place.

The next morning I stayed at home, but at about 10 A.M. the phone rang. It was Tupolev: "Please be so kind as to call on me."

I arrived at the design bureau and went up to see Tupolev. He stated decisively, "You must give Aleksei Andreyevich your best specialists in all areas of his subject area. He will form his own instrumentation department, and I will appoint a chief there. You will no longer be involved in this. Go to your department and work on the planned aircraft." I didn't hear a word of apology or excuse. Later I understood that my declaration had been perceived as an "antifamily" speech.

Tupolev's goal had been accomplished. I gave him three major specialists (A. I. Kandalov, now the deputy general designer; L. N. Bazenkov; and S. V. Svirskiy), and ten to twelve people from each team. I was never again interested in their work. . . .

A. A. Arkhangel'skiy's firing was tragic. If someone had tried to foretell what would happen after Tupolev's death, we were certain that a collegium would be formed, headed by Arkhangel'skiy, and including D. S. Markov and N. I. Bazenkov. No one could give serious consideration to Aleksei Andreyevich Tupolev. However, since he ended up as the "king" during the "game," people gradually began to shift Arkhangel'skiy to a secondary role. First of all, contacts with the Kazan' plant were given to D. S. Markov, bypassing Arkhangel'skiy. The same

thing was done for the Kuybyshev lead plant. All orders came from
N. I. Bazenkov. Thus, Arkhangel'skiy had lost two leading enterprises,
and Aleksei kept him out of the third, that is, the "R" system.

However, let us return to aviation.

Tupolev's Last Creation: The Supersonic Tu-144

It is truly impossible to stop technological progress. Having begun to
fly in heavier-than-air machines at the beginning of the century, man-
kind immediately rushed to improve their capabilities and succeeded
very convincingly. Was it really so long ago that the ANT-2 carried
two passengers aloft? Now the Il-86 can carry 350 to 400. People were
proud when Yefimov flew his Farman from St. Petersburg to Gatchina,
a distance of 40 (!) kilometers, and now a Tu-114 can easily fly nonstop
from Moscow to Havana, a distance of 11,000 kilometers. According
to newspapers at the time, the Wright brothers' biplane reached the
"unheard-of" speed of 60 km/h; the pilot Mosolov has now exceeded
3,000 km/h in the Mikoyan Ye-166. The newspaper *Rech'* wrote of the
pilot Vasil'yev, "He dared to rise above St. Isaac's Cathedral in St.
Petersburg"; the pilot Fedotov has reached 37 kilometers altitude in a
Ye-266.

In 1950, a MiG-17 exceeded the speed of sound, which is somewhat
greater than 2,000 km/h. This was first achieved in a military aircraft.
Gradually, combat aircraft became supersonic. If military pilots had
managed this, would their civilian colleagues look on this with
indifference?

Thus a serious discussion began in specialized aviation publications
on whether it was necessary to have a passenger airplane capable of
flying at supersonic speed, and what features it would have. No one had
any doubts about the ability to create such an airplane. Of course, the
problem was not a simple one, and the number of technical problems
that had to be solved were very great.

"If you think about the future of air transport of passengers from
one continent to another, you reach a single conclusion: without a doubt,
supersonic passenger airplanes are necessary. And I have no doubt," said

Tupolev, "that they will appear. Remember the skeptics who jeered at K. E. Tsiolkovskiy. 'Man will never reach other planets,' they said. But man up and landed on the moon!

"How many people laughed: Who needs passenger jets? Who would fly on them? Today, however, it is rare to meet someone who hasn't flown on a jet. The issue lies elsewhere: How can such an airplane be made as reliable, say, as our Tu-114? That is the stumbling block. How quickly can a supersonic jet be brought up to the required reliability? How can we rapidly detect all our miscalculations, which will be inevitable, and find shortcomings that may have serious consequences, and eliminate them?

"To me, personally," continued Tupolev, "it seems that, in the planning of such an aircraft, the most difficult problems, both for our design bureau and for scientists at scientific research institutes involved in solving the problems, will be:

The search for, and selection of, the optimum aerodynamic layout for the new airliner, and, in particular, its wing design.

The problem of comfort and air conditioning in the passenger cabin, when the aircraft's exterior will heat up to 100 to 150 degrees due to air friction as it rushes along at supersonic speed.

The selection, configuration, calculation, and design of air intakes capable of supporting stable engine operation over a huge range of speeds, from 300 km/h during the landing approach to 2,500 km/h during cruising.

The degree of automation of aircraft and course controls. This will require precise determination of the responsibilities of the crew and the functions of automation.

What can be done about the sonic boom that trails a supersonic aircraft?

"However"—Tupolev laughed—"the last issue smacks of scholasticism, for the laws of physics are not within the jurisdiction of you or me. Of course, it is unpleasant if windows start breaking, or pieces of tile start falling off the roof. However, we won't be flying at those speeds in Paris, London, Simferopol', or Sochi, over densely populated areas. That is simply absurd. In reality, is it so important to a tourist or resort owner whether he flies there in thirty minutes instead of one hour? It is another matter if the flight is to America, Australia, Japan, or our own Far East. Here, three hours instead of eight is essential. These routes, however, lie over oceans or uninhabited *taiga,* where there is no one to be offended by the crash of a sonic boom."

At the same time, or possibly a bit earlier, similar conversations were being held in the West. The papers subsequently announced: "General de Gaulle informed the House of Deputies that France and Great Britain have agreed to build such an airplane jointly. They have called their airplane the Concorde, that is, 'Harmony.' Many of the best-known industrial firms in these two leading European states have begun to expand their work on the Concorde. They have begun this work rather energetically, since the creation of the airplane clearly plays a role in prestige in that we will surpass America and the Soviet Union, the two most advanced aviation states."

If we, too, are to build such an airplane, then we must hurry. Many of our comrades have wondered, 'Why is he doing nothing?' And it has begun to seem to many that time is slipping away, and the Concorde will be the first to fly.

France and Great Britain did not know that for some time now the light had been lit long past midnight in the corner room on the first floor of a house on Nikolina Gora Street. The old and young Tupolevs were drawing preliminary sketches for the future airplane. On his father's instructions, Aleksei Andreyevich brought together a small group of young designers to create sketches for the Tu-144. The elder Tupolev had more than one meeting on this topic with President of the Academy of Sciences M. V. Keldysh and Minister of Aviation, P. V. Dement'yev. Although there are, as yet, no resolutions by the party Central Committee and the Council of Ministers on this topic, both these prescient state figures have instructed the scientists of the Academy of Sciences and the Central Aero-Hydrodynamics Institute to begin the necessary studies. In addition, these studies have yielded results, and in the younger Tupolev's group, the shape of the Soviet supersonic passenger liner has begun to appear—still somewhat vaguely—on the drafting boards.

Belief in the ability to create the airplane grew with each passing day. Due to experience, Tupolev gradually involved a substantial number of basic design teams and departments in his bureau, together with their leaders, in the work on the Tu-144.

On late fall evenings, the brightly lit design bureau building brought to mind an oceanliner moving inexorably toward its goal. Some of the last lights to be turned out are those on the "captain's bridge" on the third floor—Tupolev's office. The driver, who was tired of waiting, would drive him home through Moscow as it grew quiet, only to deliver him to the re-lit building the next morning.

It was necessary, first of all, to determine the nominal speed on which the airplane would be based, either 2,500 or 3,000 km/h. After learning that the Concorde had been computed for 2,500 km/h, the Americans hurried to advertise that they would produce their Boeing 2707-300 passenger airplane, with a speed of 3,000 km/h, only six months later. If you leave Moscow at 9 A.M., there is not a big difference in time if you arrive in New York at either 11:00 or 11:30 A.M. Nevertheless, the road to Sheremet'yevo Airport, and the road from Kennedy Airport to the center of Manhattan, reduce this supposed time advantage to nonsense.

Another aspect was critical, however. For heating conditions at 2,500 km/h, when the airplane's structure would heat up to 150 degrees, perhaps even several times, the airplane could be made of special dur-alumin, with titanium and steel used only at individual points or assemblies. At 3,000 km/h, however, heating rises to 250 degrees, and the airplane had to be built wholly of steel and titanium.

These materials have still not undergone serious engineering and operational testing on airliners. Relying on his old supposition—strive to obtain the very highest reliability for passenger aircraft—Tupolev formulated his thoughts thus:

"The Tu-144 must be made of duralumin, because we are counting on a speed of 2,500 km/h. It is entirely obvious to me that the Americans, in going to 3,000 km/h solely for reasons of competitiveness, will be forced to make their passenger Boeing as unreliable as they made the titanium-steel North American XB-70 Valkyrie bomber before it.

"If you recall, they struggled with the XB-70 for several years. After each flight the airplane had to be taken out of service for several weeks, and sometimes months. Only now has the reason come to light: ongoing work on the new, unproven structures made of steel and titanium."

The next decision involved the choice of aerodynamic configuration. First (and this was one of the most complicated problems), it was necessary to find the wing shape with excellent characteristics both during cruising (when the Tu-144 would be traveling at supersonic speeds) and at subsonic speeds (when it would be descending for a landing, in other words, at speeds no faster than ordinary airplanes). It would have been possible to use a variable wing, which has a straight wing at low speeds, and a swept wing at high speeds (when the wing is pulled in toward the fuselage). This type of system was developed experimentally and yielded good results. Tupolev rejected this for the Tu-144, however. In his opinion, creating a sufficiently reliable wing mechanism for an airplane such as the Tu-144 was undoubtedly possible, but only after several years of experimental work. Precisely because these years were not at his disposal, another solution had to be found. By experimenting on tens and hundreds of research models with wings of various shapes, the scientists and specialists at the Central Institute concluded that the best version would be a very thin, triangular wing with a leading edge contour similar to a long "S" and a nose with a sharp leading edge that was bent somewhat downward. Controlled wind-tunnel tests of large-scale models with this type of wing convinced us that an airplane with a tail-less design would be stable and have good control in all flight modes, if two of its characteristics could be managed. The first derived from the fact that, during transition from subsonic to supersonic speed, the aerodynamic focus of a triangular wing shifts. Preserving the balanced state of the airplane at that moment required that some type of weight be shifted along the airplane's axis. What could this weight be?

Tupolev smiled. "Naturally, not passengers and not baggage. Obviously, it is fuel. It is necessary to plan center-of-gravity tanks ahead of and behind the center of gravity. At a signal that the airplane is crossing the sound barrier, strong fuel pumps would rapidly shift several tons of fuel from one tank to another, and the problem would be solved."

While it appeared this obstacle had been conquered, another problem appeared immediately. With the wing design selected, a supersonic airplane would cruise with an angle of attack of 4 to 5 degrees, while during landing, the nose would be pulled up at an angle of 14 to 15 degrees. Even in this position, however, pilots needed an unrestricted view of the airfield and runway during landing, as if they were on a motorcycle. How could this be achieved? Flat windshields in front of the pilots worked for slow-moving airplanes. On the high-speed Tu-144, however, the airplane's nose had to be sharp and smooth, like a foil.

When Tupolev was ordered to make a clay model of the Tu-144, he recalled that, on one of their fighters, the British had started to lower the nose on takeoff and landing. He took out a pocket knife, cut off the nose on the model, and inclined it at an angle of 20 to 30 degrees. "If we attempt to glass in the entire sharp nose section, distortions in the field of vision would make it impossible to fly. Accordingly, we will try to make a folding nose."

At first, there was skepticism. No one had heard of aircraft with "folding noses." And when he told them that he had seen something similar somewhere, they began to try to convince him: "But that was on a tiny fighter, and here the nose alone is like the entire other airplane!"

"All our lives we have done nothing but things that didn't exist previously. You don't have to design a supersonic airplane if it is based on something that already exists. And in general it is best to look for new designs, and not go on about something that never was. At some time you yourselves didn't exist, but you appeared!" he responded rather irritatedly.

When the folding nose was reproduced in the model, Tupolev wanted to see how it would appear from the cockpit, from the pilots' seats. He squeezed through and sat in the seat of the left pilot. They lowered and raised the nose with a winch several times. He remained satisfied. "Well, the whiners and disbelievers groaned that it would be difficult, and wouldn't work out, but we did it and did it pretty well. Let's have a look: Where are the ejection handles?"

"We've never put them on passenger airplanes. It would be a problem here on the Tu-144. You yourself just crawled through and saw how damned tight it is. Then we have new, larger hatches, the seats are much larger, and there are the mechanisms for lowering the nose."

As they brought up more and more arguments against ejection, Tupolev's eyes grew sterner and sterner, until he burst out, "Young men, do you have a conscience? It is like darkness has descended over you. We've never made such airplanes before; everything on them is new and for the first time. I don't care, but suppose? ["Suppose" exactly what can't be said, for sure. This was a bad sign.] No, we will put seat ejection in, without fail. No one will be taking any passengers on the first flight, but the pilots have to be certain that everything has been done for flight safety."

The next fundamental question was how to position the four engines. There were two entirely opposite proposals: closer to the longitudinal axis, and several meters away from it. "Try it closer to the axis, anyway. Only this way will the rotational moment from them be minimal. Let's say we mount two powerful engines on each side, at a substantial distance from the fuselage. Let's imagine—and we have to do this—that both left or both right engines fail in flight. Haven't we made control of the aircraft more difficult without any special reason for doing so? And we have done this in such a way that managing this will be difficult. And what will pilots think of this? I would like to know their opinion."

Next was instrumentation. How could we automate the control of the airplane and navigation? "If we don't make extensive, but sensible, use of automation, it will be very difficult for the crew. On the Tu-144, the volume of information that they have to process is so great that they won't be able to manage without an onboard computer. Don't get carried away, however: remember what wiser minds have to say."

He laughed loudly and unexpectedly. He had forgotten: it seems that the very Wiener who discovered cybernetics had something to say about people who were attracted to it: "They are ready to claim that a chicken is only a means used by an egg to lay another egg." I am inclined toward the opinion of our fellow countryman, academician Kolmogorov: "Computers perform only auxiliary operations, in accordance with goals set by man." Finally, Admiral-Engineer Berg didn't put things too badly, either: "I am afraid that those who believe a milk pail is a computer and a cow a milk sensor are not entirely correct."

Tupolev said,

It is also cause for concern that everywhere you hear, "If you use an onboard computer, then make sure it is not several local computers, but only a single, global computer." I'd refer the latter "progressives" to the memoirs of Academician A. N. Krylov, to the place where he tells about a conversation between his father and M. I. Kazi, the director of the Baltiysk Shipbuilding Plant:

"You, Nikolai Aleksandrovich, point only to the need for large battleships. Just imagine that, instead of five million infantrymen, we were feeding Goliath, big as a mountain, clothing him in armor that no artillery could penetrate, and he—the son of a bitch—right on the eve of the battle, stuffs himself and gets diarrhea. . . ."

Doesn't this Goliath sound like that single computer? Therefore, automation is automation, but people—that is, the pilots—retain the decisive role. And since that is the case, don't get carried away, but set things up so that they can, when necessary, monitor the automation and, under emergency conditions, also repair it.

That is how Tupolev gradually worked out one problem after another. The time had come to begin the massive output of blueprints. The transition from concept to structure is not always smooth. A decision had to be weighed and discussed. Tupolev would gather the cream of aviation engineering thought in our country in his office. The walls in his semicircular office, where the scientists, designers, and managers from ministries would meet, were covered with diagrams of the future Tu-144. After an extensive exchange of opinions between lead designer A. A. Tupolev and his group, the basic designs in the areas of shape and layout for the airplane were approved.

The assembly shop for the plants had mezzanines all around it. From there, above, leaning on the railing, Tupolev could get a good look at the long, elegant fuselage of the 144, which brought to mind an arrow set in a bow. The exterior lines of the S-shaped wing were like the wood of the bow, and its straight trailing edge was like the bow-string. What thoughts were running through the mind of its creator? Perhaps he already saw the arrow piercing the sky.

But this would not happen very soon. While the deafening din of air hammers drifted up from below, the airplane was being assembled. Hundreds of workers were attached to the sprawled body of the airplane, while its insides were being packed with very complex instrumentation.

At the same time, they were already "flying" the Tu-144. In an adjacent building that was bright enough to bring to mind a medical examining room, a pilot cockpit packed with equipment and instruments had been mounted on the laboratory's dynamic test stand. In the seats were pilots Ye. Yelyan and M. Kozlov, fated to be the first to take the wild-looking airplane into the air. The cockpit was dark, the dials of numerous instruments glimmered. The movie screen in front of the cockpit depicted a takeoff runway at our airfield. When Yelyan increased engine revolutions, the image on the screen began to move. The runway rushed toward the airplane, faster and faster. The pilots began the liftoff, the airplane's nose rose, the ground on the screen dropped down, and the Tu-144 was climbing. Clouds appeared, sparse at first, then thicker, and then flying blind on instruments began.

While Yelyan was "flying" with Kozlov, the quality of the air conditioning was being checked in an adjacent building. Numerous heat sources heated the exterior surface of a section of fuselage to 150 degrees, the same temperature the airplane would be heated to in flight. They checked the next cooling variation, since the cabin had to be at room temperature. In a nearby insulated bay, the tapes of automatic test recorders moved; the nearly straight temperature lines deviated insignificantly from +18°C—the required temperature in the passenger cabin during flights. Hundreds of sensors measured the temperature, not just at various points in the cabin itself, but also on mannequins "seated" in seats. Each of them had a device to generate the same amount of heat and moisture as a future live passenger would radiate around himself.

Not far away, another large building was for static strength testing of structural elements. A longeron from the wing of the future airplane was slowly bent, gripped in the jaws of the huge test stand. Hydraulic jacks applied increasingly heavy loads until it broke. These strength tests were of the most important parts of the Tu-144.

The numerous buildings and laboratories contained all types of test stands.

"Our time is short. How can we achieve it so that the maximum number of questions on aircraft behavior in the air are answered before the first flight?" asked Tupolev, before answering himself: "On test stands that make the experimental conditions as close to real-life conditions as possible. Don't stint on time for these, it will be recouped. Only via all this can we produce an airplane for a first flight as if it is not a first flight, but a tenth, or, perhaps, a hundredth."

"As close to real-life as possible." This is right, of course, but how was it accomplished? When the issue was aircraft heating, or loads experienced by structures in flight, things were simple: we knew how to heat and apply loads. But what about control, for example? Its control rods permeated the entire, huge airplane, from nose to tail, passing through hundreds of brackets with bearings. Then, too, an airplane "breathes." Its long, metallic fuselage may lengthen or shorten, bend, heat, cool, or shake. Moreover, individual elements of the airplane are subject to these effects in very diverse combinations during flight.

Finally, using the cleverest of devices, it is possible to take all these factors into account. Although the stand wasn't huge, E. Yelyan and M. Kozlov could experience the same loads as during a flight by operating the control wheel and pedals.

It was similar on other test stands. Engineering skills and worker know-how have made it possible to make test conditions everywhere as close as possible to real-life conditions. In fact, there were test stands wherever you look. Among a cyclopic tangle of steel beams, landing gears were raised and lowered, and nearby the nose portion of the fuselage was raised and lowered. Far outside the city engines roared; at a test range, in order to check the airplane's antennas, radios established communications with remote subscribers. In the wooden kitchen compartment, attractive stewardesses served meals. Suitcases were loaded into the baggage compartment. Inside the wooden cabin, which simulated the passenger cabin, artists selected fabric colors for interior decor.

In the quiet of a laboratory, a computer controlled a system of instruments, and the autopilot operated the aircraft controls. Finally, at the airfield, away from everything else, surrounded by red fire trucks, they started fires artificially and then put them out in the engine nacelles.

On Tupolev's desk was a record book for each of the test stands. Heaven help anyone who did not have an entry in the record for several days running about what had been accomplished at his stand.

After their "flight," Yelyan and Kozlov reported their results in great detail. Tupolev was carping and caustic, almost to a fault. He was interested in everything: What are the loads on the control wheel and pedals? How is the autopilot working? How does the airplane behave in various flight phases? Can the instruments be seen, and are their dials easy to read? Is the cockpit laid out comfortably? Is there enough field of vision?

Once it seemed Tupolev was finished, the pilots rose to leave, but he stopped them again in the doorway. "Yelyan, did you simulate a failure of one of the control circuits?" The interrogation began anew.

The pilots left and the air-conditioning specialists entered; behind them were the engine specialists, and so he delved into all the systems for hours and days.

Tupolev belonged to that category of people who get satisfaction from performing a job well. This applied equally to the overall airplane and to individual experiments. When experiments were being conducted every day in the numerous laboratories of the design bureau, and one of them was successful, Tupolev was pained if it had occurred without him. However, when we would play a short interlude and repeat the concluding portion of an experiment in his presence, he was overjoyed, would thank the workers, and then was charged with optimism.

Everything, however, has its beginning and its end. They completed the assembly of the airplane at the plant. That evening Tupolev invited his assistants to come with him as he looked over the airplane. He sat in the pilot's seat, inspected the passenger cabin and kitchen, and peeked into the lavatories.

Just assembled, freed from its jigs, straps, hoses, and cables, as if already launched into the sky, it became possible to see the airplane for the first time in all its glory. Now it was in the hands of the testers. A feeling akin to jealousy came over the designers. They had thought, tormented themselves, missed sleep, and created, and now it had left home, like a youth beginning to live independently. How many of these experimental aircraft there were in Tupolev's life! For him, the designer, each was a living creature, with its character, inclinations, and, at times, even its caprices.

From that time onward, Tupolev's route changed. Now he went to the air base in the morning. Although his office was in a building about ten minutes away from the hangar, he felt too cut off from the airplane. He was also irritated by intolerable telephone calls. At first he moved into one of the rooms in the hangar, but even that was too far removed. His unflagging, almost youthful vigor and his desire to see everything with his own eyes, to be the first to learn the results of the work, all drove him to the hangar. Next to the airplane itself they set up a desk with telephone and chair. Now he was in the midst of events. He had only to raise his head to see the entire Tu-144 above him. But this, too, wasn't enough: he wanted to sit in the airplane, delve into everything himself, learn everything first-hand. But how can one grasp the ungraspable?

It was unfortunate when he would catch cold and remain at his dacha while we worked on the airplane. The telephone would ring constantly. He would call first one person, then another; he had to know how things were going, what had been done; and he got angry if the answer wasn't specific. People weren't offended: they understood that he was mad at himself, at his age, his indisposition, and his inability to be among them.

The working out of the aircraft's systems began. For the umpteenth time, each one of them was switched on. These weren't subjective impressions, but tapes from numerous automatic recorders and oscillographs that were figuring out the grades. The day finally came when it was no longer designers and engineers, but the crew who took their seats in the airplane; final checkout was underway. Each of them reported his conclusions to Tupolev and Aleksei Andreyevich in person. In attempting to evaluate the degree of their confidence in themselves and the airplane, he was extremely critical, but it was clear that he was satisfied.

The huge doors of the hangar slowly opened to the sides. The first snow had powdered the ground the night before and was so bright that you wanted to close your eyes. A heavy tow truck pulled the 144 onto the flightline. Fuel trucks pulled up, filling the airplane's tanks for the first time. There was a careful inspection—no, no leaks. And then the first engine was started; its powerful rumble filled the airfield. The Tu-144 had found its voice.

After trying all four engines, they began ground test runs. The airplane was now moving on its own, without outside help—its first independent steps. With its lowered nose, it brought to mind a pterodactyl. With each run, Yelyan increased the speed. Arguments were already underway: didn't it lift off?

It was freezing, without a breeze, and sunny. Dressed in their semi-spacelike armor, the crew stood alongside the airplane. Restrained, reserved, and with knitted brow, Tupolev clapped Yelyan on the shoulder and with an unexpected, warm smile, began to sing in his high voice, "So, what is it, Yelyan? Get going a little bit!" He sat in his Volga and waved to his driver: "To the runway."

And once again, on 31 December 1968, for the umpteenth time, as on 21 October 1923, almost a half-century before, when he saw his first "child" off into the air at the very spot where, in his opinion, E. Yelyan would lift the Tu-144 into the air, a figure severe in its solitude and stooped over, stood rooted to the spot—the Academician, General Designer, Hero, and prize winner, and a great and wise man—Andrei Nikolayevich Tupolev.

Tupolev's Last Days

Few people would have guessed that the Tu-144, the first supersonic passenger aircraft, would become the last major accomplishment for Tupolev.

In September 1972, at the Kremlin's Hall of Congresses, a ceremonial session was dedicated to the fiftieth anniversary of the founding of his design bureau. The welcoming speech read by G. P. Sbishchev, who had been named chief of the TsAGI, read:

In our nation's conquest of the forces of air and sea, two names are inseparable from each other: N. Ye. Zhukovskiy, and his closest disciple and assistant, Andrei Nikolayevich Tupolev. Tupolev represents an entire epoch in the establishment and development of Soviet aviation technology. He is one of the organizers of the Central Institute and a creator of the scientific base for our aviation.

Despite the fact that Tupolev was in his eighties, he retained absolute clarity of thought. He never gave up in the face of difficulties or problems, or plunged into despair. On the contrary, when he encountered such circumstances, Tupolev would rise to overcome them, full of energy from some unknown source.

But that was his mind. Having grown weary of life's agitations, and his constant stressful and difficult work with its incomparable responsibility, and having never, to the end of his life, recovered from the illness he suffered in his youth, or from his wife's untimely passing, his body gradually failed.

In late fall Tupolev again caught cold and took to his bed. At first he stayed in bed at his dacha, but then his daughter and doctor insisted on hospitalization. They admitted him to the hospital in Pokrovko-Streshnevo.

On 22 December Yuliya Andreyevich and Aleksei Andreyevich went to visit their father. He was calm, had a relaxed, ordinary family conversation about his grandchildren, and everyday matters. It was getting dark, and after saying their farewells, they went home. That night, the tired heart of Andrei Nikolayevich stopped.

He was buried, as he wished, not far from the grave of Yulenka Nikolayevna, in the historic Novodevichiy Cemetery in Moscow.

A marker has been erected on his grave. I have a photograph of it, but neither the bust of Andrei Nikolayevich, nor its setting, renders the greatness and optimism of this man.

He was a great designer and a great optimist, even in the darkest days of his own life.

AFTERWORD

Andrei Tupolev and Leonid L'vovich Kerber—their lives intersected during the turbulent decades of the 1930s. A strong bond of friendship and collaboration tied them together, beginning with the preparations for the ANT-25 for the transpolar flights of 1937. Kerber, then a young Army Signal Corps specialist, worked successfully to perfect the radio communications system for these historic flights. From that time forward, as they say, Tupolev "had his eye on him."

Kerber was born on 16 June 1903 in St. Petersburg, the son of a Russian naval officer. In 1917, Kerber joined the naval cadet corps, following the tradition of his family. That fateful year of Revolution, however, ended his naval career. Being the son of a nobleman, he found himself sealed off from any further higher education. Between 1918 and 1921 Kerber served as a soldier in the Red Army. He participated in the storming of Kronstadt, to quell the rebellion of navy mutineers against Bolshevik rule. During the Russian Civil War he fought in a series of battles with White army units led by Kolchak and Denikin.

With the end of the Civil War in 1921, Kerber began his odyssey in aviation. First, he became a mechanic in an aviation factory. Later, he joined the Red Army's signal services. Kerber's natural intelligence and capacity for hard work earned him recognition by his peers.

In the Red Army Air Force Kerber worked at the center of experimental projects in radio communications. As a specialist, he published a technical volume on radio communications, the first published in the Soviet Union. His book, with its clarity and simple explanations, was highly praised, prompting the Red Army to adopt the book as a training manual. In 1937, Tupolev asked Kerber to install radio equipment in the ANT-25, the special aircraft designed for the transpolar flights of Valery Chkalov and M. M. Gromov of that same year. During these long-distance flights, Kerber maintained radio communications with the transpolar fliers as they flew nonstop from Moscow to North America.

The year following the historic transpolar flights saw the lawlessness of Stalin's purges interrupt Kerber's career. He was arrested and sent to the Arkhangelsk region to work as a prison laborer in the forest industry. After the passing of time, however, the authorities realized that Kerber's head, not his hands, should be mobilized for work. Accordingly, the NKVD (the Soviet secret police) transferred Kerber to Moscow in 1938, where he was assigned to TsKB-29, the *Tupolevskaya sharaga* ("Tupolev's special prison"). Reunited again with Andrei Tupolev, Kerber would remain with the Tupolev Design Bureau for the remainder of his engineering career.

While in prison, and later as a free man, Tupolev attracted to his design bureau a talented group of specialists. Beyond their contributions to their area of professional work these same men displayed great leadership. To be part of this extraordinary group was not easy or simple. Yet Kerber quickly established himself as a leader in this elite group, being appointed deputy chief designer for avionics and instruments. With each successive aircraft project Kerber carried out new and more sophisticated work. These numerous accomplishments led to a series of awards and medals, including the Lenin prize and other state prizes. Kerber also received the A. S. Popov medal from the Academy of Sciences of the USSR. As a Tupolev specialist, Kerber earned the doctorate of technical science.

In 1969, at the time of retirement, Kerber turned to literary work. He published many technical and historical pieces, along with *Tupolevskaya sharaga*. Kerber's full-scale biography of Andrei Tupolev was published in 1973, *Tu—Chelovek i samolet*. His many publications in aviation, technical, and popular scientific journals brought great success and a large readership.

On 8 October 1993 Leonid Kerber died, but only after he completed his revisions of this manuscript. It is difficult to accept that we will no longer hear the stories told by this courageous and good man. In my own family Leonid Kerber persists in our memory as a man with encyclopedic knowledge who knew so much about life and his own speciality.

Maksimilian Saukke
Moscow, Russia

NOTES

INTRODUCTION

1. Howard Moon, *Soviet SST: The Techno-Politics of the Tupolev-144* (New York: Orion Books, 1989). Howard Moon provides a detailed account of the Tu-144 project and insights into Andrei Tupolev's complicated relationship with the Soviet state.
2. See Michael Scammell, *Solzhenitsyn: A Biography* (New York: Norton, 1984), 224–25. Scammell indicates that Stalin organized the first prison workshop in 1930. The first English translation of Solzhenitsyn's novel appeared in the late 1960s: *The First Circle,* trans. Thomas Whitney (New York: Harper & Row, 1968).
3. G. A. Ozerov worked with Tupolev as an aircraft structures specialist.
4. See A. Sharagin, *Tupolevskaya sharaga* (Druck: Possev-Verlag, V. Gorachek KG, Frankfurt/M., 1971).
5. Charaguine, A., *En Prison avec Tupolev,* trans. N. Krivocheine and M. Gorbov in collaboration with Nina Nidermiller (Paris: Editions Albin Michel, 1973). The French translation also contained an introduction by S. Kirsanov which appeared earlier in the West German edition.
6. Leonid Finkelstein produced the BBC excerpts on Tupolevskaya sharaga. Under the pen name L. Vladimirov, Finkelstein published *The Russian Space Bluff* (London: Tom Stacey, Ltd., 1971), which included a long foot-

note on the samizdat publication. That same year the West German magazine *Der Spiegel* serialized portions of the book, which prompted the Soviet journal *Za Rubezdom* (*Abroad*) to launch an attack on the space book. Leonid Finkelstein to Von Hardesty, 8 August 1994.

7. L. L. Kerber, *Tu-Chelovek i samolet* (Moscow: Sovetskaya Rossiya, 1973), 147.

8. L. L. Kerber wrote several articles in 1988 which provided additional detail on the special prison workshops. Kerber's nine-part series of installments in *Izobretatel' i ratsionalizator,* nos. 3–9 (1988) created a sensation. This collection of articles represented an unprecedented effort to fill in the "blank pages" of aviation history. Other articles include: "Shtrikhi k portretu," *Nauka i zhizn'* 11 (1988): 12–19; "Ne kopiya, a analog, o samolete Tu-4," *Kril'ya rodina,* no. 1 (1989): 24–26, and no. 2 (1989): 33–34; (with Maksimilian Saukke) "Vopreki vsemy . . . ," *Kril'ya rodina,* no. 6 (1988): 24–25, and no. 7 (1988): 28–29. David Gai prepared an English-language interview with Leonid Kerber; see "Performance of Duty," *Moscow News,* 29 November 1989, 13.

9. For the polemics with Yakovlev, see (with Maksimilian Saukke), "B poiskhakh istiny," *Kril'ya rodina,* no. 1 (1988): 30–31. Saukke and Kerber also prepared an essay (unpublished) prior to the Gorbachev years, "O nekotorykh spornykh vsglyadakh na istoriyu aviatsii v trudakh A. S. Yakovleva" (Moscow, 1984).

10. See Maksimilian Saukke, "Kerber-L'vovich," *Mir aviatsii,* no. 4 (1993): 9. This is a moving account of Kerber's life, written shortly after his death.

11. For the history of early Russian aviation, see Igor Sikorsky, *The Story of the Winged "S"* (New York: Dodd and Mead, 1938); Alexander Riaboff, *Reminiscences of a Russian Pilot,* ed. Von Hardesty (Washington, D.C.: Smithsonian Institution Press, 1985); and K. N. Finne, *Igor Sikorsky: The Russian Years,* eds. Carl Bobrow and Von Hardesty (Washington D.C.: Smithsonian Institution Press, 1987).

12. The centennial of Igor Sikorsky's birth was celebrated in Moscow in 1989 with an exhibition and a new book, G. I. Katyshchev and Vadim Mikheyev, *Aviakonstruktor Igor Ivanovich Sikorskiy* (Moscow: Nauka, 1989). This event and book represented the official recognition of Igor Sikorsky's many achievements in Russia and the United States. See also by the same authors *Kril'ya Sikorskogo* (Moscow: Voyenizdat, 1992).

13. Kendall E. Bailes, "Technology and Legitimacy: Soviet Aviation and Stalinism in the 1930s," *Technology and Culture* 17, no. 2 (April 1976): 55–81. Bailes provides an insightful interpretation of how Stalin used aviation and aerial feats to legitimate communism. See also Kendall Bailes, *Tech-*

nology and Society under Lenin and Stalin: Origins of the Soviet Technical In-
telligentsia, 1917–1941 (Princeton: Princeton University Press, 1978). For a
more recent study, Loren R. Graham, *The Ghost of the Executed Engineer:*
Technology and the Fall of the Soviet Union (Cambridge: Harvard University
Press, 1996).

14. For an English translation of a Soviet-era biography of Chkalov, see
Georgiy Baidukov, *Russian Lindbergh: The Story of Valery Chkalov,* ed.
with introduction by Von Hardesty (Washington D.C.: Smithsonian Institution Press, 1991). To appreciate the role of Chkalov and other aviators in Soviet society read B. Galin's "Valery Chkalov," *Pravda,* 21 June
1937, 3, as translated in *Mass Culture in Soviet Russia,* ed. James von
Geldern and Richard Stites (Bloomington: Indiana University Press,
1995), 260–66.

15. For an account of Soviet aviation in World War II, see Von Hardesty, *Red*
Phoenix: The Rise of Soviet Air Power, 1941–1945 (Washington D.C.:
Smithsonian Institution Press, 1982, 1991).

16. Maksimilian Saukke, a close friend of Kerber and collaborator for this
English translation of *Tupolevskaya sharaga,* has written a concise account
of Tupolev's prison years, "Raby sistemy, pamyati zaklyuchennykh
spetstyur'my TsKB-29 NKVD," *Mir aviatsii,* no. 4 (1993): 6–9.

17. See "Narkom Yezhov" in *Mass Culture in Soviet Russia,* ed. von Geldern
and Stites, 298–300.

18. There are many sources in the English language on Tupolev-designed aircraft, the most recent being Bill Gunston, *Tupolev Aircraft since 1922* (An-
napolis: U.S. Naval Institute Press, 1995). Another excellent source is
R. A. Mason and John W. R. Taylor, eds., *Aircraft, Strategy, and Operations*
of the Soviet Air Force (London: Jane's, 1986). Less known, but perhaps the
best source of periodical literature in reprint on Soviet aircraft, is *The*
Bulletin of the Russian Aviation Research Group of Air-Britain, edited by Ni-
gel Eastaway. The March 1996 issue of the bulletin contains sixteen separate articles and reports on Tupolev aircraft, including articles in Russian,
French, English, Czech, and German.

19. The first B-29 landed near Vladivostok on 29 July 1944, the second on 20
August 1944 at Khabarovsk, and the final two at Vladivostok, on 11 November 1944 and 21 November 1944, respectively. American bomber
crews had been ordered to stay clear of Soviet territory. Emergency landings in the Soviet Union were a tempting option, if low on fuel or facing
some other emergency. The Soviets refused to return the aircraft to the
United States. Three of the interned B-29s were used for the Tu-4 program, one to test fly, one to dismantle for study, and one to serve as a

model or template. For the role of the B-29 in the Pacific War, see Kenneth Werrill, *Blankets of Fire: U.S. Bombers over Japan during World War II* (Washington D.C.: Smithsonian Institution Press, 1996).

20. L. L. Kerber, "Ne kopiya, a analog, o samolete Tu-4," *Kril'ya rodina,* no. 1 (1989): 24–26. Kerber reported that Stalin's order to copy the B-29 was taken quite literally, even to duplicating a patch on the fuselage of the B-29 which was used as a model for the Tu-4. Kerber to Von Hardesty, 18 September 1991.

1. FORMATION OF THE MAN

1. An Ataman or Hetman is the leader of a Cossack community. There were a number of cossack communities who enjoyed autonomy as a military caste. These autonomous groups, congregated along the Don and Dnieper Rivers, played a key role in the expansion and security of the borderlands of the Russian Empire in Ukraine and the North Caucasus region.

2. L. L. Kerber notes that Tver Medical Institute is located in the *gymnasium* building. On its wall there is a memorial plate indicating that Tupolev received his secondary education here within the walls of the former *gymnasium.*

3. V. G. Belinskiy, N. G. Chernyshevskiy, N. A. Dobrolyubov, and D. I. Pisarev were members of the Russian intelligentsia in the nineteenth century, known for their radical views and opposition to tsarism.

4. *Russkoye Slovo,* no. 10 (13 January 1917): 4.

5. The Russian Civil War raged from 1918 to 1921 between the Bolsheviks and the various White armies that arose to challenge the new Soviet regime. Kerber served on the Red Army during this bloody conflict.

6. General Peter Wrangel and General Anton Denikin led White armies against the Bolsheviks. Both were portrayed as counterrevolutionaries in the Soviet period.

7. Literally, "small insect."

8. NEPmen refers to entrepreneurs who engaged in various kinds of commerical activities during the "New Economic Period" of the 1920s. This approved form of "capitalism" under the Bolsheviks would be reversed at the end of the decade with many of the so-called NEPmen purged.

9. Palaces or houses of culture were established throughout the Soviet Union to provide an institutional basis to promote cultural activity. These formal enclaves of culture were run by party functionaries and heavily influenced by political agitators, especially in the Stalin years.

10. Peterhof or Petrovorets was the palace built by Peter the Great on the Baltic. The palace, famed for its fountains, was located across from Kronstadt, the island naval base at the threshold of St. Petersburg.

11. Amtorg was the Soviet trading company in the United States. Amtorg established a number of contracts with American companies for the supply of advanced technology.

12. V. V. Khripin and A. N. Lapchinskiy were leading air theorists in the 1930s. As in the West, Soviet air theorists discussed the future role of air power, in particular the bomber, which was perceived as a potentially decisive weapon. Many Soviet air theorists were followers of Giulio Douhet (1869–1930), the Italian theorist who had promoted air power as the key to victory in any future war. With the purge of the Soviet Air Force at the end of the 1930s many of the "Douhetists" were removed from positions of influence.

13. Ivan D. Papanin was a famed Soviet Arctic explorer of the 1930s, who at the time of the Chkalov flight had established a scientific research station on a drifting ice floe near the North Pole. During this period Tupolev's four-engined ANT-6 (TB-3) flew numerous flights to the Arctic stations, establishing for the Soviet Union leadership in cold-weather flying techniques. Sigismund Levanevskiy's ANT-6, the N-209, disappeared over the Artic in a failed attempt to fly over the North Pole in August 1937.

14. The metaphor of a "Potemkin village" suggests an effort to portray a false reality. Prince Potemkin created a series of idyllic peasant villages in the eighteenth century to impress Empress Catherine the Great. These cardboard images of Russian rural life represented a studied effort to mislead the government.

15. K. Ye. Voroshilov, a long-time crony of Stalin, held the position of People's Commissar of Defense in the early 1930s.

16. OSOAVIAKhIM (*Obshchestvo sodeystviya oborone, aviatsionnomu, i khimicheskomu stroitel'stvu*) is the acronym for the Society for the Support of Defense and Aviation and Chemical Construction. The paramilitary organization for Soviet youth engaged in myriad activities, including glider training, civil defense drills, navigation instruction, model building, aircraft maintenance, and the aggressive promotion of air power. Another group, ODVF (*Obshchestvo druzey vozdushnogo flota*) or the Society of Friends of the Air Fleet numbered over a million members at branches throughout the Soviet Union. Members donated a portion of their salaries for aircraft construction, among other activities. These two mass organizations represented the systematic effort of the Communist Party to rally support for avaiation during the Stalin years.

17. The planing step is a step on the bottom of the boat. Owing to this, at speeds of 60–80 km/h the hull of the boat rises up, and it is easier to leave the water.

18. Ya. I. Alksnis was a popular commander of the Soviet Air Force who fell victim to the purges, along with Yakov V. Smushkevich and many other high-ranking Air Force personnel, between 1937 and 1941.

19. The rescue of the icebreaker *Chelyuskin* in 1934 represented one of the most dramatic examples of air rescue in the 1930s. Soviet aircraft equipped with skis flew to the trapped icebreaker, an event that prompted Stalin to establish the highest award for bravery, Hero of the Soviet Union. One of the heroic fliers who rescued the *Chelyuskin*, M. V. Vodopyanov, wrote a novel, *A Pilot's Dream*, which was later adapted as a play for the Moscow theater performance.

20. "Stakhanovite" workers were emblematic of sacrifice and herculean effort on behalf of the revolutionary goals of the Soviet Union. Aleksei Stakhanov mined 102 tons of coal in a single shift, a display of proletarian zeal that won him a place in the pantheon of revolutionary heroes. His feat led to the Stakhanovite movement, which aimed to mobilize and energize workers.

21. Andrei Nikolayevich Tupolev, *Grani derznovennogo tvorchestva* [*The Limits of Daring Creativity*] (Moscow: Nauka, 1988), 89.

22. High aspect ratio dictated that the wing be long and narrow, to allow for maximum lift. Igor Sikorsky with his four-engined *Il'ya Muromets* in 1913–14 was the first to pioneer this type of design.

23. These were magnesium lights turned on by electrical power, which replaced the landing lamps.

24. Until this time fuselages had been made of nonload-bearing wooden box frames, and the skin either of fabric or corrugated metal sheets. In monocoque fuselages, by contrast, the design was radically different with the outer stressed skin providing structural integrity. This smooth skin, fashioned of metal or plywood, was able to withstand the torsional and bending stresses.

2. IN PRISON

1. Kerber's father was a Tsarist naval officer. Under Bolshevik rule many children of the nobility and other groups deemed as class enemies were considered suspect and often denied access to educational institutions.

2. *Taiga* is a geographical region located in the northern latitudes between the Arctic tundra and forests.

3. Themis, Greek goddess of law and justice, portrayed as blindfolded and holding the scales of justice.
4. Shortly after an unpleasant conversation with Molotov (or so they said in aviation circles), M. M. Kaganovich became convinced that his older brother, Lazar Kaganovich, had offered him up as a sacrifice, and so blew his own brains out.—*L.L.K.*
5. This was KB-103, a task force, not considered to be a full-fledged design bureau.
6. That incomplete list of the most famous prisoner-specialists of the NKVD's *sharaga* would include the following: Vladimir Leont'yevich Aleksandrov, Robert Lyudvigovich Bartini, Boris Sergeyevich Vakhmistrov (armament), Amik Avetovich Yengivaryan (electrical equipment), Aleksandr Markovich Izakson (helicopters), Mikhail Minayevich Kachkachyan (hydroscopic instruments), Sergei Pavlovich Korolyov, Yuriy Aleksandrovich Krutkov, Dmitriy Sergeyevich Markov, Solomon Moiseyevich Meyerson (armament), Vladimir Mikhaylovich Myasishchev, Aleksandr Vasil'yevich Nadashkevich (armament), Aleksandr Ivanovich Nekrasov, Josef Grigor'yevich Neman, Vladimir Mikhaylovich Petlyakov, Aleksandr Ivanovich Putilov, Boris Sergeyevich Stechkin (he was soon transferred to the engine prison design bureau), Dmitriy Lyudvigovich Tomashevich, Andrei Nikolayevich Tupolev, Aleksei Mikhaylovich Cheremukhin (gyroplanes), and Vladimir Antonovich Chizhevskiy.
7. Ekaterina Furtseva was a member of the Politburo under Nikita Khrushchev.
8. "Grishka-in-Tatters."
9. Valentina Grizodubova with Paulina D. Osipenko and Marina M. Raskova flew an ANT-37, named *Rodina,* or *Motherland,* on a nonstop flight in 1938 from Moscow to the Far East, covering 6450 kilometers to establish a new record for female pilots.
10. A pun relying on the fact that the Russian word for *assembly* also means *knot.*

3. NEW CHALLENGES

1. On 13 January 1953, *Pravda* announced the arrest of a group of Kremlin doctors who were accused of terrorism and the attempted murder of Soviet officials. For many, this event signaled Stalin's desire for another great purge. The purge was stillborn because of Stalin's death on 5 March 1953.
2. A. Ya. Vyshinsky served as the presiding judge for the Shakhty trial of 1928, the first show trial under Stalin. The so-called Shakhty affair involved the purge of 52 specialists and engineers who had been accused of

sabotage in the coal mines in Shakty in the Donbass. The Shakhty trial demonstrated that the professional elite were not immune from the purge mechanism, something Tupolev would learn a decade later. Vyshinsky, a former Menshevik, presided over the huge show trials in 1936–37. A survivor, Vyshinsky outlived Stalin. Ulrich served as a court official at the show trials and is remembered as a henchman of Vyshinsky.

3. At the Twentieth Party Congress on 25 February 1956, Nikita Khrushchev delivered his "secret" speech, a long rambling address in which he attacked Stalin and the "Cult of Personality."

4. Boris Pasternak's famous novel, *Dr. Zhivago*, had been translated into English and would later be adapted for a feature-length film. At the time *Dr. Zhivago* was banned in the Soviet Union.

INDEX

Abramov (plant director), 168, 177, 208
Academy of Sciences, 84, 107, 113, 121, 255
"Aerodynamic Calculations of the Airplane" (Lukyanov, Tupolev, and Zhukovsky), 29
aerodynamics, 32, 356
Aeroflot, 16, 79, 293, 306, 338, 340; ANT-35 passenger airplane, 134; use of Tu-134, 343
Aeronautical and Seaplane Experimental Construction (AGOS), TsAGI, 64, 68, 81–82, 93, 107–8, 136
aeronautics industry. *See* aviation industry
Aeroplane (periodical), 125, 287, 293
aerosleighs, 43–44, 85
agitation aircraft, 82, 98–104
Aim of a Lifetime (Yakovlev), 215
air blowers, adaptation of, 137–39
airborne (parachute) troops, 78
air-conditioning, 268, 278, 346, 353, 362
aircraft industry. *See* aviation industry
Aircraft Radio Stations and Their Operation (Kerber), 112
Air Force, Soviet, 51, 54–55, 80, 97, 242, 331; Army Air Force, 77, 129; bomber requirements during Cold War, 279, 283; commissions for evaluation of Tu-2, Tu-14, and Tu-16, 193–95; demands on ANT-40, 132–33; dislike of Il-4, 241–42; purge, 375n12, 376n18; support for Tu-91, 333; use of Tu-2, 15.

See also Scientific Research Institute of the Air Force
Air Force Academy, Zhukovskiy, 152–54
Air Forces Directorate, 30, 31, 33, 51, 73–74
Aleksandrov, Anitoliy Petrovich (physicist), 276
Alekseyev, A. (pilot), 121, 131
Alksnis, Ya. I., 100, 107, 132, 141, 220, 262; as purge victim, 12, 376n18
alloys, 279; duralumin, 41–44, 50, 68, 86–89, 355–56
All-Union Electrotechnical Institute (VEI), 99, 102
AM-3 jet engine, 284–85, 291
AM-35 engine, 142, 227
AM-37 engine, 227
AM-38 engine, 142
AM-42 engine, 246
Amtorg (Soviet trading company), 75, 375n11
Andrews Air Base, 314, 315, 321
ANT-1 aircraft, 47–49, 50, 341
ANT-2 aircraft, 49–50, 64, 352
ANT-3 reconnaissance aircraft, 51–53, 64, 90
ANT-4 heavy bomber. *See* TB-1 heavy bomber
ANT-5 (I-4) aircraft, 181
ANT-7 aircraft, 121
ANT-6 heavy bomber. *See* TB-3 heavy bomber